CONDUCTIVE POLYMERS AND PLASTICS

in Industrial Applications

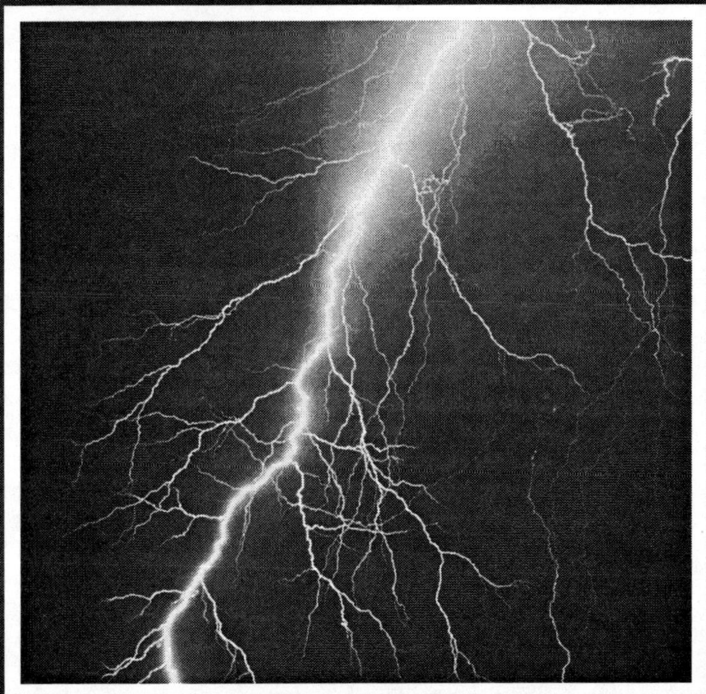

Larry Rupprecht, Editor

Society of Plastics Engineers

pdl

Plastics Design Library

ISBN 1-884207-77-4
Library of Congress Card Number 98-89320
Printed in Canada

Published in the United States of America, Norwich, NY by Plastics Design Library a division of William Andrew Inc.

Comments, criticism and suggestions are invited and should be forwarded to Plastics Design Library.

Plastics Design Library, 13 Eaton Avenue, Norwich, NY 13815 Tel: 607/337-5080 Fax: 607/33-5090

Table of Contents

Preface

The introduction of the Electromagnetic Compatibility Directive and the burgeoning use of electronic components in a wide range of manufactured goods have created interest in plastic materials designed for EMI shielding, safe packaging, corrosion protection, and other applications. Conductive plastics are positioned to play an increasingly important role in affairs of mankind, specifically in the area of electronic and electrical conductivity.

While general knowledge about conductive polymers and plastics has been available for many years, a true understanding of their application has only taken shape in the last 3 to 4 years. This is attributable to advancements in materials and processing techniques. Engineers have only begun to explore the design freedom and the economic benefits of specifying conductive polymers and plastics in industrial and business applications.

Shielding of electronic components and devices from effects of electrostatic discharge (ESD) and electromagnetic or radio frequency interference (EMI/RFI) is addressed frequently in various media. ESD problems can damage or destroy sensitive electronic components, erase or alter magnetic media, or set off explosions or fires in flammable environments. EMI can interfere with the operation of simple appliances, corrupt data in large-scale computer systems, cause inaccurate readings and output in aircraft guidance systems, and interrupt the functioning of medical devices such as pacemakers. Liability to industry from these problems totals billions of dollars each year. This book presents novel approaches and techniques in the area of electronic protection.

Beyond ESD and EMI problems lie very diverse application areas for conductive polymers and plastics. Highlighted in this book are such uses as corrosion protection of metals; as resistors, capacitors, or detectors, and improved electrostatic painting processes.

This book is a collection of papers describing efforts of many individuals - both in industry and academia - in both pure research and application development of conductive polymers and plastics. Numerous existing possibilities of material design are discussed, including intrinsically conductive polymers, polymers doped with conductive sites, ion implantation, polymers containing dispersed conductive fillers, and polymer blends technology in cost effective applications which are compared with metal plating.

Conductive fillers discussed in the book include carbon black, hollow flexible carbon fibers, nickel coated carbon fibers, other conductive fibers, and multiphase thermoplastic composites containing several fillers.

In addition to existing technology, the book discusses improvements to current plastic processing methodology that provide enhanced conductive characteristics while improving economic benefits. For instance, co-continuous phase technology in the preparation of conductive composite materials and co-injection molding techniques in forming finished articles are introduced.

Various methods of manufacture of polymer and final product are investigated, including electrohydrodynamic flow cells, transducers used as vapor-phase detectors, electrostatic paintable compounds, conductive polymer films, non-linear optical waveguides, conductive foams, thermoformed containers for electrostatic sensitive devices, disk-drive assemblies, and more.

This work is aimed at understanding the effect of processing parameters and formulation on material performance and uniform distribution of conductive components. Although, conductive additives are incorporated to change electrical properties of materials, they also affect other performance characteristics of final products. These effects are investigated and remedies proposed which allow production of defect-free finished products.

Larry Rupprecht
Winona, May 1999

Electrical Conductivity in Conjugated Polymers

Arthur J. Epstein
Department of Physics and Department of Chemistry, The Ohio State University, Columbus, Ohio, 43210-1106

INTRODUCTION

In 1977, the first intrinsic electrically conducting organic polymer, doped polyacetylene, was reported,[1] spurring interest in "conducting polymers." These polymers are a different class of materials than conducting polymers, which are merely a physical mixture of a non-conductive polymer with a conducting material such as metal or carbon powder. Initially these intrinsically conducting polymers were neither processable nor air stable. However, later generations of these polymers were processable into powders, films, and fibers from a wide variety of solvents, and also air stable.[2,3] Some forms of these intrinsically conducting polymers can be blended into traditional polymers to form electrically conductive blends. The electrical conductivities of the intrinsically conducting polymer systems now range from that typical of insulators ($<10^{-10}$ S/cm [10^{-10} $(\Omega^{-1}\text{-cm})^{-1}$]) to that typical of semiconductors such as silicon ($\sim10^{-5}$ S/cm) to greater than 10^4 S/cm (nearly that of a good metal such as copper, 5×10^5 S/cm).[2,4] Applications of these polymers, especially polyanilines, have begun to emerge. These include blends and coatings for electrostatic dissipation and electromagnetic interference (EMI) shielding, electromagnetic radiation absorbers for welding (joining) of plastics, conductive layers for light emitting polymer devices, and anticorrosion coatings for iron and steel.

The common electronic feature of pristine (undoped) conducting polymers is the π-conjugated system which is formed by overlap of carbon p_z orbitals and alternating carbon-carbon bond length.[5,6,7] (In some systems, notably polyaniline, nitrogen p_z orbitals and C_6 rings also are part of the conjugation path.[8,9]) Figure 1 shows the chemical repeat units of the pristine forms of several families of conducting and semiconducting polymers, i.e., *trans*-polyacetylene [t-$(CH)_x$], the leucoemeraldine base (LEB), emeraldine base (EB) and

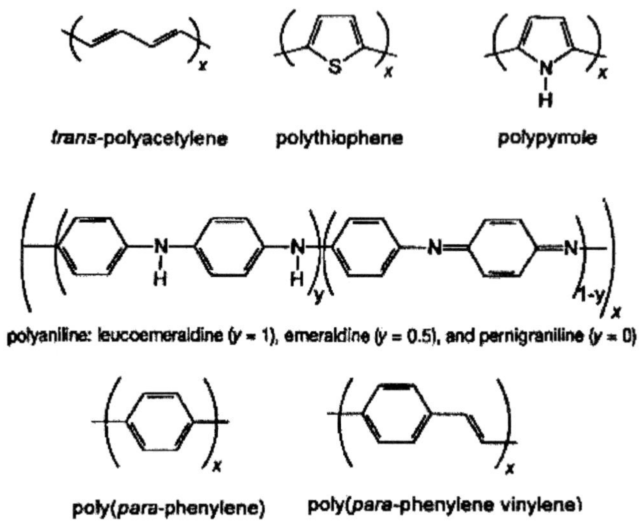

Figure 1. Repeat units of several electronic polymers.

pernigraniline base (PNB) form of polyaniline (PAN, polypyrrole (PPy) polythiophene (PT), poly(*p*-phenylene) (PPP), and poly(*p*-phenylene vinylene) (PPV).

Each of these polymers is that of an insulator, with an energy gap between filled and empty energy levels. For undoped *t*-$(CH)_x$ the energy gap arises from the pattern of alternating single (long) and double (short) bonds,[5,6,7] with an additional contribution due to electron-electron Coulomb repulsion.[5] Interchange of short and long bonds results in an equivalent (degenerate) ground state. The pernigraniline oxidation state of PAN[10] also has a two-fold degenerate ground state. The remaining polymers in Figure 1 are nondegenerate: single and double bond interchange yields electronic structures of different energy.

INCREASE IN CONDUCTIVITY WITH DOPING

The conductivities of the pristine electronic polymers are transformed from insulating to conducting through doping.[2-7] Both *n*-type (electron donating, e.g., Na, K, Li, Ca, tetrabutylammonium) and *p*-type (electron accepting, e.g., PF_6, BF_4, Cl, AsF_6) dopants have been used. The doping typically is done using vapors or solutions of the dopant, or electrochemically. (In some circumstances, the polymer and dopant are dissolved in the same solvent before forming the film or powder.) The polymer backbone and dopant ions form a rich variety of new three-dimensional structures.[11]

For the degenerate ground state polymers, the charges added to the backbone at low doping levels are stored in charged soliton and polaron states for degenerate polymers,[5-7,12,13] and as charged polarons or bipolarons for nondegenerate systems.[14] For nondegenerate polymers, high doping results in polarons interacting to form a "polaron lattice" or electrically conducting partially filled energy band.[15,16,17] Some models suggest equilibrium between polarons and bipolarons.[18] At high doping levels of *t*-$(CH)_x$, it is proposed that the soliton energy levels essentially overlap the filled valence and empty conduction bands leading to a conducting polymer.[19]

For the polyaniline emeraldine base (EB) form, the conductivity varies with proton (H^+ ion) doping level (protonic acid doping). In the protonation process, there is no addition or removal of electrons to form the conducting state.[15] Figure 2 schematically demonstrates the equivalence of p-doping of leucoemeraldine base and protonic acid doping of EB to form the conducting emeraldine salt. Both organic acids such as HCSA (camphor sulfonic acid), and inorganic acids, such as HCl, are effective,[20] with the organic sulfonic acids leading to solubility in a wide variety of organic solvents, such as chloroform and m-cresol.[21] The protonic acid may also be covalently bound to the polyaniline backbone, as has been achieved in the water soluble sulfonated polyanilines,[22] Figures 3a and 3b. Similar electronic behavior has been observed for protonic acid doped PAN as for the other nondegenerate ground state systems.[15-17] That is, polarons are important at low doping levels, and, for doping to the highly conducting state, a polaron lattice (partially filled energy band) forms. Polaron pairs, or bipolarons are formed in less ordered regions of doped polymers.[23]

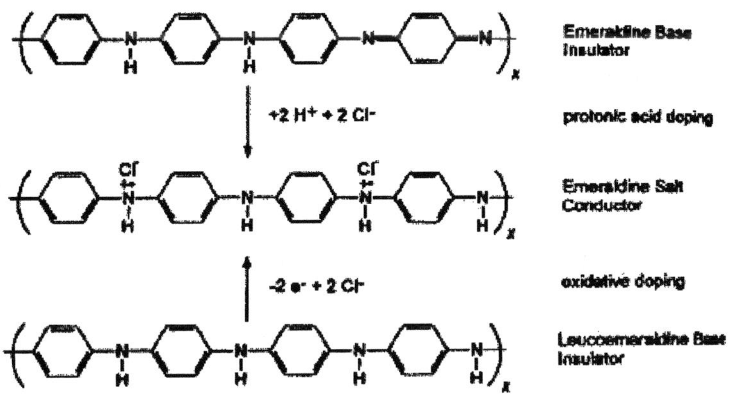

Figure 2. Illustration of the oxidative doping (p-doping) of leucoemeraldine base and protonic acid doping of emeraldine base, leading to the same final product, emeraldine salt.

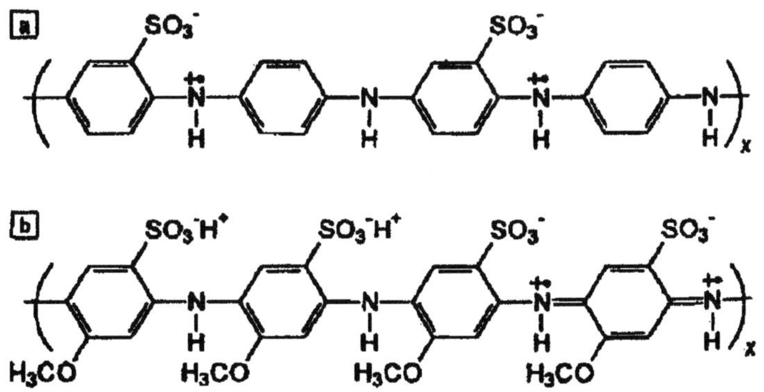

Figure 3. Schematic illustrations of (a) 50% sulfonated and (b) 100% sulfonated polyanilines (self-doped forms).

Iodine doped $(CH)_x$ was initially reported[1] with $\sigma \sim 100$ S/cm. Subsequently, $(CH)_x$ was synthesized by alternate routes that yielded higher conductivities upon doping, reportedly[24] as high as $\sim 10^5$ S/cm, rivaling that of traditional metals such as copper ($\sigma_{DC} \sim 6\times10^5$ S/cm). Recent advances in the processing of other conducting polymer systems have led to improvements in their σ_{DC} to the range of $\sim 10^3$ - 10^4 S/cm. The absolute value of the highest conductivities achieved remains controversial. Many traditional signatures of an intrinsic metallic nature now have become apparent, including negative dielectric constants, a Drude metallic response, temperature independent Pauli susceptibility, and a linear dependence of thermoelectric power on temperature. However, the conductivities of even new highly conducting polymers, though comparable to traditional metals at room temperature, generally decrease as the temperature is lowered. Some of the most highly conducting samples remain highly conducting even at millikelvin.[25,26]

As there is a great diversity in the properties of materials synthesized by even the same synthetic routes, correlated structural transport, magnetic, and optical studies of the same materials are important. The conductivity of a polymer, for example HCSA doped polyaniline, can vary greatly both in magnitude (in this case, nearly four orders of magnitude) and temperature dependence (both increasing and decreasing conductivity with decreasing temperature) as a result of processing in different solvents. The effect of solvent and solvent vapors on the structural order and subsequent electrical conductivity of intrinsically conducting polymers, especially polyanilines, is termed[20] "secondary doping."

MODELS FOR ELECTRICAL CONDUCTIVITY

Much work has focused on the nature of the charge carriers in the highly doped metallic state. They may be spatially localized by structural disorder so they cannot participate in transport except through hopping.[2,4] Figure 4 is a schematic view of the inhomogeneous disorder, with individual polymer chains passing through both ordered regions (typically 3 - 10 nm across) and disordered regions. The percent "crystallinity" may vary from near zero to 50 or 60% for polypyrroles and polyanilines, respectively, to greater then 80% for polyacetylenes. The chains in the disordered regions may be either relatively straight, tightly coiled, or intermediate in disorder.

Impurities and lattice defects in disordered systems introduce backward scattering of these electron waves with resulting[27] "Anderson localization." The ramifications, include a finite density of states $N(E_F)$ produced at the Fermi level E_F between mobility edges.[28] When the Fermi level or chemical potential lies in the localized region, $\sigma(T = 0\ K)$ is zero even for a system with a finite density of states. Mott variable range hopping (VRH) model is applicable to systems with strong disorder such that the disorder energy is much greater than the band width. For Mott's model $\sigma = \sigma_0 \exp[-(T_0/T)^{1/(d+1)}]$, where d is the dimensionality and, for

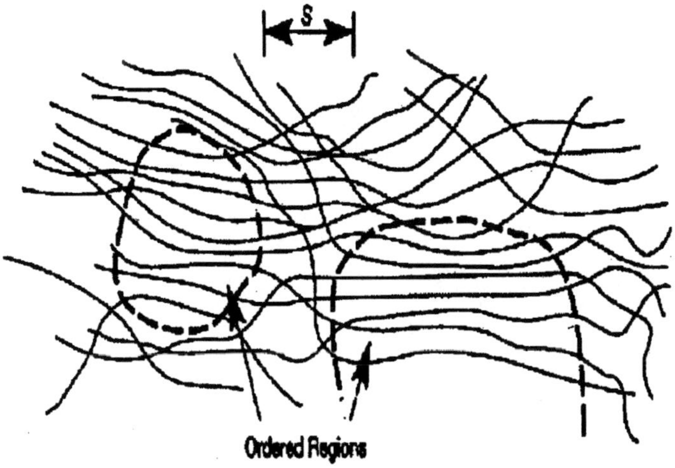

Figure 4. Schematic view of the inhomogeneous disorder in these doped polymers, with individual polymer chains passing through both ordered regions (typically 3 - 10 nm across) and disordered regions (of length 's').

three-dimensional systems, $T_0 = clk_BN(E_F)L^3$ (c is the proportionality constant, k_B the Boltzmann constant, and L the localization length). If the Fermi level is at an energy such that the electronic states are extended, then finite σ at 0 K is expected. This model assumes that the substantial disorder is homogeneous throughout the isotropic three-dimensional sample.

For isolated one-dimensional metallic chains localization of charge carriers arises for even weak disorder because of quantum interference due to static back-scattering of electrons,[28] contrasting to the strong disorder required for localization in three-dimensional systems. The localization effects in the inhomogeneously disordered (partially crystalline) conducting polymers are proposed to originate from one-dimensional localization in the disordered regions. The inhomogeneous disorder model[25,29,30] represents the doped polymer as relatively ordered regions ("crystalline islands") interconnected through polymer chains traversing disordered regions, Figure 4. Within this model, conduction electrons are three-dimensionally delocalized in the "crystalline" ordered regions (paracrystalline disorder may limit delocalization within these regions[29]). To transit between ordered regions, the conduction electrons must diffuse along electronically isolated chains through the disordered regions where the electrons easily become localized. The localization length of these electrons depends the details of the disorder (e.g., electrons traveling along tightly coiled chains are expected to have much shorted localization lengths then electrons traveling along expanded coil or relatively straight chains). Photon-induced enlargement of the localization length increases the conductivity with higher temperature. Three-dimensional crystalline order facilitates delocalization. If the localization length for some conduction electrons exceeds the separation between the ordered regions then will be substantially enhanced.

For conventional metals, many of the electrical transport properties can be described by the Drude model with a single scattering time τ. The model explains high and frequency independent conductivity of metals from dc to the microwave (~10^{10} Hz) frequencies, and a real part of the dielectric constant (ε_r) which is negative below the screened plasma frequency, $\omega_p^2 = 4\pi ne^2/m^*\varepsilon_b$; n is the density of carriers, m* is the carrier effective mass, and ε_b is the background dielectric constant. In the low frequency Drude limit ($\omega\tau << 1$), the Drude response can be deduced as $\varepsilon_r = -\omega_p^2\tau^2$ and $\varepsilon_i = \omega_p^2\tau/\omega$, where ε_i is the imaginary part of the dielectric constant.

ELECTRICAL CONDUCTIVITY OF CONDUCTING POLYMERS

The $\sigma(T)$ of heavily iodine doped $(CH)_x$ and PF_6 doped PPy down to mK range vary as a function of aging.[31] The highest σ_{dc} at room temperature reported in this study is ~5×10^4 S/cm for I_3 doped T-$(CH)_x$ and ~10^3 S/cm for the highest conducting PPy(PF_6). For both of these materials, σ decreases with decreasing temperature to a minimum at T_m ~ 10 K. Below T_m, σ increases by ~20% and then is constant to 1 mK. Some preparations of PAN-CSA show similar behavior.[25] Lower conductivity samples of doped $(CH)_x$, doped polyaniline, and doped polypyrrole become insulating at low temperatures.

Hydrochloric acid as well as CSA doped polyaniline prepared in chloroform often show quasi-one-dimensional variable range hopping (VRH), $\sigma = \sigma_0\exp[-(T_0/T)^{1/2}]$, where $T_0=16/[k_BN(E_F)Lz]$. Here L is the one-dimensional localization length and z the number of nearest neighbor chains. Generally, the higher σ samples have a weaker temperature dependence (T_0 ~ 700–1000 K, T<80 K), and lower σ samples a stronger temperature dependence (T_0 ~ 4000 K). Smaller T_0 is associated with weaker localization and improved intrachain and interchain order.

The microwave frequency dielectric constant provides a measure of the charge delocalization in individual samples. The low temperature dielectric constant, ε_{mw}, for a series of emeraldine hydrochloride samples is proportional to the square of the crystalline coherence length, ξ^2, independent of the direction of orientation of the sample with regard to the microwave frequency electric field, demonstrating that the charge is delocalized three-dimensionally within the crystalline regions of these samples.[30]

The sign, magnitude, and temperature dependence of the 6.5×10^9 Hz dielectric constant for very highly conducting T-$[CH(I_3)_y]_x$, PPy-PF_6, and m-cresol prepared PAN-CSA are quite striking.[2,4,29,30] Each of these systems has a large (10^4 - 10^6) and negative value of ε_{mw}. Using the Drude model for low frequencies ($\omega\tau < 1$), plasma frequencies of ω_p = 0.01 - 0.02 eV (~200 cm^{-1}) and room temperature scattering times of ~10^{-11} sec were calculated. The exact values correlate with the sample preparation conditions. The ω_p are much smaller than one expects from the usual Drude model, suggesting that only a small fraction of the conduction band electrons participate in this low frequency plasma response. Similarly, the value of τ is two orders of magnitude larger than usual for an alkali, noble, or transition metal, perhaps associated with the time for electrons to transit the disordered regions.[32]

For the conducting doped conjugated polymers, there are zero, two, three, or one zero crossings of the real part of the dielectric function (ε_1) as the frequency is decreased.[2,4,25,30] For the least conducting materials, ε_1 remains positive for the entire optical frequency range (50–50,000 cm^{-1}), reaching values of several hundred at microwave frequencies. For higher conductivity materials, ε_1 crosses zero between 1 and 3 eV (the all-conduction-electron plasma response) and then becomes positive again below 1000 cm^{-1}, reaching values in excess of 10^4 at microwave frequencies. For "metallic" doped PAN and PPy with $\sigma_{dc} \sim 400$ S/cm, ε_1 has the previous two zero crossings, and a third zero crossing occurs to negative values at a "delocalized conduction electron plasma frequency" of several hundred wavenumbers. For very highly conducting doped polyacetylene, ε_1 crosses zero at the all conduction electron plasma frequency and remains negative to the lowest measured optical frequencies.[33]

APPLICATIONS

Intrinsically conducting polymers are promising materials for shielding electromagnetic (EM) radiation and reducing or eliminating EMI because of their relatively high σ and ε and their ease of control through chemical processing.[34] Also, they are relatively lightweight compared to standard metals, flexible, and do not corrode. The capabilities are in the range for many commercial (~40 dB) and military (~80 - 100 dB) applications.

Intrinsically conductive polymers, especially polyanilines, can be used in welding (joining) of thermoplastics and thermosets. The conducting polymer film or blend of the conductive polymer and the thermoplastic or thermoset to be joined is placed at the interface. Exposure to microwave frequency radiation results in heating of the joint and subsequent fusing (welding).[35] The resulting joint may be as strong as that of the pure compression molded thermoplastic or thermoset.

The corrosion of steel has long been an important problem. Polyaniline has been shown to have corrosion protecting capabilities both when doped[36] and neutral.[37] The mechanism for corrosion protection was found to be anodic, i.e., the polyaniline film withdraws charge from the metal, pacifying its surfaces against corrosion. Large values (up to ~1.5 cm) of throwing power were obtained for emeraldine base protected cold rolled steel.

There is a need for low voltage, reliable operation of light emitting polymer devices. One approach is to overcoat the transparent conducting indium tin oxide electrode with a layer of nearly transparent conducting polymer, especially polyaniline,[38] or incorporating networks of conducting polymer fibers in the polymers.[39]

SUMMARY

Intrinsically conducting polymers are a broad class of (often) processable materials based upon doped π conjugated polymers. They vary from insulators through to semiconductors and even good metals. A wide variety of electronic phenomena are observed. This class of polymer is potentially of use in many technologies.

ACKNOWLEDGMENT

I thank A.G. MacDiarmid, J.P. Pouget, V. Prigodin, G. Ihas, T. Ishiguro, Y. Min, and D. Tanner for extensive and stimulating discussions and collaborations. Contributions by those at OSU, especially J. Joo, R. Kohlman, Y.Z. Wang, and Z.H. Wang. This work was supported in part by NSF DMR-9508723 and ONR.

REFERENCES

1 C.K. Chiang, C.R. Fincher, Jr., Y. W. Park, A.J. Heeger, H. Shirakawa, E.J. Louis, S.C. Gau, and A.G. MacDiarmid, *Phys. Rev. Lett.*, **39** (1977) p. 1098.
2 R. Kohlman, J. Joo, and A.J. Epstein, in **Physical Properties of Polymers Handbook**, edited by J.E. Mark, (*AIP Press*, 1996), Chapter 34, p. 453.
3 Proc. International Conference on Science and Technology of Synthetic Metals, Seoul, Korea, 24-29 July 1994, (*Synth. Met.* 69-71, 1995); Proc. International Conference on Science and Technology of Synthetic Metals, Snowbird, Utah, 28 July-2 August 1996 (*Synth. Met.* 84-86, 1997).
4 Kohlman and A.J. Epstein, in **Handbook of Conducting Polymers**, edited by T. Skothern, R.L. Elsenbaumer, and J. Reynolds (*Marcel Dekker, Inc.*, New York) p. 85 (1997).
5 D. Baeriswyl, D.K. Campbell, and S. Mazumdar, in **Conjugated Conducting Polymers,** edited by H.G. Keiss (*Springer-Verlag,* Berlin, 1992), p. 7.
6 E.M. Conwell, *IEEE Transactions on Electrical Insulation*, **EI-22** (1987) p. 591.
7 A.J. Heeger, S.A. Kivelson, J.R. Schrieffer, and W.P. Su, *Rev. Mod. Phys.*, **60** (1988) p. 781.
8 J. M. Ginder and A.J. Epstein, *Phys. Rev.*, **B 41** (1990) p. 10674; J. Libert, J.L. Brédas, and A.J. Epstein, *Phys.Rev.*, **B 51** (1995) p. 5711.
9 J. Libert, J.L. Brédas, and A.J. Epstein, *Phys.Rev.*, **B 51** (1995) p. 5711.
10 M.C. dos Santos and J. L. Brédas, *Phys. Rev. Lett.*, **62** (1989) p. 2499; J.M. Ginder and A.J. Epstein, *Phys. Rev Lett.*, **64** (1990) p. 1184.
11 J.P. Pouget, Z. Oblakowski, Y. Nogami, P.A. Albouy, M. Laridjani, E.J. Oh, Y. Min, A.G. MacDiarmid, J. Tsukamuto, T. Ishiguro, and A.J. Epstein, *Synth. Met.*, **65** (1994) p. 131.
12 S.A. Brazovskii, *Sov. Phys. JETP, Lett.*, **28** (1978) p. 606.
13 M.J. Rice, *Phys. Lett.*, **71A** (1979) p. 152.
14 D.K. Campbell and A.R. Bishop, *Phys. Rev.*, **B 24** (1981) p. 4859.
15 A.J. Epstein, J.M. Ginder, F. Zuo, R.W. Bigelow, H.-S. Woo, D.B. Tanner, A.F. Richter, W.-S. Huang, and A.G. MacDiarmid, *Synth. Met.*, **18** (1987) p. 303.
16 S. Strafstrom, J.L. Brédas, A.J. Epstein, H.S. Woo, D.B. Tanner, W. S. Huang, and A.G. MacDiarmid, *Phys. Rev. Lett.*, **59** (1987) p. 1474 .
17 J.L. Brédas, B. Thémans, J.G. Fripiat, J.M. Andréh, and R.R. Chance, *Phys. Rev.*, **B 29** (1984) p. 6761.
18 F. Genoud, M. Guglielmi, M. Nechtschein, E. Genies, and M. Salmon, *Phys. Rev. Lett.*, **55** (1985) p. 118.
19 E.M. Conwell, H.A. Mizes, and S. Javadev, *Phys. Rev.*, **B 40** (1989) p. 1630.
20 A.G. MacDiarmid and A.J. Epstein, *Synth. Met.*, **65** (1994) p. 103.

21 Y. Cao and A.J. Heeger, *Synth. Met.*, **52** (1992) p. 193.
22 J. Yue and A.J. Epstein, *J. Am. Chem. Soc.*, **112** (1990) p. 2800; W. Lee, G. Du, S.M. Long, S. Shimizu, T. Saitoh, M. Uzawa, and A. J. Epstein, *Synth. Met. in press.*
23 M.E. Jozefowicz, R. Laversanne, H.H.S. Javadi, A.J. Epstein, J.P. Pouget, X. Tang, and A.G. MacDiarmid, *Phys. Rev.*, **B 39** (1989) p. 12958.
24 J. Tsukamoto, *Adv. in Phys.*, **41** (1992) p. 509.
25 R.S. Kohlman, A. Zibold, D.B. Tanner, G.G. Ihas, Y.G. Min, A.G. MacDiarmid, and A.J. Epstein, *Phys. Rev. Lett.*, **78** (1997) 3915.
26 T. Ishiguro, H. Kaneko, Y. Nogami, H. Nishiyama, J. Tsukamoto, A. Takahashi, M. Yamaura, and J. Sato, *Phys. Rev. Lett.*, **69** (1992) p. 660.
27 P.W. Anderson, *Phys. Rev.*, **109** (1958) p.1492.
28 N.F. Mott and E. Davis, in **Electronic Processes in Non-Crystalline Materials** (*Clarendon Press*, Oxford, 1979) p. 6.
29 Z.H. Wang, A. Ray, A.G. MacDiarmid, and A.J. Epstein, *Phys. Rev.*, **B 43** (1991) p. 4373; J. Joo, V.N. Prigodin, Y.G. Min, A.G. MacDiarmid, and A.J. Epstein, *Phys. Rev.*, **B 50** (1994) p. 12226.
30 R.S. Kohlman, J.Joo, Y.Z. Wang, J.P. Pouget, H. Kaneko, T. Ishiguro, and A.J. Epstein, *Phys. Rev. Lett.*, **74** (1995) p. 773; R.S. Kohlman, J. Joo, Y.G. Min, A.G. MacDiarmid, and A.J. Epstein, *Phys. Rev. Lett.*, **77** (1996) p. 2766.
31 T. Ishiguro, H. Kaneko, Y. Nogami, H. Nishiyama, J. Tsukamoto, A. Takahashi, M. Yamaura, and J. Sato, *Phys. Rev. Lett.*, **69** (1992) p. 660; H. Kaneko, T. Ishiguro, J. Tsukamoto, and A. Takahashi, *Solid State Commun.*, **90** (1994) p. 83.
32 V.N. Prigodin and A.J. Epstein, *submitted.*
33 J. Tanaka, C. Tanaka, T. Miyamae, M. Shimizu, S. Hasegawa, K. Kamiya, and K. Seki, *Synth. Met.*, **65** (1994) p. 173.
34 N.F. Colaneri and L.W. Shacklette, *IEEE Trans. Instrum Meas.*, **IM-41** (1992) p. 291; T. Taka, *Synth. Met.*, **41-43** (1991) p. 1177; J. Joo and A.J. Epstein, *Appl. Phys. Lett.*, **65** (1994) p. 2278.
35 A.J. Epstein, J. Joo, C.-Y. Wu, A. Benatar, C.F. Faisst, Jr., J. Zegarski, and A.G. MacDiarmid, in **Intrinsically Conducting Polymers: An Emerging Technology**, M. Aldissi, ed. (*Kluwer Academic Pubs.*, Netherlands, 1993) p. 165.
36 B. Wessling, *Adv. Mat.*, **6**, 226 (1994).
37 S. Jasty and A.J. Epstein, *Polym. Mater. Sci. Eng.,* **72** (1995) p. 565.
38 J.C. Scott, S.A. Carter, S. Karg, and M. Angelopoulos, *Synth. Met.*, **86** (1996) p.1197.
39 Y. Yang, E. Westerweele, C. Zhang, P. Smith, and A.J. Heeger, *J. Appl. Phys.*, **77** (1995) p. 694.

Polyaniline as Viewed from a Structural Perspective

M. J. Winokur
University of Wisconsin-Madison
B. R. Mattes
Los Alamos National Laboratory

INTRODUCTION

Polyaniline and its related derivatives form a diverse family of conducting polymers in terms of both their electronic and structural characteristics.[3] These properties are intimately coupled to the subtle molecular level intrachain and interchain interactions. Amorphous polyaniline (a-PANI) exhibits relatively poor interchain electron transport behavior thus rending these materials quasi-one-dimensional.[4] Crystalline PANI (c-PANI) salts can, depending on the primary dopant,[5] the presence of water and/or secondary dopants[6] and the sample processing history,[7] display a vastly improved interchain ordering which results in an increase in the overall conductivity. This leads to a three-dimensional band picture and the development of a metallic-like state.[8] Understanding the central structure/property issues underlying this broad diversity is a formidable goal and requires a detailed knowledge of both the nascent polymer structure and its subsequent evolution.

The nominal main chain architecture which best describes PANI is given by

$$[(C_6H_4\text{-}NH\text{-}C_6H_4\text{-}NH\text{-})_{1-x}(C_6H_4\text{-}N{=}C_6H_4{=}N\text{-})_x]_n$$

PANI may be prepared in a range of oxidation states[9] varying from the fully oxidized (x = 1) pernigraniline form to the fully reduced (x = 0) leucoemeraldine form.

Emeraldine base, at x = 0.5, is essentially an insulator and contains a four monomer chain repeat comprised of three benzenoid rings and one quinoid ring as shown in Figure 1. These polymers can also be prepared in the conducting salt form, either during synthesis or after reaction with an appropriate acid HX, to yield

$$[(C_6H_4 - NH - C_6H_4 - NH^+-)(X^-)]_n$$

with $y = X^-/N$ where X^- is the counter-ion and $y = 0.5$ is the maximum dopant concentration found in the most heavily doped form. In the best circumstances PANI-ES's conductivity can increase (referenced to EB) by at least 14 orders of magnitude to well over 1000 1/Ω-cm. Either EB or ES can be effectively processed[10] using conventional polymer processing techniques to give useful articles such as fibers, films, and coatings that are also electronically conducting. Examples of potential applications for such PANI based articles are summarized in ref. [11]

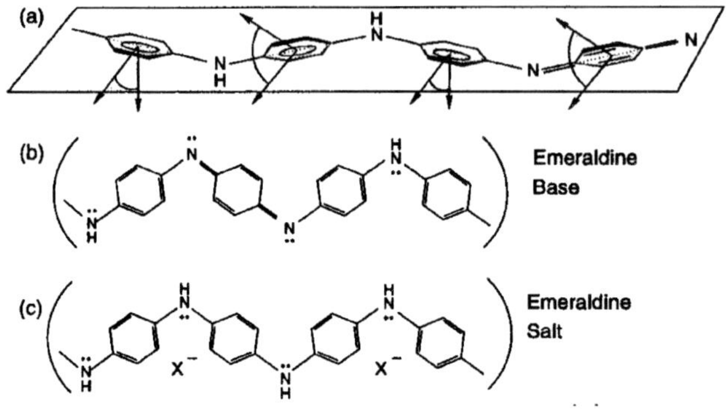

Figure 1. Schematic drawings for PANI. (a) showing the ring torsions referenced to the average molecular plane and both (b) base and (c) salt forms.

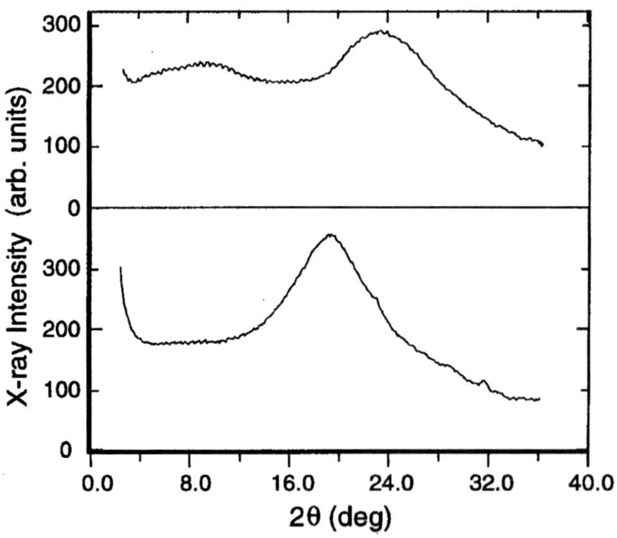

Figure 2. Typical powder diffraction spectra from fully amorphous PANI EB (bottom) and HCl-doped ES (top) in the class-II form.

The presence of modest flexibility about the alternating amine/imine linkages and the disparate torsional response of the two fundamental main chain ring units appears to frustrate the overall interchain packing. These factors, in combination with the possibility of interchain N···HN hydrogen bonding and the coinsertion of H_2O, yield a parent EB or ES polymer which can be prepared in either a partially ordered state having modest crystallinity[2,12] or in a fully amorphous form[13,14] as shown in Figure 2. There is a well-documented polymorphism in which at least two distinctly different structural forms can be recognized. Class I ES-I and EB-I compounds are typically obtained from HCl-doped PANI-ES solutions

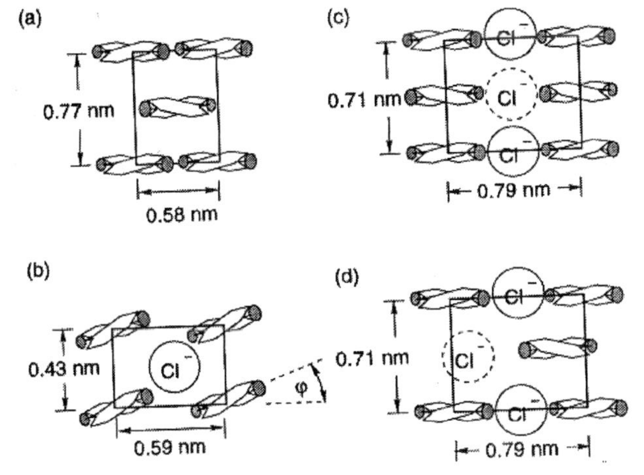

Figure 3. Proposed equatorial packings (perpendicular to the chain axis) for crystalline PANI films and powders according to ref.2 in (a) EB-II, (b) ES-I, (c) ES-II, Pc2a and (d) ES-II, $P2_122_1$.

which have been precipitated while class II structures are associated with films formed by casting of EB from organic solvents such as NMP. The proposed crystalline chain packings by Pouget and coworkers[2] for these structures are reviewed in Figure 3.

By extending these previous studies to encompass the structural evolution of c-PANI EB-I powders on doping by aqueous HF solutions and c-PANI HCl-doped PANI powder after dehydration and rehydration we are able to further clarify the nature of structural ordering within the class I family of compounds. The nominal ES-I structure of Figure 3 must be strongly modified to incorporate the existence of local chain rotations and lateral displacements which lower the overall symmetry of the unit cell and effectively generate one-dimensional channels, parallel to the polymer chain axis, which enhance the uptake of water into crystalline regions. Moreover these studies indicate a qualitatively different structural evolution during HF-doping to yield a more extensive family of class I structures.

EXPERIMENTAL DETAILS

All PANI powder samples were prepared by oxidative polymerization of aniline in an aqueous HCl solution according to the methods of ref.[14] and this synthesis yields class-I ES with a Cl^- concentration, y, approaching 0.5. Emeraldine base (EB-I) was obtained by immersion of this HCl ES-I in an excess of 0.1 M NH_4OH for a minimum of 3 h. Individual portions of this EB-I were thereafter immersed in various aqueous HF solutions with HF concentrations ranging from 25 mM to 995 mM and then dried under dynamic vacuum. These class-I ES powders were then transferred in air to glass or mica walled x-ray capillaries for further diffraction studies. A quantity of the HCl-doped ES-I powder was also transferred into the first chamber of a special dual chambered *in situ* x-ray cell and then dried under dynamic vacuum at ca. 100°C for 6 h. After drying, a small quantity of degassed water was placed in the second chamber of the evacuated *in situ* cell. The amorphous PANI films (EB-II) were cast from NMP using the methods of ref.[14] One film was then treated with 1 M aqueous HCl solution to yield the respective a-PANI ES-II sample shown in Figure 2. All films and powders were typically handled in air. Therefore all PANI samples discussed hereafter are expected to contain appreciable quantities of absorbed water (especially the ES samples) with the sole exception of the dehydrated sample.

The XRD PANI powder studies used either one of two available general purpose x-ray diffractometers employing a 15 kW rotating anode generator fitted with a copper target (λ=1.542 Å). The specific details are available elsewhere.[15] In general individual 2θ scans for the two x-ray diffractometers required between 15 m and 12 h depending on the specific requirements of the experiment.

RESULTS AND CONCLUSIONS

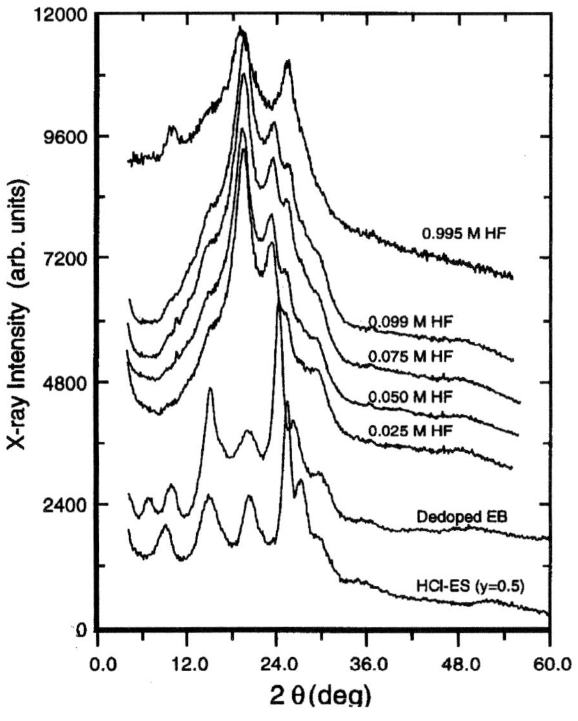

Figure 4. XRD spectra from a series of emeraldine class I powder samples after synthesis (to yield HCl-ES at bottom), treatment with 0.1 M NH_4OH (to give dedoped EB) and then exposure to increasingly stronger aqueous HF solutions. Note that all curves except the bottom one have been vertically offset for clarity.

Figure 4 displays typical powder diffraction spectra from the various indicated class-I PANI samples. The as-prepared HCl-doped ES profile (at the bottom) exhibits the largest proportion of sharp scattering features and is qualitatively similar to other XRD studies. This suggests a significant crystalline fraction. Unlike the EB-I results of Pouget *et al.*,[2] removal of HCl reduces the relative crystalline fraction of material only slightly. There are noticeable shifts in the positions and relative peak intensities indicative of the distinct changes in both the unit cell dimensions and the local structure. In addition, a pronounced new peak centered near $2\theta \approx 6.5^{\circ}$ is clearly resolved. The subsequent XRD spectra of the HF redoped samples display a number of marked charges. While the four profiles spanning the 25 mM to 99 mM HF ES-I samples are qualitatively similar to one another,[16] they are profoundly different from those of either the as-prepared HCl-doped ES sample or the dedoped EB powder. There are, however, at least two subtle systematic variations throughout this intermediate HF-doping sequence. The $2\theta = 30^\circ$ shoulder becomes much less pronounced while the $2\theta = 26^\circ$ shoulder is ultimately identified as a distinguishable peak. The 995 mM sample scan is clearly different from all the preceding curves and indicates that addition HF uptake (to give $y \approx 0.5$) is associated with a discrete change to another structural phase. In this case the scattering profile bears a strong resemblance to the H_2SO_4-doped PANI-ES results of Moon *et al.*[12] We note that throughout the entire processing sequence of samples in Figure 4 there appears to be a monotonic decrease in the proportion of scattering which can be attributed to crystalline regions of the films. Moreover these remaining peaks also appear to broaden somewhat. This general trend suggests that c-PANI is "fragile" and that the continued structural variations irreversibly lower the relative crystallinity.

There are other important scattering features that can be resolved. The HCl-ES scan of Figure 3 contains distinctive variations in the widths and shapes of the resolved peaks. Naively one expects that simple crystalline polymers tend to produce scattering peaks whose full-width at half-maximum are nearly independent of the crystal orientation and broaden only slightly with increasing 2θ. In this sample the two peaks located near $2\theta = 26^\circ$ and 28° are much narrower than any other resolved peaks including those at lower angles. While an anisotropic crystal habit may play a role in this result, a more likely possibility is that these other peaks are comprised of at least two or more superimposed scattering peaks from a low-symmetry unit cell. Hence a simple d-spacing identification is somewhat deceptive although we include this in Table 1.

Table 1. Summary of observed d-spacings

Sample	2θ, °	d, nm	Sample	2θ, °	d, nm
HCl-ES	8.9	0.99	Dedoped EB	6.5	1.40
	15.0	0.59		9.8	0.90
	20.4	0.44		15.1	0.59
	25.4	0.35		20.0	0.44
	27.7	0.32		24.3	0.37
	30.5	0.29		26.4	0.34
HF-ES (99 mM)	9.4	0.94		30.0	0.30
	14.7	0.60		36.5	0.25
	19.4	0.46	HF-ES (995 mM)	10.0	0.88
	23.4	0.38		14.6	0.61
	25.5	0.35		19.0	0.47
	29.6	0.30		25.5	0.35

Before introducing possible structural models for the aforementioned results it is first necessary to demonstrate the significance of incorporating water uptake[17] into any comprehensive discussion of the unit cell structure.[18] Hence the results of the *in situ* scattering experiment during water uptake in a dehydrated

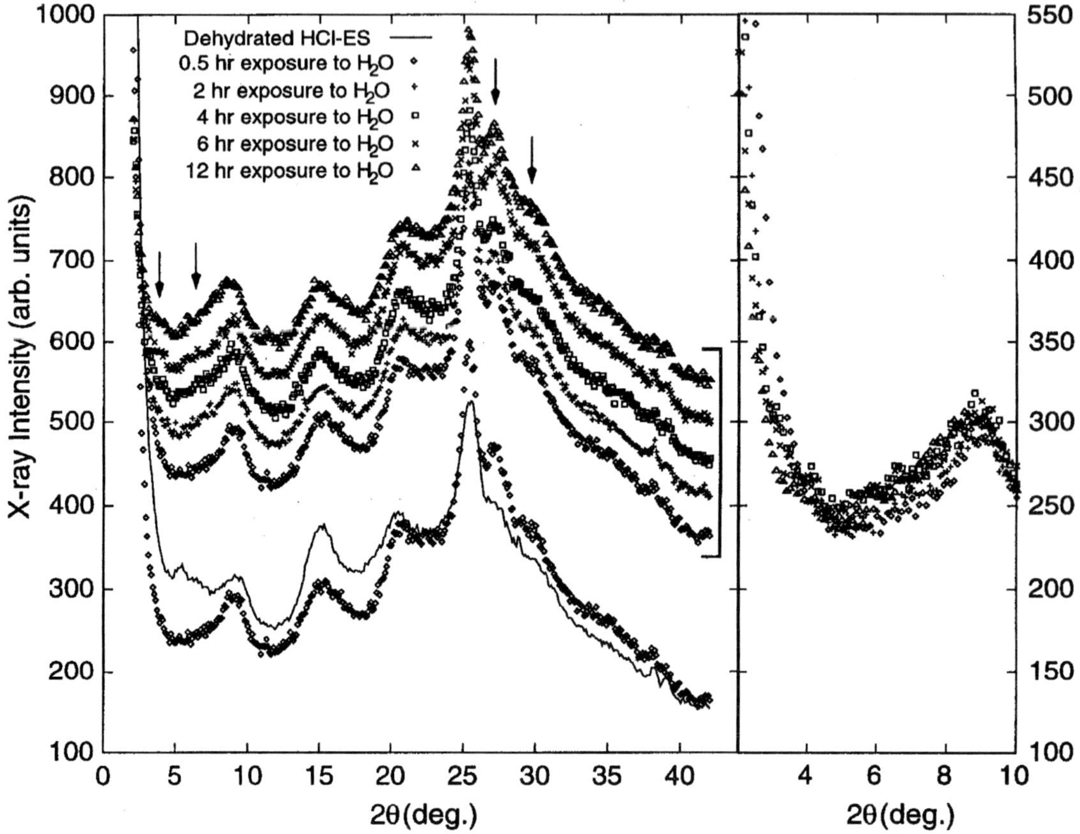

Figure 5. XRD spectra recorded *in situ*, from a dried HCl-ES (class I) sample, during continuous exposure to water vapor. The left panel is arranged so that only the upper five bracketed curves have been vertically offset. The right panel shows the low angle 2θ behavior in greater detail without offsets.

HCl-doped ES-I samples are shown in Figure 5. The bottom two spectra of the left side panel show in direct relief a comparison of the dehydrated powder and the same powder after exposure to water vapor for just 30 m. While the specific crystalline "peak" positions remain relatively unchanged, the dehydrated sample data is significantly different in a variety of important ways. Much of the scattering intensity shifts to lower angle and the relative proportion of scattering by crystalline regions of the power is sharply diminished. Moreover the relative peak intensity ratios are seen to shift strongly. There is an exceptionally large increase in the scattering intensity of the dehydrated sample at the lowest accessible 2θ regions. There are also noticeable changes between the two HCl-ES profiles representing a sample before (Figure 4) and after dehydration. In particular the rehydrated sample exhibits a profile shape closer to those reported elsewhere and it contains a measurable decrease in the relative fraction of crystalline material.

These features suggest that water has a profound impact on the crystal structure, the relative crystalline/amorphous proportions and the overall structural homogeneity. Removal of water from HCl-doped ES seems to produce three main effects: The unit cell packing becomes altered without any dramatic changes in the major d-spacings, the level of local disorder within the unit cell is significantly increased and, finally, the degree of structural inhomogeneity at larger scales is also increased. Since the small-angle scattering results of Annis *et al.*[19] identify changes primarily along the meridional direction, the increases in intensity of the scattering background seen in this work are also expected to occur likewise (along the meridional direction) and are associated with ordering by both water and halogen counter-ions within any identifiable channels. Closer inspection of the *in situ* profiles, obtained at selected times after constant exposure to water vapor, shows continued evolution of the HCl-doped ES structure. In addition to the rapid recovery of the original ES-I unit cell structure, albeit with a loss in crystallinity, the low-angle scattering in the $2\theta = 2^{\circ}$ to 4° region smoothly decreases over time. This suggests a gradual return to a more uniform water/halogen-ion ordering along the c-axis. Coupled with this are gradual increases in the scattering intensity in the vicinity of 4°, 7°, 27° at 30° (and denoted by arrows in Figure 5). Hence there are slow changes in the unit cell structure itself.

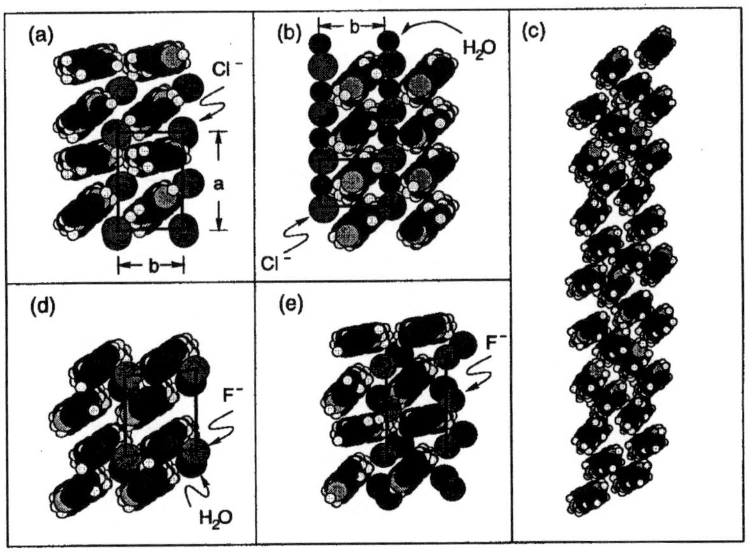

Figure 6. Various proposed new structural models for the studied emeraldine class I powders appropriate in (a) dehydrated HCl-ES, (b) HCl-ES containing two water molecules per nitrogen, (c) dedoped EB, (d) redoped HF-ES [from 25 mM to 99 mM HF aqueous solution treated powders] and (e) fully redoped HF-ES [using a 995 mM HF aqueous solution].

While comprehensive modeling studies are currently underway, it is still possible to provide preliminary structural models which reproduce many of the aforementioned scattering features. These are displayed in sequential order in Figure 6. All of the doped structures have some characteristics in common with the nominal model proposed by Pouget [shown in Figure 3(b)] but there are notable differences. The model for dehydrated HCl-ES requires that the PANI chain axis rotation alternates along the a-axis. This doubles the effectively equatorial unit cell dimensional area and creates two different PANI interchain nearest neighbor spacings along the b-axis direction. The larger of these two may serve to facilitate water diffusion upon reexposure to water vapor. In panel 6(b) the rehydrated structure is displayed with two H_2O molecules per N-atom. In this structure all PANI chains now have equivalent chain rotations thus halving the a-axis repeat. To accommodate the pronounced water uptakc the PANI chain axis rotation is large (relative to the a-axis) and the Cl^- ions are laterally displaced from the high-symmetry position of Figure 3(b). Modeling the dedoped EB sample requires a large, disordered unit cell but the overall ES-I PANI chain packing remains. Finally on HF-doping there is a sequential two-step process whereby only half the available F^- channels site are filled initially. The final HF-doped ES sample most resembles the dehydrated HCl-ES structure although the former requires water. In sum total this structural response is far richer than originally imagined.

ACKNOWLEDGMENTS

The financial support by NSF Grant No. DMR-9631575 (MJW) is gratefully acknowledged.

REFERENCES

1 M. E. Jozefowicz *et al.*, *Phys. Rev.*, **B 39**, 12958 (1989).
2 J. P. Pouget et al., *Macromolecules*, **24**, 779 (1991).
3 A. J. Epstein et al., *Synth. Met.*, **65**, 149 (1994).
4 Z. H. Wang et al., *Phys. Rev. Lett.*, **66**, 1745 (1991).
5 M. Reghu, Y. Cao, D. Moses, and A. J. Heeger, *Phys. Rev.*, **B 47**, 1758 (1993).
6 A. G. MacDiarmid and A. J. Epstein, *Synth. Met.*, **69**, 85 (1995).
7 Z. H. Wang, J. Joo, C.-H. Hsu, and A. J. Epstein, *Synth. Met.*, **68**, 207 (1995).
8 N. S. Sariciftci, A. J. Heeger, and Y. Cao, P*hys. Rev.*, **B 47**, 1758 (1994).
9 W. S. Huang, B. D. Humphrey, and A. G. Mac-Diarmid, *J. Chem. Soc., Faraday Trans.*, **82**, 2385 (1986).
10 A. Andreatta et al., in **Science and Applications of Conducting Polymers**, edited by W. R. Salaneck, D. T. Clark, and E. J. Samuelsen (*Adam Hilger*, Bristol, 1991), p. 105.
11 A. G. MacDiarmid and A. J. Epstein, **Science and Applications of Conducting Polymers** (*Adam Hilger*, Bristol, England, 1990), p. 141.
12 Y. B. Moon, Y. Cao, P. Smith, and A. J. Heeger, *Polymer*, **30**, 196 (1989)
13 M. Laridjani *et al.*, *Macromolecules*, **25**, 4106 (1992).
14 J. Maron, M. J. Winokur, and B. R. Mattes, *Macromolecules*, **28**, 4475 (1995).
15 T. J. Prosa *et al.*, *Phys. Rev.*, **B 51**, 150 (1995).
16 The absolute F^- concentrations were not ascertained.
17 M. Angelopoulos, A. Ray, A. G. MacDiarmid, and A. J. Epstein, *Synth. Met.*, **21**, 21 (1987).
18 B. Lubentsov *et al.*, *Synth. Met.*, **47**, 187 (1992).
19 B. K. Annis, J. S. Lin, E. M. Scherr, and A. G. MacDiarmid, *Macromolecules*, **25**, 429 (1989).

Processability of Electrically Conductive Polyaniline Due to Molecular Recognition

Terhi Vikki
Department of Technical Physics, Helsinki University of Technology, FIN-02150 Espoo, Finland
Olli Ikkala
Department of Technical Physics, Helsinki University of Technology, FIN-02150 Espoo, Finland and Neste Oy, P.O. Box 310, FIN-06101 Porvoo, Finland
Lars-Olof Pietilä
VTT Chemical Technology, P.O. Box 1401, FIN-02044, Finland
Heidi Österholm, Pentti Passiniemi, Jan-Erik Österholm
Neste Oy, P.O. Box 310, FIN-06101 Porvoo, Finland

INTRODUCTION

The electrically conductive emeraldine salt form of polyaniline[1] has long been regarded as an intractable material, i.e. infusible and poorly soluble, due to the aromatic structure, the interchain hydrogen bonding, and the charge delocalization effects. Emeraldine salts are known to dissolve only in certain amines, and hydrogen bonding solvents, in particular in strong acids. Melt and solution processability can be improved if PANI is protonated with specific bulky protonic acids. Well-known examples of such acids are *p*-dodecyl benzene sulphonic acid (DBSA),[2] camphor-10-sulphonic acid (CSA)[2] and methyl benzene sulphonic acid (TSA).

PANI(DBSA)$_{0.5}$-complex is soluble in excess DBSA,[3] probably because its highly acidic -SO_3H-groups are able to make a particularly strong hydrogen bonding to the aminic sites of PANI. Less acidic compounds lead to lower solubility due to smaller strength of hydrogen bonding. For example, aliphatic alcohols, long chain aliphatic carboxylic acids, phthalates and most other carboxylic acid esters and ketones are not solvents for electrically conductive PANI. However, in spite of their low acidity, phenols are good solvents for emeraldine salt, if the protonation has been made using CSA.[2,4]

The above considerations show that strong specific interaction between the emeraldine salt and an organic compound is important to achieve high solubility. Here we point out a novel concept to achieve high solubility of emeraldine salt where increased specific interaction to the solvent is obtained by sterically matching several small interactions[5,6] i.e., molecular recognition.[7] Examples of solvents fulfilling these conditions are dihydroxy benzenes and phenyl phenols. In this work solubility of PANI(DBSA)$_{0.5}$ in resorcinol i.e., 1,3-dihydroxy benzene is studied. We also show that PANI(CSA)$_{0.5}$/*m*-cresol is a limiting case of the concept.[5]

EXPERIMENTAL METHODS

PANI(DBSA)$_{0.5}$-complex was prepared by conventional methods.[4] PANI(DBSA)$_{0.5}$ and resorcinol were dried and mixed using a 3 g miniature mixer at constant temperature in N_2 atmosphere for 10 minutes. The mixing temperatures were 160, 180, 200, 220 and 240°C, and the weight fraction of resorcinol was 100, 90, 80, 70 and 60 wt%. FTIR was used to verify that no chemical reactions or major thermal degradation had occurred.

Optical microscopy in combination with a hot stage was used to study the solubility of PANI(DBSA)$_{0.5}$ in resorcinol. A small amount of mixture was inserted between two microscope glass slides and kept for two minutes at the temperature were the mixing had taken place. The morphology of the mixture was simultaneously inspected with a microscope. If a distinct “two-phase” structure containing dispersed PANI particles in a solvent rich medium was observed, it was concluded that PANI(DBSA)$_{0.5}$ was not dissolved in resorcinol. On the contrary, a green transparent “one-phase” morphology without a dispersed phase suggests solubility. Note, however, that based on optical microscope alone, one cannot unambiguously conclude whether a true solution or colloidal dispersion is obtained.

DSC measurements were conducted with a Perkin Elmer DSC 7 equipment at a heating rate 10°C/min.

COMPUTATIONAL METHODS

In order to model PANI(DBSA)$_{0.5}$/resorcinol systems, the long alkyl tail of DBSA was excluded, as it was not expected to qualitatively effect bonding. Therefore, the binding of resorcinol molecules to sulphonic acid doped PANI-complex was studied using TSA as the counter-ion. UHF/AM1 optimized structure of PANI chain consisting of three rings and doped with two TSA molecules was studied. Eight resorcinol molecules were added to the system and 200000 steps (time step 1 fs) of molecular dynamics were performed at 300 K. The resulting structure was saved after each 1000 steps and the 200 structures were optimized. The Insight/Discover software with the pcff force field by Biosym Technologies was used in these calculations.

Conformations of CSA-protonated PANI chains and the PANI(CSA)$_{0.5}$/*m*-cresol system were modeled using the semiempirical quantum chemical method AM1 implemented to the MOPAC software package. The models were limited to PANI compounds consisting of three rings and checked with eight rings.

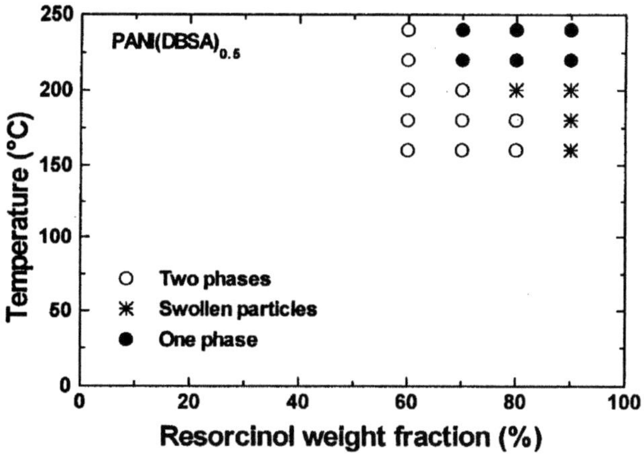

Figure 1. Dissolution phase diagram of PANI(DBSA)$_{0.5}$ and resorcinol mixtures.

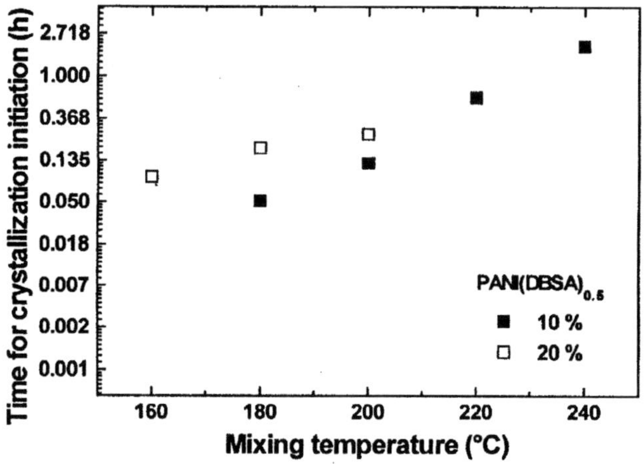

Figure 2. Induction time for resorcinol crystallization as a function of the mixing temperature.

RESULTS AND DISCUSSION

Solubility of PANI(DBSA)$_{0.5}$ in resorcinol depends both on temperature and PANI-complex weight fraction. Figure 1 depicts the morphologies of PANI(DBSA)$_{0.5}$/resorcinol mixtures at elevated temperatures by optical microscopy. High temperatures and low PANI-complex weight fractions promote dissolution, manifested as a one-phase morphology. PANI(DBSA)$_{0.5}$ can be dissolved in resorcinol up to 40 wt% at temperatures below 240°C. This behavior suggests one branch of phase boundary corresponding to the upper critical solution behavior with a high critical temperature.

The same morphologies as in Figure 1 are observed also at room temperature immediately after rapid cooling. No crystallinity is observed in PANI(DBSA)$_{0.5}$/resorcinol mixtures. However, after an induction period spherulitic crystals start to emerge, see Figure 2. This is in contrast to pure resorcinol which crystallizes immediately after cooling to room temperature. Long induction time is observed for samples with high mixing temperature, i.e., for samples that have been well

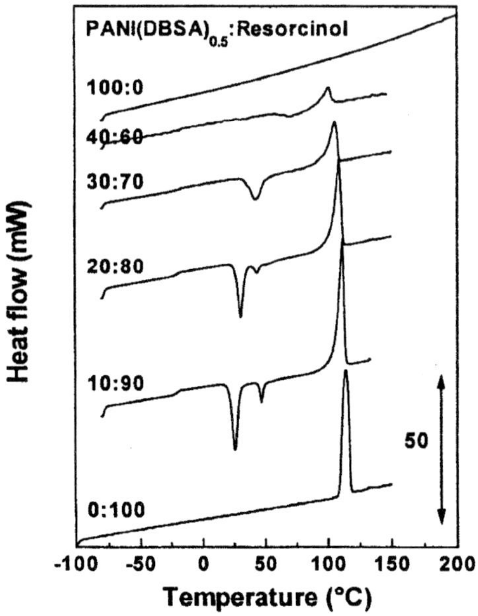

Figure 3. DSC traces of PANI(DBSA)$_{0.5}$/resorcinol samples mixed at 200°C.

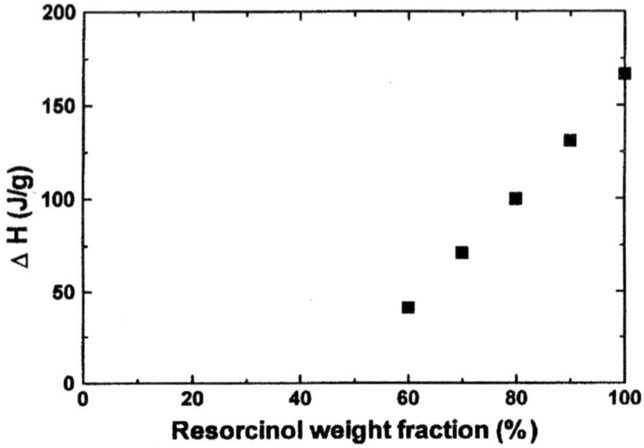

Figure 4. Resorcinol heat of fusion in PANI(DBSA)$_{0.5}$/resorcinol samples mixed at 200°C.

dissolved according to Figure 1. This observation suggests that the dissolved PANI(DBSA)$_{0.5}$ molecules delay the crystallization of resorcinol. A similarly slow development of crystallinity was also observed for mixtures of PANI(CSA)$_{0.5}$ and resorcinol by WAXS in a related study.[6]

The DSC traces for the second heating of the samples mixed at 200°C are shown in Figure 3. The mixtures were aged a few weeks at room temperature before measurement. By comparing different aging times, it was concluded that resorcinol was fully crystallized. Melting point depression of resorcinol is observed suggesting interaction between the components (Figure 3). Pure resorcinol crystallizes at about 115°C and the melting point is depressed to 98°C as 40 wt% PANI(DBSA)$_{0.5}$ is mixed with resorcinol at 200°C. Also the heat of fusion shows interaction between the components (Figure 4). The heat of fusion determined from the first heating thermogram depends linearly on the weight fraction of resorcinol. It vanishes for mixtures with less than 2.8 moles of resorcinol associated per PhN repeat unit of PANI. This suggests that only part of resorcinol is able to crystallize as the rest is strongly associated with PANI(DBSA)$_{0.5}$.

The association of 8 resorcinol molecules to the system comprising three PANI repeat units doped by two TSA molecules is shown in Figure 5, i.e., there are 2.7 moles of resorcinol vs. 1 mol of PhN repeat unit of PANI. The first 4 resorcinol molecules form strong hydrogen bonds directly to the two sulfonate groups of TSA. The strong dipole moment of the sulfonate

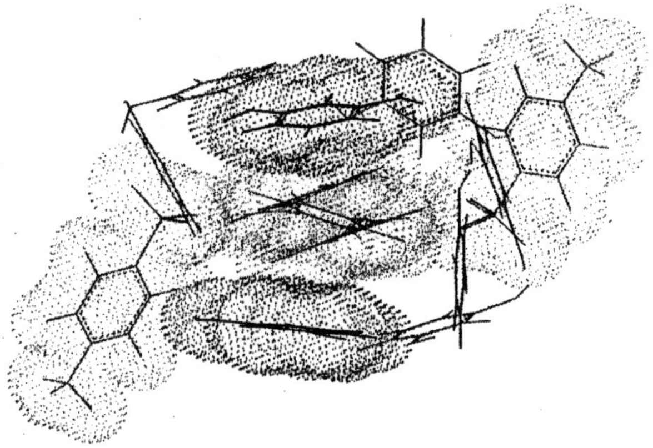

Figure 5. Association of 8 resorcinol molecules with PANI protonated by TSA.

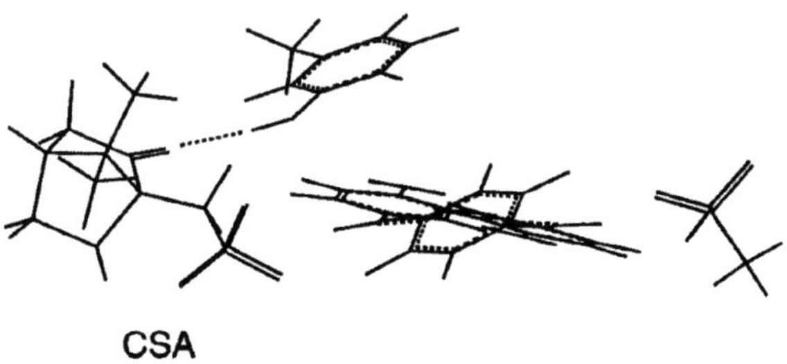

Figure 6. Association of *m*-cresol molecules with PANI protonated by CSA.

groups is able to orientate these "first-layer" resorcinol molecules due to the hydrogen bonding OH-groups. The "first-layer" resorcinol molecules effectively shield the sulfonate groups. The nature of the available hydrogen bonding to additional resorcinol molecules is therefore changed, and the additional 4 resorcinol molecules are bonded both by two hydrogen bonds and one phenyl/phenyl interaction on top of the PANI rings. There are several specific reasons that allow the phenyl/phenyl stacking of the "second-layer" molecules. Firstly, the stacked structures are possible because the distance of the OH-groups of resorcinol matches the corresponding distances of the hydrogen bonding moieties of the PANI(DBSA)$_{0.5}$, thus allowing steric fit of two hydrogen bonds and one phenyl/phenyl interaction, i.e., molecular recognition. Secondly, resorcinol is a rigid structure, for which the thermal movements do not change the distances. Thirdly, the phenyl/phenyl interaction plays an important role, as further manifested by phenyl phenols and bisphenols which are examples of other solvents. In these cases also the periodicities of the phenyl rings within the solvents approximately match the periodicity of PANI chains, allowing steric fit of the successive phenyl rings in combination with the hydrogen bonds.

Finally it is shown that PANI(CSA)$_{0.5}$ dissolved in *m*-cresol is a limiting case of the above molecular recognition concept.[5] In this case there are three possible sites for the association of *m*-cresol molecules. First, there is the sulfonate anion of CSA, secondly the PANI amine group and finally the carbonyl group of CSA. The last bonding site is specific to CSA and does not exist in DBSA, for example. Figure 6 demonstrates the optimized structure showing >C=O··HO hydrogen bonding between CSA and *m*-cresol and the stacking of the *m*-cresol phenyl ring on top of the PANI phenyl ring. In this case the net interaction of *m*-cresol consists of one hydrogen bond and one phenyl/phenyl interaction, leading to a cyclically associated species. This observation is in agreement with the observed high solubility of $PANI(CSA)_{0.5}$ in *m*-cresol, while the solubility of $PANI(DBSA)_{0.5}$ in *m*-cresol remains poor.[4,5]

CONCLUSIONS

We suggest that molecular recognition can be systematically applied to identify a large class of novel low acidic solvents for PANI protonated by essentially any organic acid. In this concept the phenyl rings of PANI are considered as potential sites of phenyl/phenyl interaction with a periodicity of ca 6 Å. At the same periodicity there are also hydrogen bonding sites, consisting of amines and sulfonates due to protonating sulfonic acids. The first requirement for low acidic solvents is that the solvent has to comprise phenyl rings and sufficiently strong hydrogen bonding functional groups at the same periodicity. Secondly, for PANI protonated by generic sulfonic acid such as DBSA, TSA, or methane sulfonic acid an additional requirement is that at least one hydrogen bond and at least one phenyl/phenyl interaction is made, the total number of such interactions being ≥ 3. Suitable compounds are dihydroxy benzenes, phenyl phenols, bisphenols, hydroxy benzoic acids. In the special case where the counter ion itself allows a suitable hydrogen bonding, such as CSA, the critical number of the interactions is reduced to 2. An example of this case is $PANI(CSA)_{0.5}$ dissolved in *m*-cresol.

In order to demonstrate the feasibility of the concept, dissolution of $PANI(DBSA)_{0.5}$ in resorcinol is illustrated in more detail.

REFERENCES

1 J.-C. Chiang, A.G. MacDiarmid, *Synth. Met.*, 1986, 13, 193.
2 Y. Cao, P. Smith, A.J. Heeger, *Synth. Met.*, 1992, 48, 91.
3 T. Kärnä, J. Laakso, E. Savolainen, K. Levon, **European Patent Application EP 0 545 729** A1, 1993.
4 Y. Cao, J. Qiu, P. Smith, *Synth. Met.*, 1995, 69, 187
5 O.T. Ikkala, L.-O. Pietilä, L. Ahjopalo, H. Österholm, P.J. Passiniemi, *J. Chem. Phys., in press.*
6 T. Vikki, L.-O. Pietilä, H. Österholm, L. Ahjopalo, A. Takala, A. Toivo, K. Levon, P. Passiniemi, and O. Ikkala, submitted.
7 For a review, see Rebek, J. Jr., **Topics in Current Chem.**, 1988, 149, 189.

Crystallinity and Stretch Orientation in Polyaniline Camphor-Sulphonic Acid Films

L. Abell, P. Devasagayam, P. N. Adams and A. P. Monkman
Department of Physics, University of Durham, England

BACKGROUND

Conceptually, being able to replace metals with conductive polymers is a very attractive proposition. Practically, the last decade has shown that although there is a great deal of promise, we still have a way to go before commercial products are realized. To this end polyaniline (PANI) has been shown to be the most promising material to fulfill such applications[1] being air stable, cheap to produce in large scale and most importantly processible to some degree.[2] One problem with this polymer has always been how to efficiently dope it once it has been processed. This problem tends to rule out using 'base' polymer processing, i.e. solution processing unprotonated or base PANI in N-methyl-2-pyrrolidone.[3] as such material requires post process acidification to render the polymer conductive, which is difficult to achieve homogeneously in a dense film.[4] This problem has, however, been circumvented by the discovery of acid solution processing routes. These were first described by Cao *et al.*[5] using two specific functional acids, camphor-sulphonic acid (CSA) and dodecyl benzene sulphonic acid (DBSA) in various organic solvents.

One such system has proven to be of considerable interest however. This is PANI:CSA cast from *m*-cresol solution. Using this particular system enables films to be produced which show metallic transport, i.e. an increase in conductivity with decreasing temperature above some critical temperature below which the conductivity drops again with lowering temperature.[6] Intriguingly, it seems that only this particular system yields such well defined transport properties. To try to find out why this is, we have performed a range of measurements on PANI:CSA films aimed at probing the role which CSA and *m*-cresol play in both microstructure and electrical transport and to understand why it is that only this combination gives such desirable physical properties. To do this we have employed x-ray crystallography along with various film processing conditions, film orientation *via* application of uniaxial stress and temperature dependent transport analysis to monitor how both microstructure and transport are controlled by CSA and *m*-cresol content along with film processing conditions.

EXPERIMENTAL SECTION

High molecular weight polyaniline was prepared at -25°C using the standard Durham procedure.[7] Gel permeation chromatography in N-methyl-2-pyrrolidone solvent (+ 0.1% LiCl) with polyvinylpyridine molecular weight standards[8] indicated that the polymer had a weight average molecular weight, M_w, of 174000 and number average molecular weight, M_n, of 21000 Daltons. The polydispersity was therefore 8.3.

To obtain stretch oriented CSA doped polyaniline films, good quality isotropic feedstock with a thickness variation of less than 5% is imperative. This was achieved by solvent casting in the usual way, i.e. a 50% CSA doped, 1.6% PANI in *m*-cresol polymer solution was poured onto a polished silicon wafer and dried at 60°C under a dynamic vacuum for 20 hours. A quantity of chlorobenzene was added to the solution to prevent gelation. This process gives a sample thickness variation of less than 5%. Dumbbell shaped specimens with a gauge aspect ratio of 6 were then guillotined from the feed stock. Differential scanning calorimetry studies have shown that the glass transition temperature of PANI:CSA (*m*-cresol) is approximately 145°C,[9] thus the deformation temperature was chosen to be 150°C. Attempts to draw samples at lower temperatures have all failed at low strains (< 25% extension).

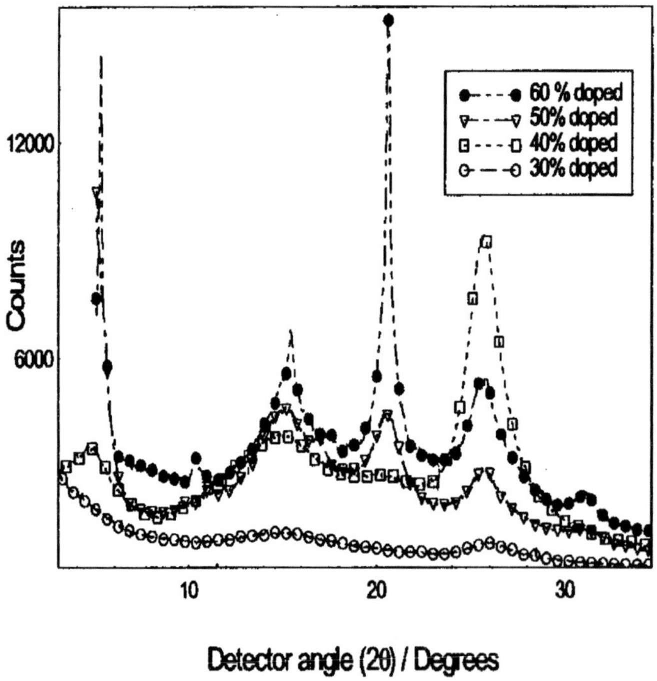

Figure 1. 2θ scans of PANI:CSA with various doping levels.

The specimens were deformed using an Instron 4505 tensile testing machine which had been preheated to 150° C for an hour. Fiducial markers were drawn onto the surface of each specimen at 3 mm intervals so that the amount of deformation parallel to the draw direction could be determined. The specimens where then placed, in turn, into the oven where they were left until the oven reached thermal equilibrium. Thermogravimetric analysis of a typical sample shows that (5±2)% of the sample's weight is lost during this period due to loss of *m*-cresol and chlorobenzene. As 15-20% of the as cast sample's weight is due to the retention of *m*-cresol and chlorobenzene, sample 'hardening' due to removal of the plasticisers during the thermal emersion before stretching should not be a significant problem. The specimens were then deformed uniaxially at a rate of 5mm/min, a strain rate of 0.2 min^{-1}.

To probe the transition to the "metallic" conductive state upon protonation with CSA, the temperature dependence of the electrical conductivity of films was measured as a function of doping level. Films were prepared with CSA contents intended to yield 30, 40, 50 and 60% protonation levels. Conductivity was measured under vacuum ($< 10^{-3}$ Torr) over the temperature range 10 to 300 K using a four in line constant current technique.[10] Samples were held for 5-6 hours in the vacuum environment of the cryostat prior to the measurement process to ensure that any volatiles were extracted. The conductivity of PANI:CSA after such treatment was typically 80 to 85% of that measured in normal atmospheric conditions. The 2θ WAXS diffraction patterns of the various films were collected using a Philips diffractometer and CuK_α radiation in reflection (see Figure 1), air scattering and line broadening due to slit effects were corrected for. A careful study of diffraction patterns obtained from several forms of CSA show that the features we observe in PANI:CSA are not due to CSA crystallites within the films. To measure the effect of solvent removal from films, samples were placed in high vacuum ($<10^{-6}$ torr using a turbo molecular vacuum pump) for 48 hours prior to examination.

Texture induced in the drawn samples was studied using a Huber 4 circle goniometer with CuK_α radiation. This technique relies on the fact that the measured intensity after corrections is proportional to the number of crystallographic plane normals or 'poles' which are parallel to the scattering direction, a reference direction which bisects the angle between the source and detector and lies in the plane formed by the source, detector and sample. Thus the orientation distribution of a chosen set of (hkl) crystallographic planes can be measured by fixing the detector at the appropriate 2θ angle so that it receives radiation scattered from these planes and then rotating the sample, the scan is parameterized by defining two spherical angles. The angle between the scattering direction and the sample plane normal is defined as α, and the angle between the projection of the scattering direction onto the sample and the symmetry axis of the sample perpendicular to the draw direction and in the plane of the film (the transverse direction) is defined as β.

The effect of varying the amount of CSA dopant on crystal structure can clearly be seen in Figure 1. We are currently in the process of fitting and refining this data to yield crystal structures, however trends present in the 2θ scans are very informative. The main features

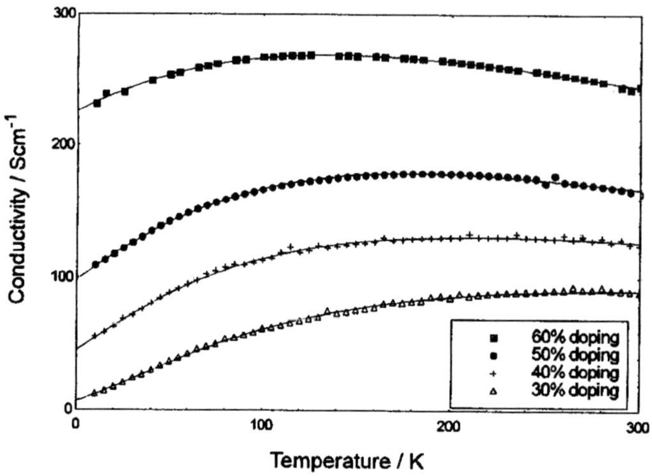

Figure 2. Conductivity data from PANI:CSA as a function of doping level.

which will be seen to be the most sensitive to *m*-cresol content are a very sharp peak at ca. 2°-4° corresponding to a d-spacing of 20A. This feature is strongly dependent on the doping level and is most pronounced at 60% CSA level corresponding to the most metallic samples that we have produced and scales directly with thc metallic conductivity contribution (see Figure 2). The half width of this 2 θ peak implies a coherence length within the film of many hundreds of Angstroms and so must be a feature associated with crystal regions, not amorphous regions as implied by others.[11]

RESULTS

Further features which are clearly dependent on the level of CSA in the sample are seen at ca. 6A, 4.5A and 3.5 A d-spacing. The widths of these peaks vary quite markedly. This is in fact due to anisotropy within the films even though they were produced simply by casting. This has been shown clearly by Minto and the group at the University of Reading.[12] In the reflection geometry that we have used to obtain our spectra a peak at 9.2A is not very visible whereas others observe such a feature for example.[12] Conductivity data for films at all doping levels are presented in Figure 2. It is clear that as the level of protonation is increased, the conductivity increases rapidly, also the conductivity becomes a weaker function of temperature in the more heavily protonated films. Each data set in the 30% to 60% doping range possesses a characteristic maximum in σ(T) somewhere in the measured range. The temperature at which this peak occurs has an inverse relation to the level of doping of the sample in question. The trend revealed by samples prepared with protonation levels in the range between 30% and 60% are consistent. At low temperature, the conductivity displays an activated behavior. As the level of protonation is increased from 30% the decrease in conductivity at the lowest temperatures is much reduced, compared to each films room temperature value. At the theoretical 50% maximum doping point[13] the peak in conductivity is observed at approximately 180 K. However, at 60% doping the magnitude of the conductivity is increased significantly, this is consistent with the x-ray data, the peak value of conductivity also occurs at a temperature significantly lower than that of 50% doped samples. The reduction in conductivity at temperatures below the peak is also less pronounced. This evidence implies that doping above the 50% level is capable of driving the PANI:CSA system far closer to true metallic behavior which is mirrored by the increase in the enhanced crystallinity of the 60% films.

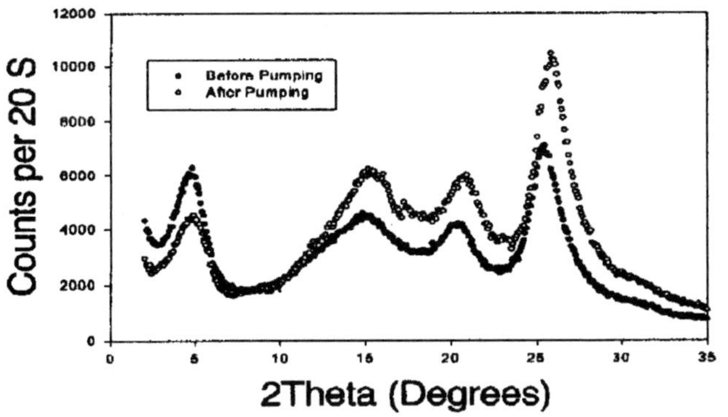

Figure 3. WAXS scans of pumped and unpumped CSA doped polyaniline.

To probe where the *m*-cresol may be residing in the system a nominally 50% doped film was taken and after recording a 2θ scan, subjected to 48 hours in high vacuum. After this treatment another 2 θ was recorded. The before and after scans are shown in Figure 3. Removal of *m*-cresol is seen to have a marked effect on the crystallinity of the sample. Peak positions were accurately determined with respect to Al peaks emanating from the sample holder (not shown in the figure for clarity). After pumping all peaks are seen to shift to higher 2θ angles indicating decreasing d-spacing. The relative intensities of the features also changes markedly. The 20A feature losses intensity whereas all other features gain intensity with some small line shape changes and line broadening also being noted. Such pumped films also suffer loss of conductivity, and in most cases the turn over temperature moves to higher temperatures. Importantly, when we have measured 60% films in a liquid He cryostat with sample in exchange gas we find that the magnitude of low temperature conductivity increases and the turn over temperature is observed at 70K,[14] observations consistent with the pumping experiment. A comparison of the 2θ diffractometer scans of the isotropic and drawn samples (Figures 1 and 4), which have been corrected for thickness and any geometrical effects, shows that there is a decrease in scattering from <u>both</u> the amorphous phase and the equatorial crystalline reflection. at 2θ=25.5° (the poles of which lie in the a-b plane of the unit cell) of the drawn sample. Little change is observed in the crystalline peak at 2θ=20.75°, and no new diffraction peaks are observed within the range 2θ=2-100° due to the production of a new crystal structure during deformation.

DISCUSSION

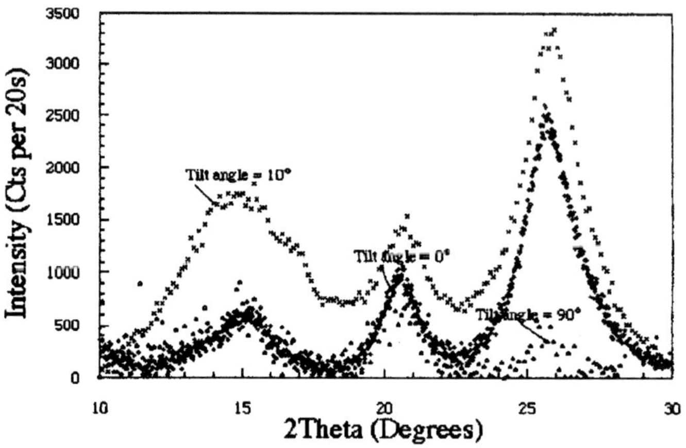

Figure 4. Comparison of 2θ scans from a drawn (67% elongation) PANI:CSA film.

From the transport data it is evident that PANI:CSA is an example of a conductor close to a metal-insulator (MI) transition. As the level of doping is increased from 30% to 60% the trends in conductivity imply that there is a transition in the nature of charge transport from hopping/tunneling due to localization of carriers on some scale, to partial 'metallic' diffusive transport. The exact mechanism by which this transition occurs must be related to the crystal structure of the sample. If we view PANI-CSA as a composite material and invoke the heterogeneous conductor model with two charge transport mechanisms at work, these being metallic diffusion within crystalline regions and temperature activated transport in the disordered regions. A detailed discussion of these processes is given elsewhere.[10] It is clear that the "metallic" conductivity is not only controlled by the doping level (which is controlling the crystal structure) but also by the content of *m*-cresol in the sample (again this is also controlling the crystallinity). From our x-ray analysis the optimum dopant level for maximum crystallinity is 60%, this agrees very well with the transport data. In both cases the degree of crystallinity scales well with the magnitude of electrical conductivity and the strength of the metallic signature. The role the solvent, *m*-cresol plays in all of this is obvious to see but at present it is difficult to interpret exactly what is going on. To fully elucidate the subtle structure property correlation's of this system requires extensive modeling and fitting of the wide range of experimental data now available. However from the trends presented here a few processes are apparent. On removal of *m*-cresol from a film the magnitude of its conductivity drops, as does the metallic contribution to the overall conductivity. Looking at the crystal structure of the two states we see that the 20A feature reduces in intensity, since this feature scales so well with metallic conductivity in the doping studies, it is consistent with the fact that in the pumped films the metallic nature is much reduced. The increase in the amorphous background on removal of the *m*-cresol indicates that the solvent is required within the crystalline phase in order to prevent some process which would otherwise disrupt the unit cell. We do not understand why the 3.5 A feature behaves as it does unless it is a signature of "effective" dedoping, or rather a shift in the CSA dissociation equilibrium such that less charge appears on the PANI chain. This would have many effects including changes to the backbone geometry and a reduction of coulombic repulsion between adjacent chains. On a molecular level we have to assume that the *m*-cresol is involved with the CSA to aid dissociation and thus drive protonation of the PANI chain, and it prevents the carbonyl group of the CSA from H-bonding to the PANI chain. What exactly occurs upon its removal is impossible to say with the data we have at present. However, as the CSA is a bulky counter ion, to achieve efficient close packing of chains, neighboring CSA counter ions have to interdigitate, removal of the *m*-cresol will allow the CSA to H-bond to the chains disrupting this efficient packing. This will cause shrinkage of the unit cell, consistent with the observed decreases in d-spacing, breaking of long range order and most importantly weakening of cofacial interchain ring overlap destroying metallic charge transport. On a macroscopic scale it is assumed that the polymer structure can be described by a fringed micelle model.[15] Removal of the *m*-cresol from this type of structure will very readily lead to loss of long range order and local shrinkage of the unit cell, consistent with what is observed upon pumping out the solvent. A great deal more work is required to fully understand this problem note that the x-ray peaks shown here are few and planes of electron density within the sample, many processes will lead to "peaks" being absent and the random alignment of crystallites will confuse relative peak intensities, only modeling and fitting can resolve these problems effectively. Once solved however we shall be in a position to start to be able to design truly metallic conjugated polymers.

The effects observed upon drawing are not simple. The behavior of the crystallites in particular are unusual if uniaxial orientation is occurring, with the chain axes orienting towards the draw direction. In this case one would expect orientation of the a-b crystallographic axes (which are themselves randomly oriented about the c crystallographic axis) into the plane formed by the transverse direction and normal to the plane of the film (normal-transverse plane). Some orientation and crystallization of the amorphous regions would also be expected. Thus for uniaxial orientation an increase in scattering from the crystalline reflections accompanied by a decrease in scattering from the amorphous phase is expected. Alternatively any decrease in crystallinity should be accompanied by an increase in scattering from the amorphous phase due to destruction of crystalline order during processing. A decrease in scattering from both phases, which a comparison of Figures 1 and 4 shows, suggests the development of preferred orientation away from $\alpha=0^\circ$ and crystallization\orientation of the amorphous phase. To check this, an α scan with the scattering direction in the normal-transverse plane ($\beta=0^\circ$), and a β scan with the scattering direction in the plane of the film, $\alpha=90^\circ$, of the $2\theta=25.54^\circ$ peak, the equatorial reflection showing the biggest effect, were performed.[15] This work showed that the poles of this reflection are orienting into the normal-transverse plane, which is consistent with the crystallographic c (or chain) axes orienting towards the draw direction, and the development of a preferred direction at $\alpha=10^\circ$. This is supported by the 2θ scans shown in Figure 4 which where taken at $\alpha=10^\circ$ and 90°. It can be seen from this that the intensity of scattering from the crystal phase increases and then decreases, which supports the idea of non-uniaxial orientation of the crystallites developing. The increased intensity at $\alpha=10^\circ$ corresponds to scattering from crystal planes which are oriented at 10° to the plane of the film. This we feel may represent a plane formed by the phenylene rings along the polymer backbone which are inclined at 10°.

SUMMARY

Our initial work on the PANI:CSA system indicates that by careful sample preparation very consistent results can be achieved. By varying the percent of CSA in the films, one starts to get some insight into the role played by the CSA. X-ray data reveals that sample crystallinity depends on the doping level. Transport measurements mirror this. We observe the onset of a metallic conductivity component at ca. 30% doping. This becomes more dominant on increased doping level to 60% doping. At this CSA doping level the samples remain "metallic" to 135 K. In agreement with the development of maximum sample crystallinity. Thus we observe a good correlation between crystallinity and metallic conductivity. On removal of *m*-cresol we observe changes in the crystalline phase, especially the loss of the 20 A feature. Such pumped samples are much less metallic and have lower overall conductivity. Stretch orientation further enhances the physical properties of the films, giving room temperature conductivities up to ca. 960 S cm^{-1} at 120% elongation.

ACKNOWLEDGMENTS

This work is sponsored by BICC Cables and EPSRC via a ROPA award (GRK 35433). We wish to thank all our collaborators for their support.

REFERENCES

1 See for example, **Science and Applications of Conducting Polymers**, ed. W. R. Salaneck, D. T. Clark and E. J. Samuelsen, *Adam Hilger, IOP Publishing Ltd.*
2 A. P. Monkman and P. N. Adams, *Solid State Comm.*, **78** (1991) 29.
3 A. P. Monkman and P. N. Adams, *Synth. Met.*, **41/2** (1991) 891.
4 A. P. Monkman and P. N. Adams, *Synth. Met.*, **41/2** (1991) 627.
5 Y. Cao, P. Smith and A. J. Heeger, *Synth. Met.*, **48** (1992) 91.
6 M. Reghu, Y. Cao, D. Moses and A. J. Heeger, *Phys. Rev. B*, **47** (1993) 1758.
7 P. N. Adams, P. J. Laughlin, A. P. Monkman and A. Kenwright, *Polymer*, **37**(15) (1996) 3411.
8 P. N. Adams, D. C. Apperley and A. P. Monkman, *Polymer*, **34**(2) (1993) 328-332.
9 L. Abell and A. P. Monkman, *Synth. Met., Accepted for publication.*
10 E. Holland *et al, J. Phys.Cond. Mat.*, (1996) 2991.
11 O. T. Ikkala *et. al, J. Chem. Phys.*, **103** (1995) 4855.

12 C. Minto and A.S. Vaghan, *Polymer, in press.*
13 J-C. Chiang *et al, Synth. Met.*, **13** (1986) 193.
14 K. Chow et al, *Synth Met. Accepted.*
15 L. Abell, P. N. Adams, and A. P. Monkman, *Polymer Comm., in press.*

Structure-Property Characteristics of Ion Implanted Syndiotactic Polystyrene

Chang-Meng Hsiung and Caiping Han
Louisiana Productivity Center, Chemical Engineering Department/University of SW Louisiana, Lafayette, LA 70504-4172
Y. Q. Wang, W. J. Sheu, and G. A. Glass
Acadiana Research Laboratory /University of SW Louisiana, Lafayette, LA 70504
Dave Bank
The Dow Chemical Company, Midland, Michigan 48667

INTRODUCTION

Syndiotactic polystyrene (sPS) is a newly developed engineering semi-crystalline polymer that is based on metallocene technology. It has some attractive physical characteristics including high melting point (270°C), low specific gravity (1.045), excellent hydrocarbon resistance, a high degree of dimensional stability, enhanced mechanical performance at elevated temperature, very good electrical properties, and low viscosity at typical shear rates.[1,2] This combination of properties opens a wide variety of applications including automotive, appliance, medical, electrical/electronic, fibers, and films. The structure-property characteristics of injection molded neat and glass fiber reinforced sPS parts have been studied by Hsiung and Cakmak.[3,4,5]

Ion implantation is a well-known technique in the electronic industry to modify the electronic and physical properties of materials. In the recent years, this technique has been widely applied to many new materials, including organic polymers. Applications of ion implantation to polymers is of growing interest mainly because polymers are inexpensive, light and can be easily shaped into various forms. However, most polymers exhibit poor surface properties. Ion implantation can significantly modify the surface, mechanical, and electrical performance of polymers.[6]

We studied the structure-property characteristics of carbon ions implanted sPS under different doses and energy levels.

EXPERIMENTAL

SAMPLE PREPARATION

Neat s-PS resin with an average molecular weight of 400,000 was kindly donated by the Dow Chemical Company. The raw material was dried in a vacuum oven at 85°C for 12 hours before molding. sPS disks with a diameter of 20 mm and thickness of 0.8 mm were made by a Carver compression molding machine. Mold temperature was set at 280°C. To ensure smooth surfaces of the samples, two glass plates and a template made of copper were used to mold the samples.

IMPLANT PROCEDURE

Ion implantation experiments were performed at the Acadiana Research Laboratory with a National Electrostatics Corporation 5SDH-2 1.7 MV Tandem Pelletron Accelerator. The accelerator system has two ion sources: the Source of Negative Ions by Cesium Sputtering (SNICS) to produce heavy ions, and the Radio Frequency (RF) sources to produce helium ions. Carbon ions were produced from a graphite pellet inside the SNICS source of the accelerator. The pressure in the system was maintained at 10^{-7}Torr. A PPT Residual Gas Analyzer (RGA) was attached to the chamber to monitor the gas emission from the sPS samples during the implantation. In the first part of this research, the carbon ions were kept at a constant energy of 1.0 MeV but the dose was varied from 10^{11} ions/cm^2 to 10^{15} ions/cm^2. In the second part of this research, the implanted dose was kept to be 10^{13} ions/cm^2 but the energy changed from 0.5 MeV to 4.0 MeV. Low current densities (around 25 nA/cm^2) were used in both cases to minimize the effects of beam heating.

CHARACTERIZATION OF SURFACE STRUCTURE AND PROPERTIES

Surface composition was analyzed by a Residual Gas Analyzer (RGA) and a Elastic Recoil Detection Analysis (ERD). Surface morphology and roughness were measured by Atomic Force Microscopy (AFM). Surface hardness was studied by a Nanoindenter. Wear resistance and friction coefficient were investigated by a Tribometer. Surface wettability and contact angles were characterized by a Kruess Processor Tensiometer. Solvent resistance was measured by the weight change of samples immersed in toluene and chloroform. Surface electrical conductivity was measured by a Keithley Electrometer.

RESULTS

COMPOSITIONAL AND MORPHOLOGY ANALYSIS

The RGA results showed that during ion implantation of sPS samples volatile species including H_2 and C_2H_2 were released. This is caused by the irreversible cleavage of covalent bonds

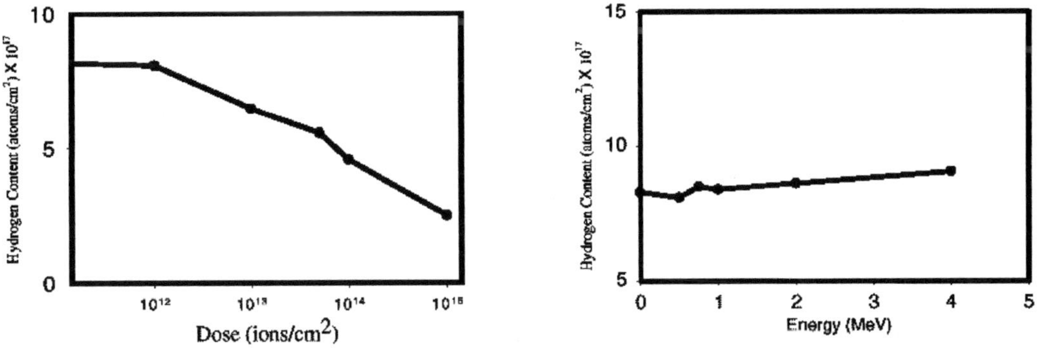

Figure 1. (left) Effect of dose on the hydrogen content in implanted sPS samples. (right) Effects of energy on the hydrogen content in implanted sPS samples.

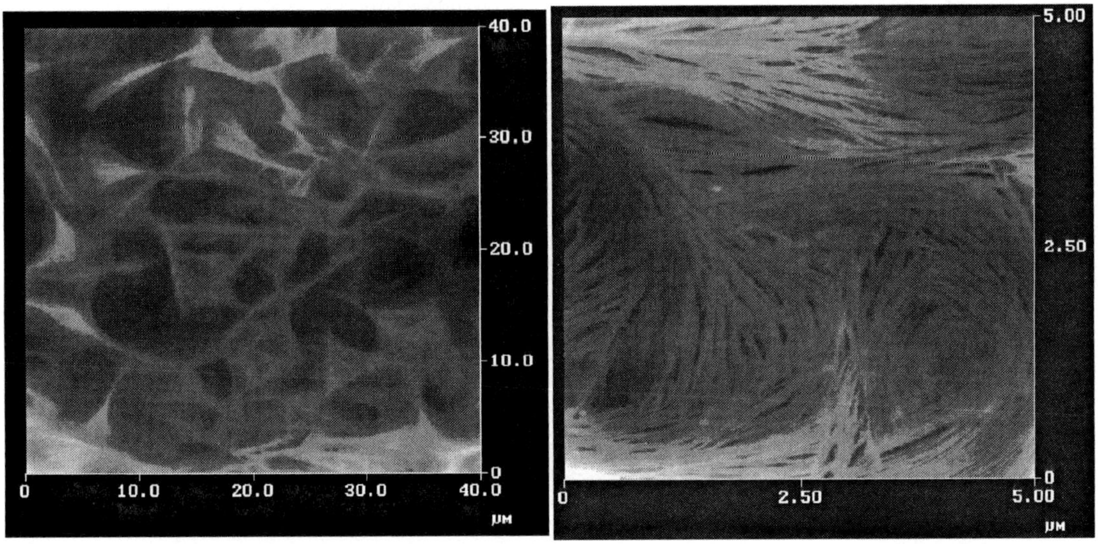

Figure 2. AFM pictures of sPS before treatment.

within the polymer chains. The results from ERD study confirm that the hydrogen content in the surface of implanted sPS is reduced by increasing dose but implantation energy seems to cause little change (Figure 1).

Figure 2 shows the surface morphology of untreated sPS samples in two magnifications. In these AFM pictures spherulitical structure can be seen very clearly and the average size of these spherulites are about 5 ~ 10 μm. Figure 3 shows the AFM pictures of implanted sPS

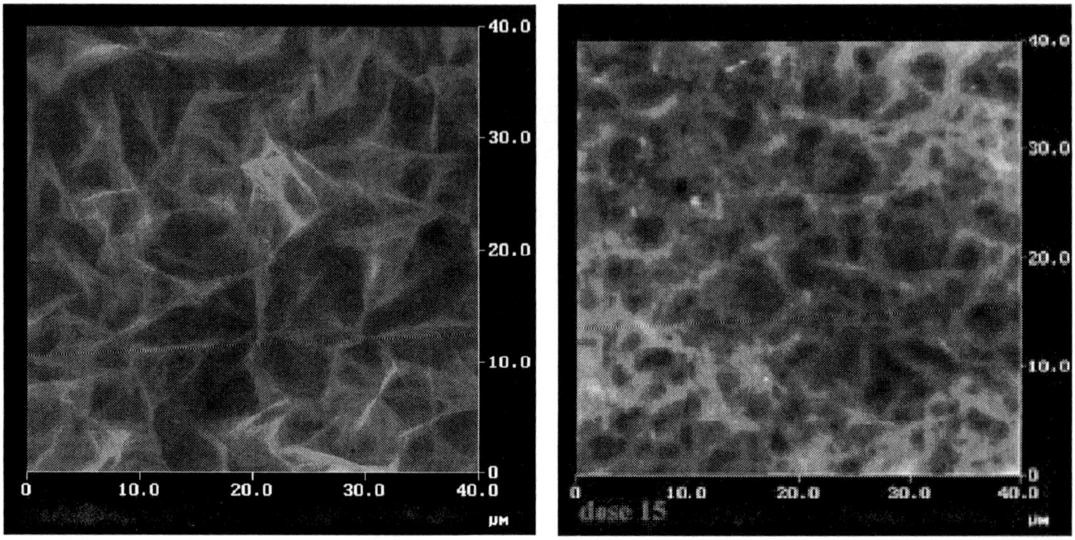

Figure 3. AFM pictures of sPS after treatment with different dose of ion beam. (left) dose = 10^{11} ions/cm^2, (right) dose = 10^{15} ions/cm^2.

samples. At lower dose of 10^{11} ions/cm^2, there seems to be only minor visual changes. But the surface structure shows melted regions at the highest dose of 10^{15} ions/cm^2.

SOLVENT RESISTANCE AND WETTABILITY

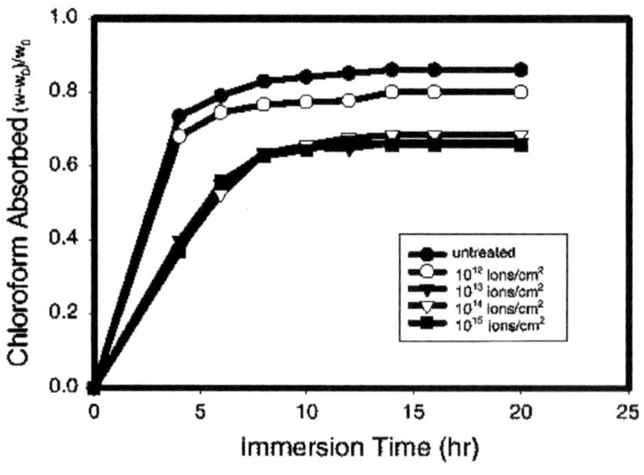

Figure 4. Effects of ion implantation on the solvent resistance of sPS samples.

The solvent resistance of the implanted sPS samples were studied by monitoring the amount of solvent absorbed when these samples were immersed in various solvents. The higher the amount of solvent absorbed means the poorer the solvent resistance. In general, if the dose is not too high, ions bombardment can cause crosslinking of polymer chains on the surfaces and this can improve the solvent resistance. Figure 4 shows that ion implantation can improve the solvent resistance. However, this effect saturated at dose of 10^{13} ions/cm^2. Further increase in doses beyond that will

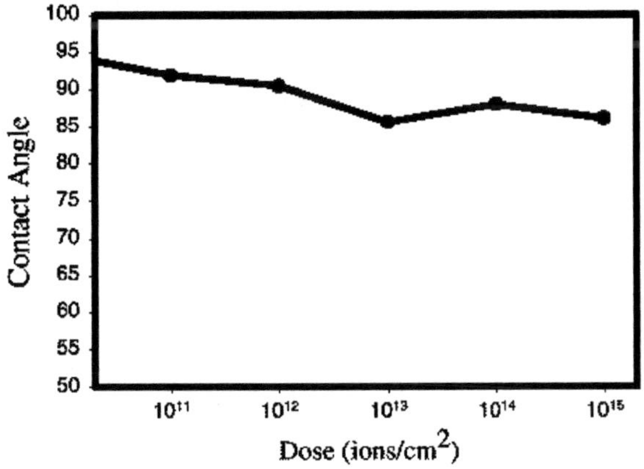

Figure 5. Effect of dose on the contact angle of ion implanted sPS samples.

not improve the solvent resistance. Similar trend was found on the effect of implantation energy but to a less extent.

The wettability of ion implanted sPS samples were studied by measuring the contact angles with respect to distilled water. The lower these numbers means the better the wettability. Figure 5 shows that wettability improves slightly with increasing dose. The effect of energy was also studied but it shows no influence.

MECHANICAL AND ELECTRICAL PROPERTIES

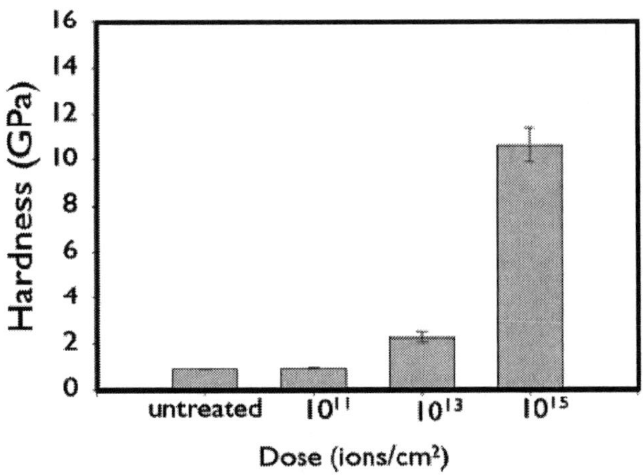

Figure 6. Effects of dose on the hardness of implanted sPS samples.

Hardness is ultimately a manifestation of three-dimensional bond strength, which can be altered by ion implantation. During ion implantation, rapture of C-H bonds occurred and gaseous elements lost, leaving dangling C bonds, which then might link together forming a rigid three dimensional carbon structure. The hardness of sPS samples treated with ion implantation is shown in Figure 6. Compared to the untreated sample, ion implantation can dramatically improve the surface hardness by more than 10 times. Samples implanted at dose of 10^{15} ions/cm^2 are even harder than stainless steel which typically has a hardness of 7 GPa.

In general, surfaces with smaller coefficient of friction have better wear resistance. Figure 7 shows the coefficient of friction of both untreated and ion implanted sPS samples. It seems that implantation dose of 10^{15} ions/cm^2 is needed to improve the wear resistance dramatically while implantation with 10^{13} ions/cm^2 and below have little effect. Friction and wear are very complex phenomena, which depend upon load, speed, humidity, mechanical interlocking, molecular interactions, heat generation, and electrostatic force.[7] The reasons for the enhancement of wear resistance of implanted polymers might be:

- Change in the structure and composition of the near surface region produced a tough new surface that forms a long lasting barrier to wear.
- High concentration of carbon ions in the near-surface region produces compressive stress that close up the microcracks inherent in the implanted surface.
- The formation of lubricate graphite-like structure on the implanted surface.

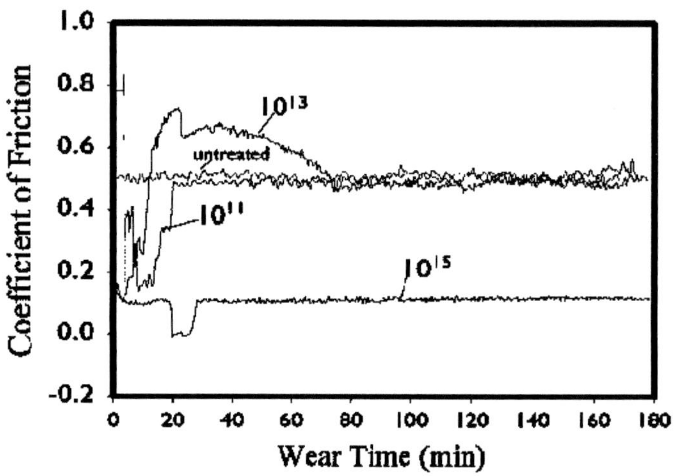

Figure 7. Coefficient of friction of sPS samples showing the effect of dose.

When ions bombard on the polymer, they lose energy, release hydrogen, and form a carbon-enriched structure. This carbon-enriched cluster is more conductive than the untreated polymer region. When the dose increases, many of these clusters will start to contact each other and finally overlap to form a continues carbon rich conductive surface, which contributes to the measurable electrical conductivity. Ion implantation typically increases surface electrical conductivity of polymers. However, due to the fact that the neat sPS polymer has a very high resistivity, the measurement and analysis is comparatively difficult. In this study, the electrical conductivity of samples implanted with

Table 1. Electrical conductivity of ion implanted sPS samples

Dose, ion/cm^2	V, V	I, A$\times 10^{-12}$	R, $\Omega\times 10^{12}$	$R_{\square}$ Ω/sq$\times 10^{12}$	Resistivity, ρ, Ω-cm	Conductivity, σ, s/cm
untreated	92.4	-	-	-	$>10^{16}$	$<10^{-16}$
10^{11}	92.4	-	-	-	-	-
10^{12}	92.4	-	-	-	-	-
10^{13}	92.4	-	-	-	-	-
10^{14}	92.4	0.48	193.0	2600	6.1×10^{11}	1.64×10^{-12}
10^{15}	92.4	550	0.168	2.35	5.2×10^{8}	1.92×10^{-9}

different doses was studied and the conductivity enhancement of the sPS samples as a result of high dose implantation is remarkable (Table 1). Electrical conductivity caused by the ion implantation with doses lower than 10^{14} ions/cm^2 could not be measured. The conductivity of the sample with dose of 10^{15} ions/cm^2 is several orders higher than the sample with dose of 10^{14} ions/cm^2.

CONCLUSIONS

1. It was found that C-H bonds broke and several volatile species (especially hydrogen) were released during the ion implantation process.
2. Ion implantation improved the solvent resistance of sPS samples. Especially, increased dose had a definite effect on the improvement of solvent resistance. However, ion implantation performed at different energy levels showed less effect.
3. The wettability of sPS samples was improved slightly by ion implantation.
4. Increased dose of ion implantation will improve the surface hardness of the sPS samples. The sPS surface as hard as stainless has been created by the implantation at a highest dose of 10^{15} ions/cm^2.
5. Implantation dose up to 10^{15} ions/cm^2 was needed to improve the wear resistance of these sPS samples.
6. Implantation dose up to 10^{14} ions/cm^2 was required to show increases in electrical conductivity. Further increase in ion dose should improve the electrical conductivity.

ACKNOWLEDGMENTS

This work was supported by Louisiana Education Quality Support Fund (Grant # LEQSF(1997-00)-RD-B-15 and LEQSF(1995-98)-RD-B-99) and the Department of Energy/Louisiana Education Quality Support Fund in Cooperative Agreement Number DE-FC02-91ER75669. sPS material donation and financial support from the Dow Chemical Company is highly appreciated.

REFERENCES

1 J. H. Schut, *Plastics Technology*, **2**, 26 (1993).
2 D. Bank and R. Brentin, *Plastics Technology*, **43**(6), 52 (1997).
3 C. M. Hsiung, J. Miao, Y. Ulcer, and M. Cakmak, *SPE Annual Technical Papers*, **1788**, 1798 (1995).
4 Y. Ulcer, M. Cakmak, J. Miao, and C. M. Hsiung, *Journal of Applied Polymer Science*, **60**, 669 (1996).
5 X. Zhang, C. M. Hsiung, and D. Bank, *SPE Annual Technical Papers*, **2339** (1997).
6 H. Ryssel and I. Ruge, **Ion Implantation**, 1986, *John Wiley & Sons.*
7 E. H. Lee, M. B. Lewis, P. J. Blau, and L. K. Mansur, *J. Mater. Res.*, **6**(3), 610 (1991).

Carbon Black Filled Immiscible Blend of Poly(Vinylidene Fluoride) and High Density Polyethylene: Electrical Properties and Morphology

Jiyun Feng and Chi-Ming Chan
Department of Chemical Engineering, The Hong Kong University of Science and Technology, Clear Water Bay, Kowloon, Hong Kong

INTRODUCTION

In recent years, conductive polymer composites with a low percolation threshold have received increasing attention.[1-7] One important approach to prepare the composites is to selectively localize a conductive filler in one polymer phase or at the interface of an immiscible polymer blend. The advantage of this approach is that the composite may achieve a high electrical conductivity at very low CB contents and retain reasonable mechanical properties. In addition, they can be manufactured at lower costs and with simpler processing procedures.

The reason for the high electrical conductivity of the composites at low CB contents is an uneven distribution of CB in immiscible polymer blends. Several examples have been found.[1-7] Narkis *et al.* studied CB-filled immiscible blends of polypropylene(PP)/Nylon and PP/polycarbonate(PC) and found that CB has stronger affinity to Nylon and PC than to PP, resulting in its preferential localization in the former phases.[3,4] These results are due to the higher surface tension and high polarity of Nylon and PC in comparison to PP. Sumita *et al.* investigated CB filled HDPE/PP blends and discovered that CB is in the HDPE phase.[6,7]

It is known that past research on the composites is focused on the CB distribution and the relationships between their electrical conductivity and morphology. Double percolation model is used to predict the electrical behaviors of the composites. However, the effect of morphology on the PTC and NTC effects of the composites is absent in the literature.

In the present work, the electrical conductivity, PTC, and NTC effects of CB filled PVDF/HDPE composites were studied. Morphology of the composites was observed. The relationships between electrical behaviors and morphology are also discussed.

EXPERIMENTAL

The polymers used in this study were PVDF (Hylar 460 from Ausimount Co. USA) and HDPE (HMM 5502 from Philips International Petroleum Inc.). The CB used was V-XC72 from, Cabot. The CB-filled PVDF/HDPE composites were prepared using a Haake mixer at 200°C and 30 rpm for 15 min. The materials obtained were further compressed into 2 mm thick sheets using a hot press at 200°C. Two group of samples were prepared. One group of samples contains a fixed PVDF/HDPE ratio (1/1) but different CB contents. Another group of samples contains a fixed CB content (10 wt%) but different PVDF/HDPE volume ratios.

The resistivity of the composites were measured with a multimeter. Before measurements, the sample surfaces were coated with silver paint to eliminate the contact resistance. The resistivity of the composites as a function of temperature was measured using a computerized system, which comprises a multimeter, a computer, and a programmable oven. The heating rate was 2°C/min. The morphology of the composites was determined using optical microscopy and the transmission mode was used. Thin sections of 1 μm in thickness were obtained by a cryomicrotome at -100°C.

RESULTS AND DISCUSSION

ELECTRICAL CONDUCTIVITY

The electrical conductivity of CB-filled PVDF/HDPE composites with a fixed PVDF/HDPE volume ratio versus CB volume fraction is illustrated in Figure 1. Apparently, the electrical conductivity of the composites increases dramatically when the CB content attains the percolation threshold approximately at 0.035 volume fraction of CB. According to the percolation theory, the electrical conductivity can be correlated with the volume fraction of the conductive filler by the scaling law as follows.

$$\sigma = \sigma_o \left(\Phi - \Phi_c \right)^t \quad [1]$$

By using a log-log plot of the electrical conductivity versus the excess of conductive filler volume fraction of $(\Phi - \Phi_c)$, as shown in Figure 2, the best fit was obtained with $\Phi_c = 0.037$ from the slope and the intercept of the straight line, the values of t and σ_o were determined to be 2.75 and 93.3, respectively. The linear correlation coefficient was 0.998.

In addition to the CB content, the PVDF/HDPE volume ratio also affects the electrical conductivity of the composites. Figure 3 displays the electrical conductivity versus PVDF/HDPE volume ratio. Clearly, the electrical conductivity of the composites increases rapidly after the PVDF/HDPE volume ratio is greater than 0.17. The increase becomes more

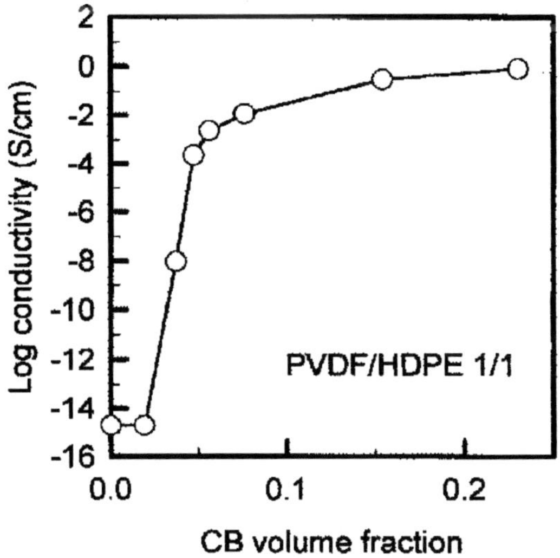

Figure 1. Plot of log conductivity vs. CB volume fraction.

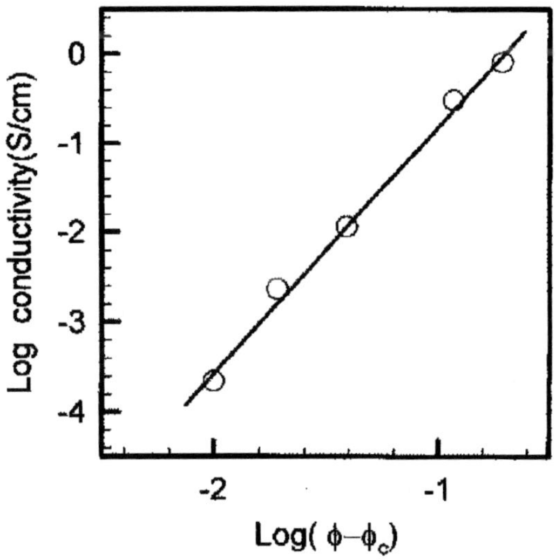

Figure 2. Plot of log conductivity vs. $(\phi - \phi_c)$.

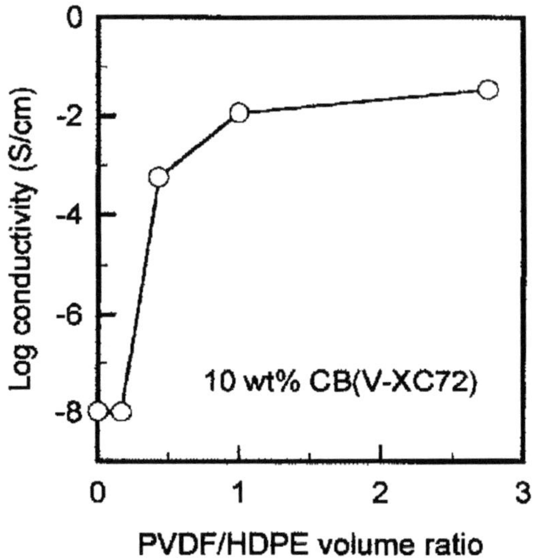

Figure 3. Plot of log conductivity vs. PVDF/HDPE volume ratio.

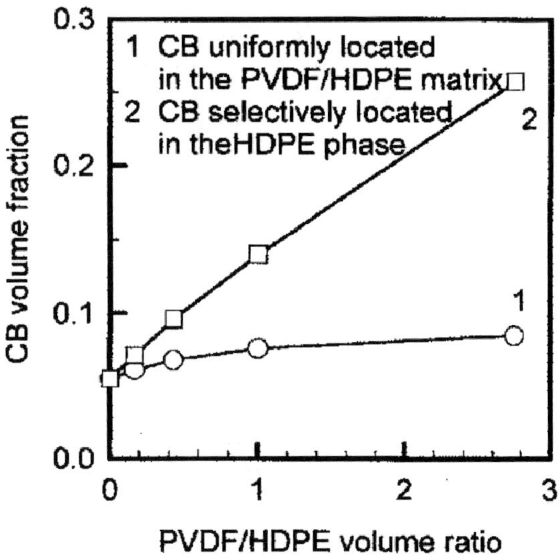

Figure 4. CB volume fraction vs. PVDF/HDPE volume ratio.

gradual when the PVDF/HDPE volume ratio is greater than 0.43. The results suggest that a decrease in HDPE content significantly increases the conductivity of the composites. Hence, it can be concluded that the distribution of CB in the PVDF/HDPE composite is uneven and CB is just located in the HDPE phase. Figure 4 shows the CB volume fraction versus PVDF/HDPE volume ratio in two different situations. If the CB is evenly distributed in the PVDF/HDPE matrix, the CB volume fractions at different PVDF/HDPE volume ratios do not show any significant differences as shown in Figure 4. Obviously, this is not a correct model when compared with the experimental results depicted in Figure 3. However, if we assume that the CB is totally localized in the HDPE phase, the CB volume fraction in the HDPE phase increases when the PVDF/HDPE volume ratio increases, resulting in a large increase in electrical conductivity. There is no doubt that this model is consistent with the experimental data in Figure 3.

PTC AND NTC EFFECTS

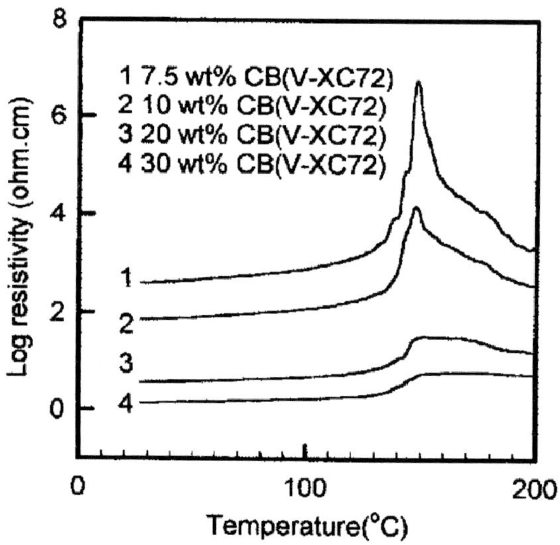

Figure 5. Plot of log resistivity vs. temperature.

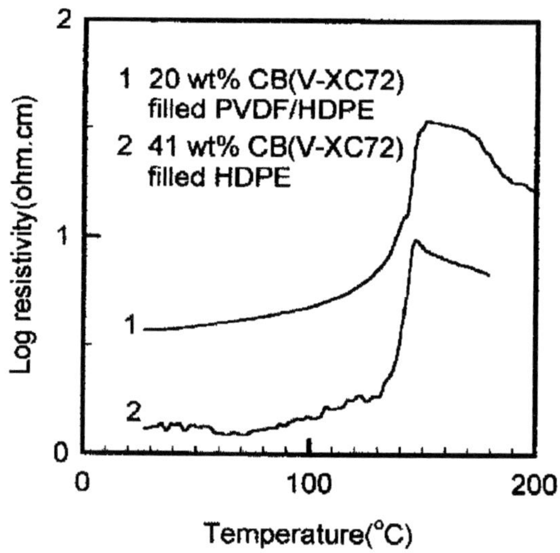

Figure 6. Plot of log resistivity vs. temperature.

Figure 5 depicts the resistivity of the CB-filled PVDF/HDPE composites versus temperature. The resistivity peak of the composites is observed at about 145°C which is a little higher than that of the melting point of HDPE. However, at the melting point of PVDF, no resistivity increase is observed. These results reveal two important facts. First, the PTC effect of the composites is caused by the thermal expansion by the melting of the HDPE phase in the

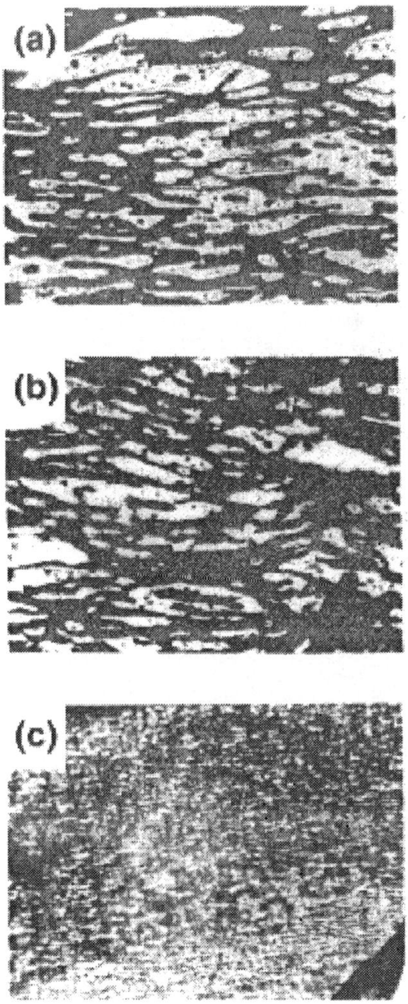

Figure 7. Optical micrographs of CB filled PVDF/HDPE composites (PVDF/HDPE 1/1). (a) 7.5, (b) 10, (c) 20 wt% CB.

composites. Second, the CB is only located in the HDPE phase in the composite, indicating that the distribution of the CB is uneven. This result provides a strong support for the conclusion that the CB is mainly located in the HDPE phase. In addition, the PTC intensity and room temperature resistivity decrease as the CB content increase.

It should be noted that in the case of 7.5 and 10 wt% CB composites, the NTC effect is observed when the temperature is over the melting point of HDPE which is very similar to that of CB-filled neat HDPE composite. However, in the case of 20 wt% CB composite, the material shows a PTC effect first, then a zero temperature coefficient (ZTC) effect but not a NTC effect until the temperature attains the melting point of PVDF, indicating that a delayed NTC effect occurs.

In order to detect the cause for the delayed NTC effect, a comparison between a CB-filled neat HDPE and the composite was made as shown in Figure 6. The CB content in the neat HDPE and in the HDPE phase of the composite is the same. The CB filled neat HDPE composite show an obvious NTC effect when the temperature is above the melting point of HDPE, indicating that the delayed NTC effect is caused by the PVDF phase but not the CB concentration in the HDPE phase. The local restriction provided by the PVDF phase on the CB filled HDPE phase delays the NTC effect to higher temperature. When the temperature reaches the melting point of PVDF phase, the PVDF phase melts and the restriction is suddenly released resulting in a sharp decrease in resistivity. This is a new physical phenomenon and it can be used to develop new polymer thermistors without crosslinking.

MORPHOLOGY

Figures 7(a), (b), and (c) show the optical micrographs of the cryomicrotomed sections of the composites with different CB contents. It is very clear that a two-phase structure is present and the CB is just located in one phase. Based on the forgoing conclusions, the light areas are identified as the PVDF phase, while the dark areas are the CB-filled HDPE phase. At a fixed PVDF/HDPE volume ratio (1/1), an increase in the CB content leads to an significant decrease in the domain size of the PVDF phase, possibly due to the increase in the viscosity of the CB-filled HDPE phase as a result of the increase in CB content. In the case of 7.5 and 10 wt% CB, the CB-filled HDPE phase forms a continuous phase and the PVDF phase is the dispersed phase. However, when the CB content reaches 20 wt%, the composite exhibits an interlocking structure which is significantly different from those of lower CB content composites. These difference in morphology is the reason for the observed difference in the NTC effect.

In the case of lower CB content composites, the CB-filled HDPE phase forms a continuous phase, the composites show a PTC effect first, then a NTC behavior, indicating that the PTC and NTC effects of the composites are very similar to those of a CB-filled neat HDPE. For the 20 wt% CB composite, its morphology is an interlocking structure and it shows a normal PTC effect then a delayed NTC effect. It is the interlocking structure that makes the NTC effect delay to a higher temperature. This is a new physical observation and has some important potential applications in industry. This material can be used as polymer thermistors without crosslinking. Hence it is important to understand the mechanism of the delayed NTC effect.

Consider the CB filled HDPE phase in an interlocking structure, when the temperature is above the T_m of the HDPE and further increases, the viscosity of the CB-filled HDPE phase is basically controlled by the two factors: the temperature and local pressure provided by the surrounding PVDF phase. It is known that when the temperature increases, the viscosity of the CB-filled HDPE phase decreases. On the other hand, when the temperature increases, the HDPE phase expands, because the structure is an interlocking structure and the PVDF phase is still hard, the PVDF phase restricts the expansion of the HDPE phase. In fact, this restriction produces a pressure on the local CB filled HDPE phase and this pressure greatly affects the viscosity of the HDPE phase. It is also known that for the CB filled HDPE phase, when the temperature is above the viscous flow temperature of the polymer, if a pressure is applied to the polymer, the free volume of the polymer decreases, and molecular interaction increases, resulting in an increase in viscosity. In the case of CB-filled HDPE phase, it is believed that the viscosity increases by the local pressure due to the surrounding PVDF phase is larger than or almost equals the viscosity decrease caused by the temperature increase. The viscosity of the CB-filled HDPE phase is high and the movements of the CB particles is greatly restricted. Hence, a flocculated structures can not form, resulting in a delayed NTC effect. But when the temperature attains the T_m of the PVDF phase, the PVDF phase melts, the local pressure on the CB filled HDPE phase is suddenly released. In this case, the viscosity of the CB-filled HDPE phase suddenly decreases, leading to a formation of the flocculated structures immediately, producing a sharp NTC effect occurs. This mechanism successfully explains the delayed NTC effect observed in this research.

CONCLUSIONS

CB is selectively localized in the HDPE phase due to the stronger affinity to HDPE than to PVDF. For the composites with a fixed PVDF/HDPE volume ratio of 1/1, a percolation threshold of 0.037 volume fraction of CB is observed. At a fixed CB content (10 wt%), the electrical conductivity increases when the PVDF/HDPE volume ratio increases, indicating that the CB is mainly located in the HDPE phase. In addition, only the PTC effect that is associated with the melting of HDPE phase is observed, confirming that the CB is localized in the HDPE phase. An increase in the CB content can greatly decrease the domain size of the PVDF phase due to the viscosity increase of the CB-filled HDPE phase. When the CB-filled HDPE forms a continuous phase and the PVDF forms the dispersed phase, the PTC and NTC behaviors of the composites are very similar to those of a CB-filled neat HDPE composite. But when the composite exhibits an interlocking structure, a delayed NTC effect is observed and the delayed NTC effect is caused by the local restriction provided by the PVDF phase.

ACKNOWLEDGMENT

This work was supported by UPGC Research Infrastructure Grant under Grant No. RI93/94EG.

REFERENCES

1 G. Genskens, E. De Kezel, S. Blacher and F. Brouers, *Eur. Polym. J.*, **27**,1261 (1991).
2 B. Wessling, *Kunststoffe*, **76**, 930 (1986).
3 F. Gubbels, R. Jermore, Ph. Tessie, E. Vanlathem, R. Deltour, A. Calderone, V. Parente, and J. L. Breads, *Macromolecules*, **27**, 1972(1994).
4 M. Narkis, R. Tchoudakov, and O. Breuer, *ANTEC'95*, P1343-1346.
5 R. Tchoudakov, O. Breuer, M. Narkis and A. Siegman, *Polym. Eng. Sci.*, **36**, 1336(1996).
6 M. Sumita, K. Sakata, S. Asai, K. Miyasaka, and H. Nakagawa, *Polym. Bull.*, **25**, 265(1991).
7 M. Sumita, K. Sakata, H. Nakagawa, S. Asai, K. Miyasaka and M. Tanemura, *Colloid Polym Sci.*, **270**, 134(1992).

Conductivity/Morphology Relationships in Immiscible Polymer Blends: HIPS/SIS/Carbon Black

R. Tchoudakov, O. Breuer, M. Narkis and A. Siegmann
Department of Chemical Engineering, Haifa 32000, Israel

INTRODUCTION

It is generally known, that CB-loaded polymer blends often exhibit a higher conductivity level than either of the component polymers at the same low CB content. This phenomenon is due to the preferential CB location within one of the phases and/or at the interfaces, resulting in segregated phase structures of the blend leading to the double percolation concept.[1,2] Usually, block copolymers are added to immiscible polymer blends to function as alloying agents or surfactants which contribute to the stabilization of the multiphase structure.[3] There are numerous publications concerning the morphology and properties of blends such as polystyrene (PS) with its copolymers. However, almost nothing is reported on such blends loaded with carbon black (CB), especially the conductive CBs. In addition, utilization of a block copolymer as the dispersed phase within a PS matrix is also uncommon. Various publications concerning conductive elastomer/CB compositions reveal fundamental parameters determining their conductivity level, among them the CB-polymer interaction.[4,5] A thermoplastic elastomer mixed with CB, being a part of the multiphase conductive systems, shows a blend conductivity affected by the thermoplastic elastomer intrinsic microstructure.[6,7] The subject of this study is the correlation between the morphology and resistivity of CB-loaded compositions of a high impact PS (HIPS) and a tri-block copolymer, styrene-isoprene-styrene, (SIS).

EXPERIMENTAL

The polymers used in this study were HIPS, Galirene HT 88-5, MFI - 4,5; Carmel Olefins, Israel and SIS, Quintac 3421, MFI - 11, 14% PS, Japan. The carbon blacks were CB-EC, Ketjenblack EC-300 Akzo, Netherlands and CB-MT, Thermal black N990, Vanderbilt, characterized by surface area (BET) 950 and 9 m^2/g and particle diameters ~ 30 and (285-500) nm correspondingly.

HIPS/SIS/CB blends containing 5-45 wt% SIS and 0-4 phr CB were produced by melt mixing in a Brabender plastograph at 190°C and subsequently compression molding. Their volume resistivity was measured as previously reported.[8] The blend morphology was studied using a SEM Jeol 5400. Freeze-fractured and microtomed surfaces (both prepared in liquid nitrogen) were investigated.

RESULTS AND DISCUSSION

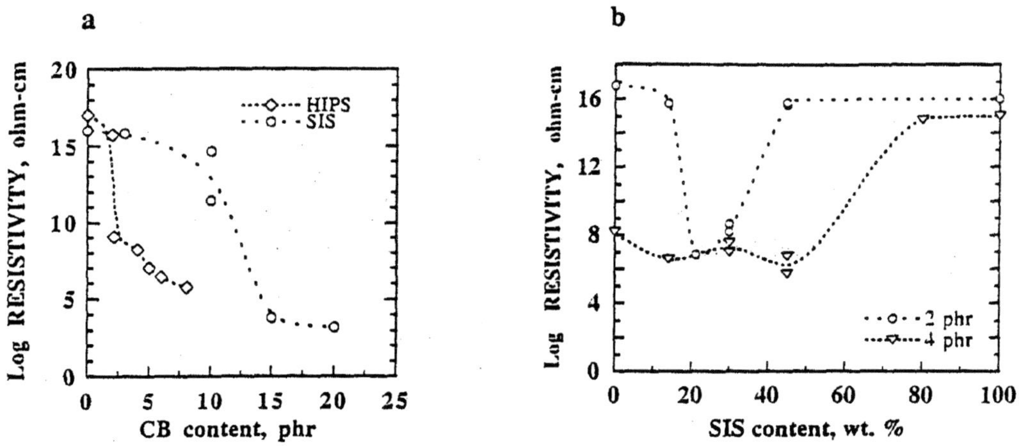

Figure 1. Resistivity vs. composition. (a) individual polymers, (b) blends.

The resistivity of polymer/CB-EC compounds as a function of the CB content is presented in Figure 1a. It is clearly seen that the SIS/CB percolation occurs at a rather high critical CB concentration, about 10 phr. This value is much higher than the corresponding value for HIPS/CB compounds, about 2 phr. Figure 1b depicts the effect of SIS content on the blend resistivity at a constant CB concentration in the blend. Interestingly, blends with 2 phr CB depict a stronger change in resistivity with blend composition compared to the blends containing 4 phr CB. It should be noted, however, that the resistivity of HIPS containing 4 phr CB is 9 orders of magnitude lower than that of HIPS containing just 2 phr CB. The resistivity of the former is practically unaffected by the addition of SIS up to 45 wt%, while the 2 phr CB containing blends exhibit a large resistivity reduction for SIS content of 20-30 wt%. It is well established that the critical concentration of a given CB at the percolation threshold in different polymers depends on the polymer/CB interaction.[9] Such interactions are related to some polymer charac-

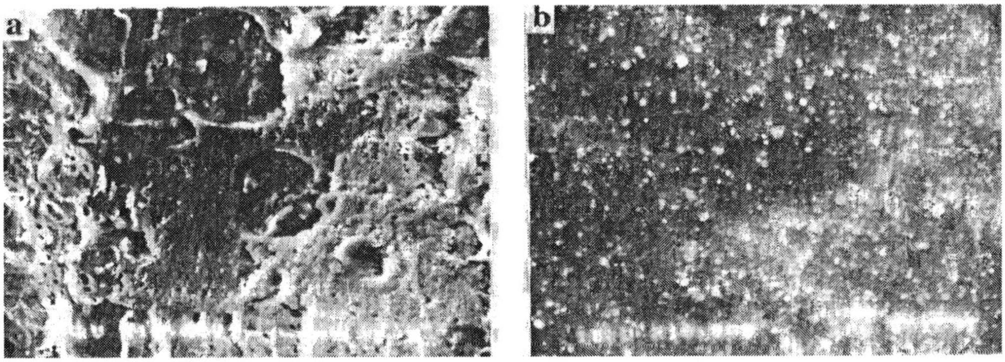

Figure 2. SEM micrographs of the individual polymers with CB-EC. (a) HIPS + 2phr CB, microtomed surface; (b) SIS + 15 phr CB, freeze-fractured surface.

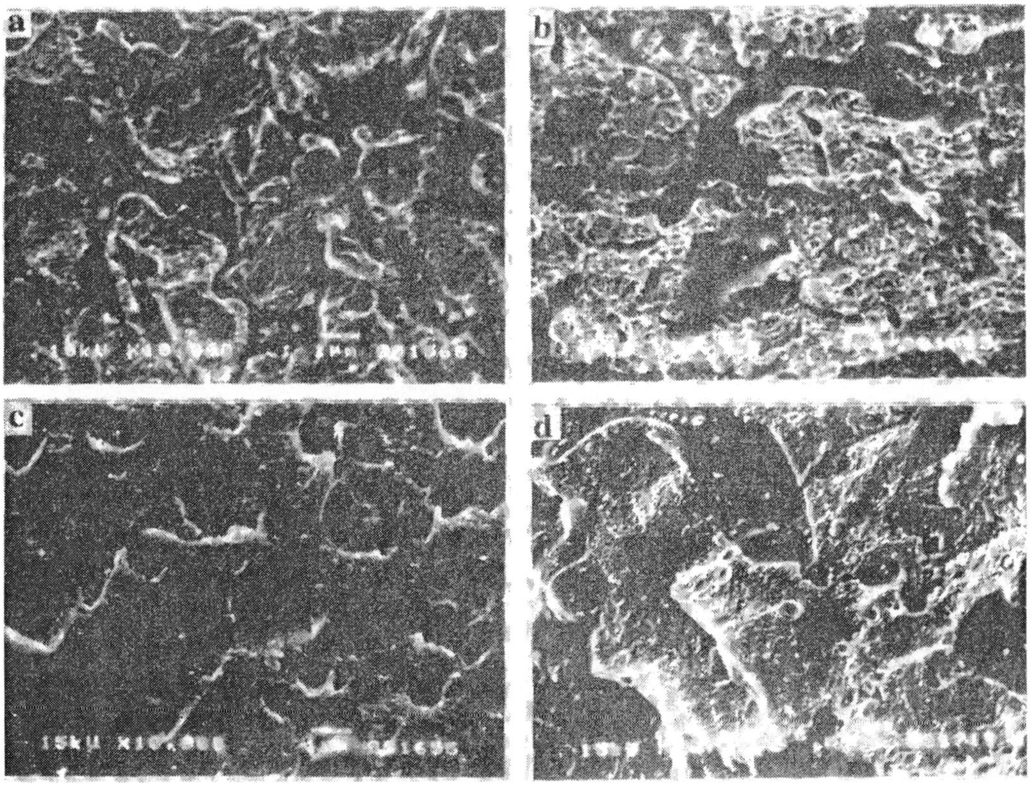

Figure 3. SEM micrographs of HIPS/SIS blends. (a) 80/20, without CB; (b) 70/30, 2 phr CB-EC; (c) 85/15, 4 phr CB-EC; (d) 70/30, 4 phr CB-MT; (a), (b), (d) - freeze-fractured surfaces; (c) microtomed surfaces.

teristics, such as surface tension, polarity, crystallinity and viscosity, where surface tension often seems to be the dominant parameter. SEM micrographs of the HIPS/CB compounds enable to observe the distribution of CB EC-300 only of microtomed surfaces. Their micrographs clearly show that the rubber inclusions 1-3 μm in diameter, do not contain CB. Carbon black addition to SIS is manifested by the appearance of bright particles increasing in quantity when CB concentration increases. It may be noted that appearance of the CB dispersion within SIS after the percolation threshold resembles a chain-like particles distribution.

The morphology of neat and CB-containing blends is shown in Figure 2. The structure of HIPS/SIS blends without CB consists of very fine SIS dispersion at 5 wt% content (particle size 0.2 to 1 μm) up to the co-continuous morphology obtained at about 20 wt% SIS in the blend. The two components are strongly inter-connected as there are no visible gaps at the interface between the phases, characteristic of low interacting components in immiscible polymer blends. It is important to point out again that the bright tiny details present in HIPS are also visible in SIS. Therefore to determine the genuine location of CB particles within the blend is not an easy task. Rubber inclusions are clearly seen in the HIPS without any traces of CB-EC present. A clear phase structure, together with a well-defined CB-EC location, are seen only in a microtomed surfaces of the blends. The SIS component appears smooth, without an evident presence of CB-EC agglomerates, whilst HIPS contains numerous well-distinguished CB agglomerates (Figure 3c). Hence, preferential CB-EC location in HIPS occurs, contrary to the case of CB-MT addition, which does not demonstrate a preference to either phase. The CB-MT large size particles added to these polymers are clearly seen penetrated in both phases (Figure 3d).

Returning to Figure 1b, the conductivity of the CB-filled HIPS/SIS blends is determined by the conductivity of the CB-rich HIPS component and its continuity. SIS addition to HIPS/CB-EC blends increases the CB effective concentration in HIPS, transforming the insulative HIPS/2phr CB compound, in the absence of SIS, to relatively conductive upon about 30 wt% SIS addition. When the continuity of the CB-rich conductive HIPS is disrupted, the blend reverts insulative. HIPS compounds with 4 phr CB-EC are conductive in the absence of SIS, therefore no significant change in resistivity with SIS addition is observed.

The presently reported finding that CB-EC is preferentially located in the polymer of lower percolation threshold is exceptional since the opposite is true for most known immiscible polymer blends.[8] It is assumed that the particular SIS structure, where PS domains are dispersed in the rubber matrix,[10] may cause the unusually high CB percolation found in SIS. Taking into account the dimensions of CB-EC particles and PS domains within SIS, a model of CB-EC dispersion in SIS is suggested (Figure 4). The model is based on the preferential location of CB-EC in PS rather than in polyisoprene which is described by the engulfing of the CB particles by the PS blocks until their "saturation". This engulfing of CB-EC with the PS blocks isolates the CB particles and therefore does not contribute to the material's conductiv-

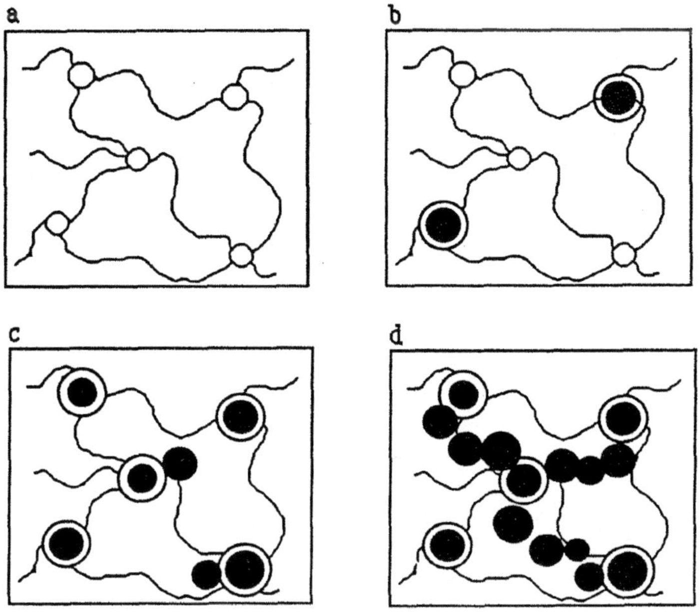

Scheme of CB-EC distribution in SIS. (a) neat SIS; (b) CB encapsulation in the PS blocks; (c) PS blocks "saturated" with CB, excess CB starts to appear in the rubber phase; (d) CB percolation within rubber is realized.

ity. Only when the engulfing process is completed, the excess CB starts its distribution within the polyisoprene phase. Therefore SIS becomes conductive when CB-EC percolation the polyisoprene phase has been achieved. This explains the relatively high CB-EC content, necessary for percolation in SIS and converting it into a conductive form. Thus for HIPS/SIS blends, the added CB is first located in the PS phase of HIPS, abundant in PS, and only thereafter it may be distributed in the SIS component as well. Considering the suggested model, the absence of preference of CB-MT location within the studied blends is understandable. The large size of CB-MT particles hinders its mobility and selective interaction with either the plastic or the rubber phase of the thermoplastic elastomer. Thus, CB-MT particles are immobilized within the phase in which they have been initially incorporated during the mixing procedure.

CONCLUSIONS

Low concentration of CB-loaded HIPS/SIS blends demonstrates interesting conductive properties and unexpected morphology. The preferred CB-EC location in HIPS occurs in spite of the fact that the percolation threshold of HIPS/CB is much lower than that of SIS/CB. The blends are conductive as long as the HIPS component is continuous and the CB contained in it exceeds its percolation value. A difference in the distribution of CB-EC and CB-MT within the blends was observed, depicting the significance of both CB size and properties and CB/polymer interaction. A physical model of CB-EC distribution within the SIS triblock copolymer has been proposed, explaining the observed unique morphology and conductivity as a function of composition.

ACKNOWLEDGMENT

The authors wish to thank the US-Israel CDR Program, Agency for International Development, Washington D. C., for supporting the work presented in this publication, Grant No. HRN-5544-G-00-2-66-00. R. Tchoudakov is also grateful to the Giladi Fund and Israel Ministry of Science for supporting her work.

REFERENCES

1 M. Sumita, K. Sakata, S. Asai and K. Miysaka, The Sixth Annual Meeting, PPS, Nice 1990, Francc.
2 F. Gubbels, E. Vanlathem, R. Lerome, R. Deltour and Ph. Teyssie, The Second International Conference on Carbon Black, Mulhouse (F), 1993.
3 M. J. Folkes, **Processing, Structure and Properties of Block Copolymer**, *Elsevier Eds.*, London - New York, 1985, p.14.
4 A. Medalia, *Rubber Chem. Technol.*, **59**, 432 (1986).
5 N. K. Dutta, N. Roy Choudhury, B. Haidar, A. Vidal, J.-B. Donner, L. Delmotte and J. M. Chezear, *Polymer*, **35**, 4293 (1994).
6 S. Radhakrishnan and D. R. Saini, *Polymer International*, **34**, 1, 111 (1994).
7 J. Sakamoto, S. Sakurai, K. Doi and S. Nomura, *Polymer*, **34**, 4837 (1993).
8 R. Tchoudakov, O. Breuer, M. Narkis and A. Siegmann, Polym. *Polym. Networks Blends*, **6**, 1-8 (1996); *Polym. Eng. Sci.*, **36**, 1336 (1996).
9 K. Miasaka, K. Watanabe, E. Jojima, H. Aida, M. Sumita and K. Ishikawa, *J. Mater. Sci.*, **17**, 1610 (1982).
10 J. A. Manson and L. H. Sperling, **Polymer Blends and Composites**, *Plenum Press,* New York, 1981, p. 132.

Rheological Characterization of an Electrically Conductive Composite

Allen C. Nixon
Raychem Corporation, 300 Constitution Drive, Menlo Park, CA 94025, USA

INTRODUCTION

Dynamic mechanical properties of carbon black composites based on rubber or elastomers have been studied for many years. An extensive review was written by A.I. Medalia.[1] Earlier work by Payne[2] studied the strain dependence of the moduli. A more recent review by Gerspacher[3] summarizes the dynamic viscoelastic properties of carbon black filled elastomers. In comparison, very little research has been done on the dynamic mechanical properties of carbon black filled polyethylene composites. Below the melt, crystallinity dominates the mechanical properties. Above the melt, however, the behavior is not unlike that of elastomeric composites.

EXPERIMENTAL

We tested composites based on two types of carbon black. One was a commercial grade as received. The other was heat treated at 1100°C to modify its surface composition. Loadings of 2.5 to 40v% of the modified and unmodified carbon black were compounded into a commercial high density polyethylene (HDPE) using a Brabender mixer. Samples were then granulated, pressed into 1 mm thick slabs in a hot press, and quenched in a cold press. 25 mm diameter disks were punched out of the slab using an arbor punch.

Rheological testing was conducted on a Bohlin VORM controlled rate rheometer, Bohlin CSM controlled stress rheometer, and a Rheometrics Mechanical Spectrometer RMS-705. The 25 mm disks were loaded into a preheated rheometer using parallel plate geometry and held for 10 minutes before testing. All isothermal testing was done at 190°C. Oscillatory measurements were made at a frequency of 10 radians/second.

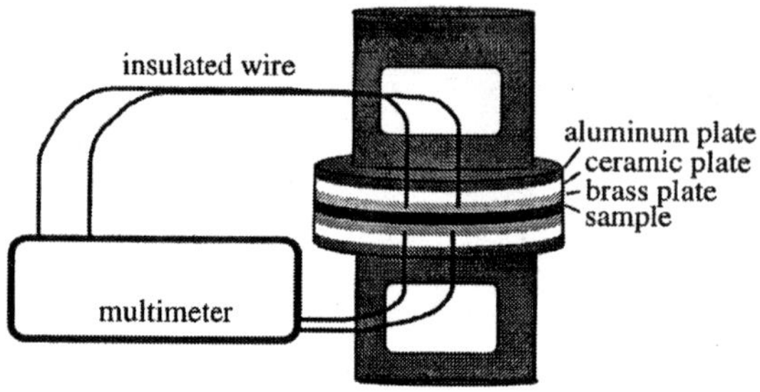

Figure 1. Electrical-rheological plate fixture for Rheometrics RMS-705 rheometer.

Custom rheometry plates were made to enable electrical resistivity and shear modulus to be measured simultaneously as seen in Figure 1. Ceramic plates were sandwiched between standard aluminum plates (that fit in the RMS disposable plate fixture) and brass plates which contact the sample. Steel screws were used to secure the plates together and make electrical contact with the leads. These screws were insulated from the disposable plate fixture by ceramic spacers. The leads were connected to a Fluke multimeter with which resistance measurements were made.

RESULTS

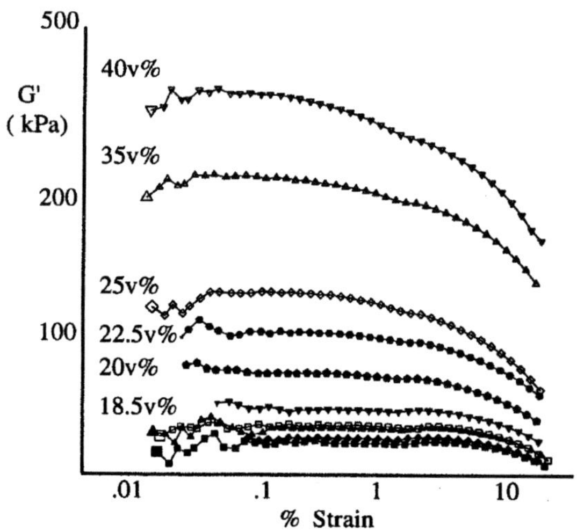

Figure 2. Effect of unmodified carbon black loading on G' strain dependence. Bohlin VORM, 10 radians/sec., 190°C.

The elastic shear modulus (G') as a function of dynamic strain amplitude is shown for composites of 2.5 to 40v% unmodified carbon black in Figure 2. Figure 3 is the corresponding graph for composites of 10v% to 40v% surface modified carbon black. G' values for the composites with less than 17v% carbon black are independent of strain up to about 5% strain and all lie within experimental error of each other. The noise seen at the lower strains was due to low torque values. Above about 18v% carbon black, G' becomes increasingly dependent on strain amplitude as the carbon black loading increases. At the low strain limit (about 0.5% strain) G' is at a maximum and independent of strain. Both the unmodified and modified carbon black composites show increased G'(low strain) at about 18v% carbon black which is consistent with a percolation theory.[4] At 25v% carbon black and

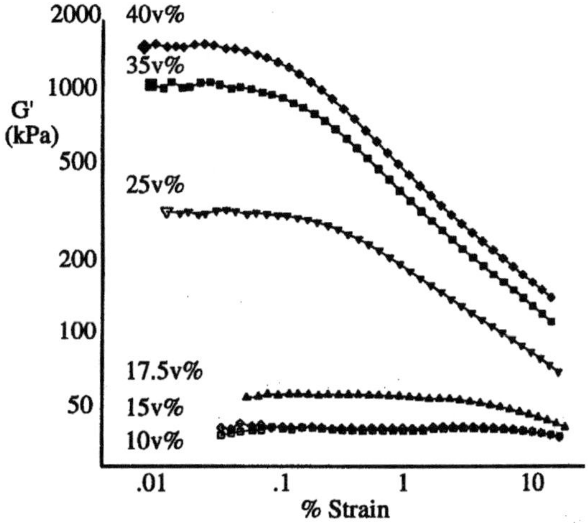

Figure 3. Effect of surface modified carbon black loading on G' strain dependence. Bohlin VORM, 10 radians/sec., 190°C.

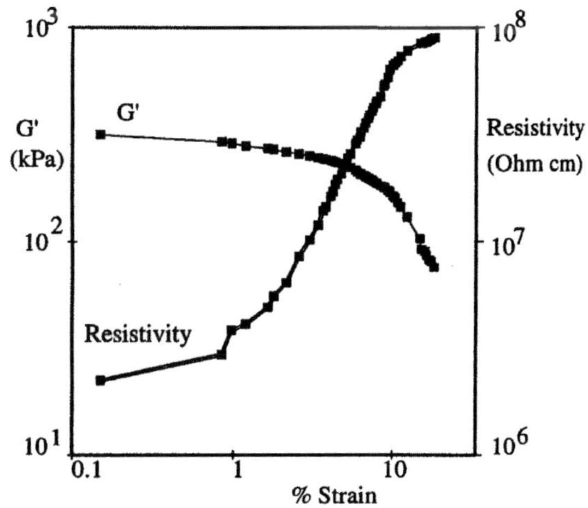

Figure 4. Simultaneous G' and resistivity measurements of 35v% unmodified carbon black composite with increasing strain amplitude. RMS-705, 10 radians/sec., 190°C using electrical-rheological fixture.

above, the G'(low strain) and the strain dependence of G' were much greater for the surface modified carbon black composite than for the unmodified carbon black composites.

Resistance measurements were made concurrent with G' measurements (using the RMS custom plates) as a function of strain amplitude. In Figure 4, the resistivity increases and G' decreases with increasing strain amplitude for the unmodified 35v% carbon black composite. This result is consistent with the findings of Voet and Cook[5] for carbon black in rubber composites.

The resistivity decreased with decreasing temperature while G'(low strain) increased. The sharp change below 130°C was due to crystallization of the composite.

Shear creep compliance for the unmodified 35v% carbon black composite for the curves from 32 to 128 Pa are within experimental error. The creep compliance increases between shear stresses of 128 Pa and 256 Pa. This indicates yield stress behavior. The exponential shape of the curves is as expected for an uncrosslinked entangled polymer.[6]

No yield stress behavior is seen and there is an apparent linear relationship between the compliance and time. This linear relationship was unexpected. Exponential curves are typical for entangled polymers.[6]

CONCLUSIONS

Above a critical volume fraction of carbon black, the G' of the composite is very strain dependent. The critical volume fraction dependence is consistent with percolation theory and the strain dependence agrees with that recorded in the literature for other systems. These results support the theory that a carbon black network (possibly secondary agglomeration) is being measured.

Simultaneous electrical and rheological measurements show a correlation between G'(low strain) and resistivity. Increased strain amplitude or temperature results in decreased G'(low strain) and increased resistivity. This correlation supports the theory that the strain dependence of G' is due to the breaking of a carbon black network. The surface modified carbon black composite showed significantly greater G'(low strain) and strain dependence of G' than did the unmodified carbon black composite. This indicates a stronger interaction between the surface modified carbon black particles.

The linear relationship between shear creep compliance and time at constant stress for the surface modified carbon black composite also indicates a strong influence of the carbon black network. The low compliance at short times (low strain) is consistent with the high G' at low strains. With time (and strain), the carbon black network breaks down, and the creep compliance increases linearly. The unmodified carbon black composite shows more typical polymeric behavior. The creep compliance increases exponentially with time due to the relaxation of chain entanglements. This increased "polymeric" behavior may be due to lower carbon black network interaction, or a greater polymer interaction with the carbon black network.

ACKNOWLEDGMENTS

The composites were provided by Mark Wartenberg, Larry Smith, Art Lopez and Joe Pachinger of Raychem Corporation in conjunction with their research projects.

REFERENCES

1 A.I. Medalia, *Rubber Chem. Technol.*, **60**, 45-61 (1987).

2 A.R. Payne, *J. Appl. Polym. Sci.*, **8**, 2661-2686 (1965).

3 M. Gerspacher, in **Carbon Black Science and Technology**, 2nd ed., Edited by J-B. Donnet, R.C. Bansal, and M-J. Wang, *Marcel Dekker, Inc.*, New York, Chap. 11 (1993).

4 R.D. Sherman, L.M. Middleman, S.M. Jacobs, *Polym. Eng. Sci.*, **23**(1), 36-46 (1983).

5 A. Voet, F.R. Cook, *Rubber Chem. Technol.*, **41**, 1207-1215 (1968).

6 J.D. Ferry, **Viscoelastic Properties of Polymers**, 3rd ed., *John Wiley and Sons*, Inc., New York, 37-39 (1980).

Estimation of the Volume Resistivity of Conductive Fiber Composites by Two New Models

Mark Weber
Research and Technology Centre, Calgary, Alberta, Canada
M. R. Kamal
Department of Chemical Engineering, McGill University, Montreal, Quebec, Canada

INTRODUCTION

Polymers reinforced with electrically conductive particles can be used in applications where electromagnetic interference (EMI) shielding is required.[1-4] The modelling of the electrical properties of conductive composites increases the understanding of the relationship between fiber properties and composite behavior. There exist several models and theories which predict electrical conductivity. Foremost among these is the percolation theory. The essence of percolation theory is to determine how a given set of sites, which may be regularly or randomly positioned in some space, is interconnected.[5] Inherent to the theory is the fact that at some critical probability, called the percolation threshold, a connected network of sites is formed which spans the sample, causing the system to "percolate". The fraction of fibers or fillers required to achieve percolation can be modelled by a Monte Carlo method.[5-10] Predictions have shown agreement with experimental data,[9-11] but the results are dependent on many variables, including lattice size, particle-particle penetration, tunneling effects, and particle dimensions.

Although the percolation theory has received the greatest attention as a predictor of electrical conductivity, other models have also been proposed. Bueche[12] considered the problem of conductive particles in a nonconducting matrix as analogous to the concept of polymer gelation, as proposed by Flory. D'Ilario and Martinelli[13] attempted to fit experimental data for poly(*p*-phenylene sulfide) reinforced with iron and graphite particles with the Bueche model. The calculated thresholds did not agree with the experimental results. Nielsen[14] extended the equations from the theory of elastic moduli to calculate the electrical and thermal conductivities of two-phase systems. Fiber and matrix conductivities are required, and the maximum packing fraction of particles must be estimated. Bigg[15] and Berger and McCullough[16] used the Nielsen model to predict the resistivity of aluminum particle composites. The discrepancy between model predictions and experimental data was very large. McCullough[17] modified a generalized combining rule for transport properties for application to percolation transport mechanisms. The equation predicts the composite conductivity in either the longitudinal, transverse, or normal directions Berger and McCullough[16] found that the generalized combining rule equation showed good agreement with experimental data for an aluminum powder-polyester composite. Ondracek[18] derived a model for field properties of multiphase materials which are at equilibrium and whose microstructure is homogeneous. A model structure is assumed to be similar to the real structure. Excellent agreement was obtained between model predictions and experimental data when the matrix and particle conductivities were of similar magnitudes. When the fiber conductivity was much greater than the matrix conductivity, the agreement was much worse. Another proposed model is the effective medium theory, which replaces the inhomogeneous medium found in an actual composite with a homogeneous "effective" medium. For a conductive composite, the original lattice consisting of randomly distributed conductances is replaced by a lattice of similar symmetry. The conductances are regularly placed so that the electrical properties are, on average, identical in each case.[19] The model does not predict a percolation threshold and is insensitive to changes in the fiber aspect ratio.

In general, predictions from the above models are only in agreement with experimental data when the fiber and matrix conductivities are similar. The percolation theory is able to accurately predict the percolation threshold in a conductive composite, but cannot predict the actual conductivity of such a sample. In this paper, two models are proposed which predict the electrical properties of conductive fiber composites using microstructural data. Background information on each is provided, along with the relevant equations. Predictions from the models are compared to experimental data for nickel-coated graphite fiber-polypropylene composites.

THEORETICAL

END-TO-END MODEL

Starting from basic principles, a relation between electrical conductivity and microstructural parameters is derived for a sample consisting of connected "strings" of fibers in a polymer matrix. The fibers are assumed to be connected end-to-end. The matrix conductivity is very small, so the composite conductivity is determined by the fiber conductivity. For more information regarding these derivations, refer to the thesis of Weber.[20]

Figure 1 shows conductive fibers embedded in a polymeric matrix. The fibers have a length l and a diameter d, and are aligned at an angle θ which describes their orientation rela-

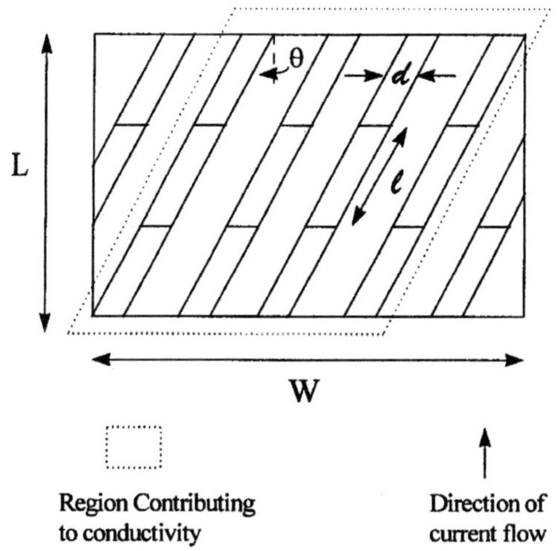

Figure 1. Sample containing connected strings of fibers, oriented at angle θ to the test direction.

tive to the test direction. They are contained in a composite sample with an overall length L, width W, and thickness T. Since the matrix does not contribute to the conductivity of the composite, the latter is determined by the fiber contribution. When the fibers have such an alignment, the sample size becomes important. If the sample is too small or narrow, it is possible that no fibers will traverse the test direction, and the composite conductivity will be zero. For the present derivation, it is assumed that the sample is always large enough so that it is conductive. It should be emphasized that, in many composite samples, the fibers will have a range of orientations. Therefore, there will generally be a significant volume fraction of fibers that contribute to the conductivity. Thus, the composite conductivity in the x-direction, $\sigma_{c\ long}$, is proportional to the fiber conductivity, σ_f, and is dependent on the number of conductive strings. By relating the number of strings of conductive fibers to the volume fraction of fibers in the sample, ϕ, the following relationship for the resistivity in the longitudinal direction can be obtained:

$$\rho_{c\ long} = \frac{\rho_f}{\phi \cos^2 \theta} \qquad [1]$$

Using a similar derivation, the resistivity in the transverse direction can be calculated as:

$$\rho_{c\ trans} = \frac{\rho_f}{\phi \sin^2 \theta} \qquad [2]$$

When the fibers are perfectly aligned parallel to the test direction, θ is zero, the longitudinal resistivity becomes inversely proportional to the fiber volume fraction, and the resistivity in the transverse direction (Equation 2) is zero. Similarly, when the fibers are perfectly aligned parallel to the sample width, θ is 90°, and Equation 2 becomes $1/\phi$. These equations give the lower bounds for resistivity in the longitudinal and transverse directions.

FIBER CONTACT MODEL

Most of the existing resistivity models do not accurately predict the volume resistivity of a composite because they do not account for particle-to-particle contact. It is usually assumed in these models, as well as in the model described in the previous section, that the resistivity of the connected string is equal to the resistivity of the fiber. However, this is an idealized case and only gives a lower bound for the resistivity. The contacts between fibers are rarely end-to-end; they are usually end-to-body or, most likely, body-to-body. The area of contact for these situations is much smaller than in perfect end-to-end alignment, and thus will have an effect on composite resistivity.[21,22] Therefore, a model which accounts for realistic contacts is needed. Batchelor and O'Brien[23] have derived a model for the thermal or electrical conduction through a granular material consisting of conducting particles in a matrix. The conductivity of the particles is very high, and the ratio of particle to matrix conductivity is much greater than one. The model derived in this paper applies the Batchelor and O'Brien model, which was developed for thermal conductivity, to electrical conductivity and extends its application to fiber-filled composites. Refer to the work of Weber[20] for further details regarding the mathematical derivation of the equations.

For composites, where the conductivity of the inclusion is large compared to that of the matrix, essentially all of the current flows through the inclusions. The potential gradient within a particle is very small, except near points of contact with other particles. In the vicinity of these points, the magnitude of the current density and the gradient of potential are large compared to values far from a contact point. Therefore, the conditions near the contact points determine the total current through the particle. The following model assumes a small, flat circle of contact between the fibers, and accounts for the percolation threshold. The resistivity in the longitudinal and transverse directions are derived as:

$$\rho_{c\,long} = \frac{\pi d^2 \rho_f X}{4\phi_p d_c l \cos^2\theta} \qquad [3]$$

$$\rho_{c\,trans} = \frac{\pi d^2 \rho_f X}{4\phi_p d_c l \sin^2\theta} \qquad [4]$$

where: d = fiber diameter, ρ_f = volume resistivity of fiber, X = factor related to fiber contacts, ϕ_p = volume fraction of fibers participating in conductive strings, d_c = diameter of circle of contact, l = fiber length, θ = average angle of orientation

Equations 3 and 4 show the dependence of the resistivity on fiber length, orientation, and volume fraction, as well as the area of contact. The predicted relationship between composite volume resistivity and fiber orientation and volume fraction is identical to that found in the simple model derived in the previous section. These equations represent a percolation type of model, and include quantitative parameters which can account for the orientation, length, and concentration of the fibers, as well as the nature of particle-particle contact.

RESULTS AND DISCUSSION

END-TO-END MODEL

The end-to-end model predicts that the resistivity is dependent on test direction and fiber orientation. Predictions from the model are compared to experimental data for nickel-coated graphite fiber-reinforced polypropylene, processed by compression molding, extrusion, and injection molding.[20] The effect of sample size is again omitted. For the compression molded and extruded samples, the experimental data has shown that the resistivity is independent of the size of the samples. The injection molded specimens are assumed to be large enough relative to the length of the fibers so that sample size effects will be negligible. The orientation of the fibers in the composites is needed in the Equations and was determined as follows. An average orientation parameter is calculated using the data from all the fibers in the sample.[20] From this, an average angle of orientation, is found. Therefore, the actual microstructure in a sample is replaced by an "effective" microstructure. Each fiber in the sample is assumed to be connected end-to-end and has the same effective fiber orientation, length, and diameter. The average fiber length is for the entire sample. Table 1 summarizes the effective microstructure in the composites.

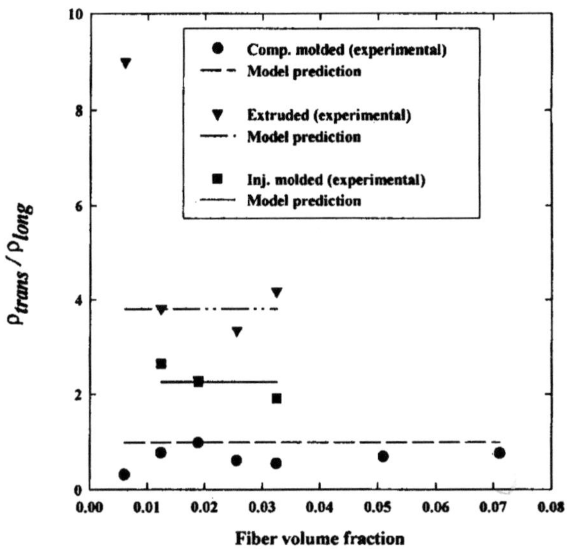

Figure 2. Comparison between aligned fiber model predictions of sample anisotropy and experimentally determined anisotropy.

The anisotropy of the composites can be determined by a ratio of transverse to longitudinal resistivity. Figure 2 compares experimental data to model predictions for compression molded, extruded, and injection molded samples. The model predictions are obtained by dividing Equation 2 by Equation 1, to give

$$\frac{\rho_{c\,trans}}{\rho_{c\,long}} = \cot^2\theta \qquad [5]$$

The ratio given by Equation 5 is independent of volume fraction. As seen in Figure 2, the ratio of the experimental resistivities is also independent of fiber concentration. Agreement between the predictions and experimental results is good for all processing methods. The simple model derived from basic principles predicts the general behavior of the conductive composites and gives the correct relation between resistivity and fiber concentration and orientation.

FIBER CONTACT MODEL

Volume resistivity predictions in the longitudinal and transverse directions are made using Equations 3 and 4. The fibers in the composites are assumed to have a flat circle of contact. Table 1 presents the values of the parameters used in the equations for the compression molded, extruded, and injection molded nickel-coated graphite fiber composites.[20] For all cases, the number of contacts, m, was assumed to vary from a minimum of 2 to a maximum of 15. The ratio of the contact diameter to the fiber diameter, d_c/d, was held constant at 4×10^{-5}. Figure 3 compares the experimental resistivity values and model predictions in the longitudinal direction, while Figure 4 gives similar results in the transverse direction. The prediction of the longitudinal and transverse resistivities by the fiber contact model are in agreement with the experimental results. The difference between the predictions and experimental data is very small in both the extruded and injection molded composites. The predictions for the compression molded plaques are also similar to the experimental data, but disagree slightly near the percolation threshold. The assumptions in the model are only valid above the critical

Table 1. Values of parameters used in fiber contact model

Sample	Fiber length, μm	Fiber diameter, μm	Angle of orientation, θ	ϕ_c	β
Compression molded					
Longitudinal alignment	394	7.6	45.2	0.01	25
Transverse alignment				0.01	25
Extruded					
Longitudinal alignment	819	7.6	27.2	0.003	34
Transverse alignment				0.005	22
Injection molded					
Longitudinal alignment	278	7.6	33.6	0.009	24
Transverse alignment				0.011	25

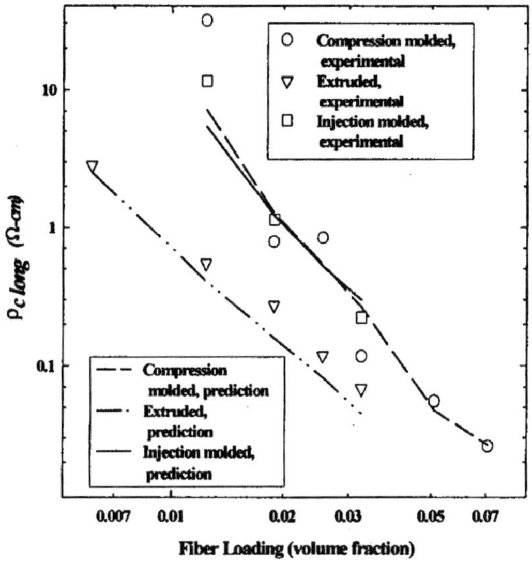

Figure 3. Comparison between fiber contact model predictions and experimental data (longitudinal direction).

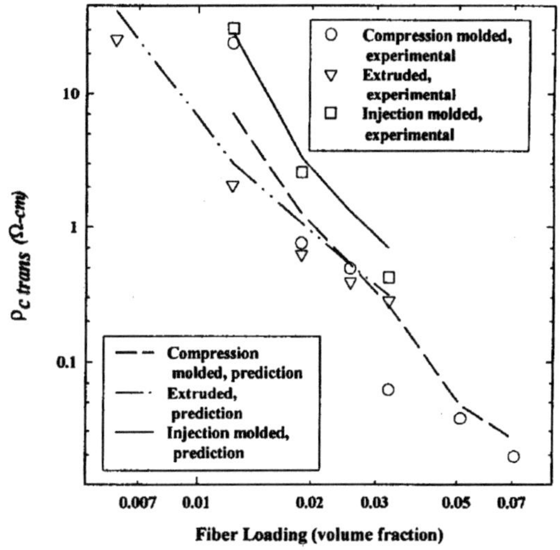

Figure 4. Comparison between fiber contact model predictions and experimental data (transverse direction).

concentration, so the model predictions in this vicinity are expected to show the greatest deviation. Using microstructural data in the model equations, along with the concepts of a percolation threshold and fiber contact, produces excellent results.

CONCLUSIONS

The ability of existing theories to predict electrical properties of conductive fiber composites has been shown to be lacking. Few models account for microstructural details, percolation threshold, and fiber-fiber contacts. Two models which predict the resistivity of a composite from microstructural data are presented. Starting from basic principles, the first model predicts the general behavior and anisotropy of the composite. Fiber orientation and concentration are accounted for. Due to the assumptions made, this model provides lower bounds for the composite resistivity. To obtain more realistic predictions, the effect of fiber-fiber contact must be considered. The second model accounts for these contacts, and includes the effect of fiber length. Predictions from this model are in excellent agreement with experimental data. Both models extend our understanding of the relationship between electrical properties and microstructure of conductive composites.

REFERENCES

1 D.M. Bigg, *Adv. Polym. Technol.*, **4**, 255 (1984).
2 T. Katsura, M.R. Kamal, and L.A. Utracki, *Adv. Polym. Technol.*, **5**, 193 (1985).
3 B.A. Luxon and M.A. Murthy, *SPE ANTEC Tech. Papers*, **32**, 233 (1986).
4 L. Li and D.D.L. Chung, *Composites*, **25**, 215 (1994).
5 G.E. Pike and C.H. Seager, *Phys. Rev. B: Solid State,* **10**, 1421 (1974).
6 N. Ueda and M. Taya, *J. Appl. Phys.*, **60**, 459 (1986).
7 M.T. Kortschot and R.T. Woodhams, *Polym. Compos.*, **9**, 60 (1988).
8 I. Balberg and N. Binenbaum, *Phys. Rev. B: Condens. Matter*, **28**, 3799 (1983).
9 E.A. Holm and M.J. Cima, *J. Am. Ceram. Soc.*, **72**, 303 (1989).
10 S.F. Wang and A.A. Ogale, *Compos. Sci. Technol.*, **46**, 93 (1993).
11 A. Dani and A.A. Ogale, *SPE ANTEC Tech. Papers*, **40**, 1392 (1994).
12 F. Bueche, *J. Appl. Phys.*, **43**, 4837 (1972).
13 L.D'Ilario and A. Martinelli, *J. Mater. Sci. Lett.*, **10**, 1465 (1991).
14 L.E. Nielsen, *Ind. Eng. Chem. Fundam.*, **13**, 17 (1974).
15 D.M. Bigg, *Polym. Eng. Sci.,* **17**, 842 (1977).
16 M.A. Berger and R.L. McCullough, *Compos. Sci. Technol.*, **22**, 81 (1985).
17 R.L. McCullough, *Compos. Sci. Technol.*, **22**, 3 (1985).
18 G. Ondracek, *Rev. Powder Metall. Phys. Ceram.*, **3**, 1205 (1987).
19 R.D. Sherman, L.M. Middleman, and S.M. Jacobs, *Polym. Eng. Sci.*, **23**, 36 (1983).
20 M.Weber, **Ph.D. Thesis**, *McGill University*, Montreal, Quebec, Canada (1995).
21 C. Rajagopal and M. Satyam, *J. Appl. Phys.*, **49**, 5536 (1978).
22 B. Bridge, *J. Mater. Sci. Lett.*, **7**, 663 (1988).
23 G.K. Batchelor and R.W. O'Brien, *Proc. R. Soc. London, Ser. A*:, **355**, 313 (1977).

Effect of Thermal Treatment on Electrical Conductivity of Polypyrrole Film Cast from Solution

J.Y. Lee, D.Y. Kim, and C.Y. Kim
Polymer Materials Lab., Korea Institute of Science and Technology, Seoul, Korea
K.T. Song, S.Y. Kim
Dept. of Fiber and Polymer Science, Seoul National Univ., Seoul, Korea

INTRODUCTION

Polymers containing conjugated double bonds are very attractive since the materials show the characteristics of a semiconductor and become electrically conductive on doping. The discovery of electrically conductive polymers[1-4] generated excitement to show a great potential in applications. Polypyrrole (PPy) is successfully used as a key material in an electrolytic capacitors.[5,6] Polyaniline has been tried as an electrode in a polymer battery[7] and as an electromagnetic interference shielding material in polyvinyl chloride.[8] Polyphenylenevinylene is semiconductive and capable of emitting light.[9]

Despite many attractive properties, electrically conductive polymers had a shortcomings in processability since the materials were infusible or insoluble. Soluble polymers with conjugated double bonds have been developed by modification of the monomers but showed poor electrical conductivities.[10,11] Polyaniline soluble in organic solvents has been finally developed to produce a film or sheet.[12-15] Soluble PPy is also introduced without sacrificing the conductivity.[16,17]

Electrical conductivity of the conducting polymer must be stable upon heating since the polymer is used or processed at elevated temperatures in the most cases. The present work examined the effect of thermal treatment on electrical conductivity of the PPy film cast from the solution in chloroform. Electrical conductivities of the films with various thickness were monitored upon heating in air or nitrogen. UV-VIS-NIR spectroscopy was carried out to correlate the electrical property with changes in the chemical structure of the polymer.

EXPERIMENTAL

Pyrrole (Aldrich) was dried with CaH_2 for 24 hours, followed by distillation under reduced pressure. Ammonium persulfate (APS, Kanto Chemical Co. Inc.) as an oxidant and dodecylbenzene sulfonic acid (DBSA, Janssen Chimica) as a dopant were used as received.

A typical polymerization process is as follows. 0.15 mol DBSA and 0.3 mol pyrrole were dissolved in 500 ml distilled water with vigorous stirring. To the above solution of which temperature was controlled with an accuracy of ±0.1°C, 0.03 mol APS in 100 ml of distilled water was slowly added. Reaction was carried out for 24 hours at -2°C and then terminated by pouring excess methanol. The resultant PPy powder was filtered and washed sequentially with distilled water, methanol and acetone, followed by filtering and drying in a vacuum oven at 25°C for 12 hours.

PPy powder obtained (1 g) was completely dissolved by ultrasonification in 25 ml of chloroform with additional 1 g of DBSA, and filtered through a 1 μm Teflon membrane filter. The solution was transferred onto a glass plate and the solvent was dried, resulting in an excellent quality of free-standing films with the thickness of 35 and 100 μm. A thin film with the thickness of 4 μm was obtained by using the spin-casting technique. The film cast from the chloroform solution was washed with acetone to leach out the extra DBSA. Intrinsic viscosity of the PPy solution in chloroform was measured by using a Ubbelohde viscometer. Electrical conductivity of the free-standing film was measured by the four-probe method for the bulk conductivity and by the two-probe method for the surface resistivity.

A Perkin-Elmer System 2000 NIR FT-Raman spectrometer was used to investigate the absorption characteristics of the resulting PPy in the NIR range. UV-VIS spectrum up to 820 nm was recorded by a Hewlett Packard HP 8452A diode array spectrometer. The two spectra were combined at 800 nm, leading to UV-VIS-NIR spectrum from 300 nm to 3000 nm. TGA was carried out using a DuPont 2950 Thermogravimetric Analyzer under nitrogen or air with the scan rate of 20°C/minute.

RESULTS AND DISCUSSION

Molecular weight of PPy is known to be dependent on the polymerization conditions such as the concentration of the oxidant and the polymerization time. There seems to be the critical molecular weight for a good electrical conductivity as shown in Figure 1. The conductivity increases by three decades with an increase of the intrinsic viscosity from 0.072 to 0.094 dl/g. A further increase in the viscosity raises the conductivity by less than one decade. The conductivity of the pellet pressed from PPy powder generally increases with an increase of the oxidant concentration. However, PPy formed chemically with a higher concentration of the oxidant failed to be soluble.

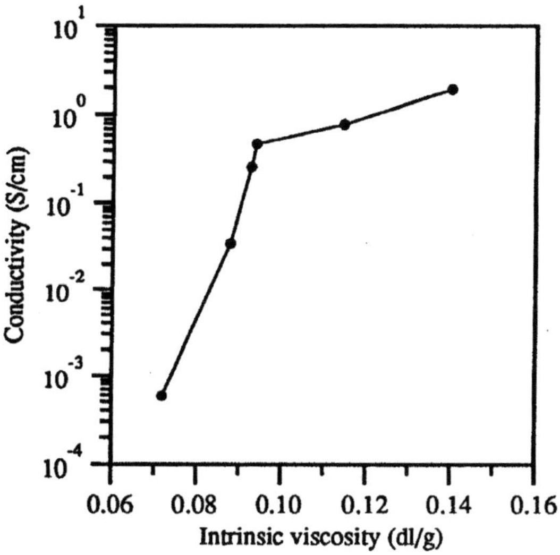

Figure 1. Electrical conductivity of PPy film as a function of intrinsic viscosity.

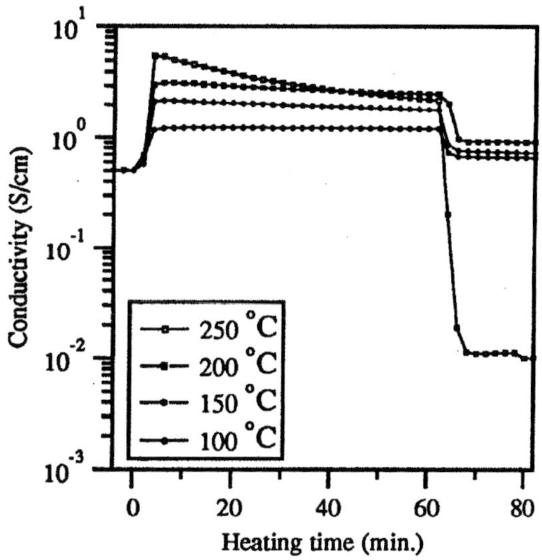

Figure 2. Electrical conductivity changes of 100 μm thick PPy film cast from chloroform solution with heating time at various heating temperatures.

Changes in electrical conductivity of a 100 μm thick PPy film upon isothermal heating at different temperatures are shown in Figure 2. The conductivity jumps rapidly by up to one decade in the beginning of the heating to the set temperature. The increase in the conductivity is larger when the set temperature is higher. The polymer shows no decrease in the conductivity during isothermal heating at up to 200°C for one hour. The conductivity of PPy slowly decreases when heated at 250°C.

It is interesting to note that the conductivity is improved when the polymer is cooled to room temperature after heating and the higher the heating temperature the higher the conductivity unless the conductivity decreases during the heat treatment. Higher conductivity at room temperature after heating than the original value indicates that there must be rearrangement of the polymer molecules on heating to show an improvement in the conductivity.[18] It would be improved stacking in molecules to enhance the planarity and the conductivity. The shrinkage of the polymer on cooling to room temperature disturbs just a little the planarity improved in the bulk on heating to decrease the conductivity but not completely. Heating the polymer to 250°C probably causes chemical reaction in it to break the conjugation in the molecules. The polymer loses the conductivity by about three decades after being heated at 250°C for one hour and then cooled to room temperature.

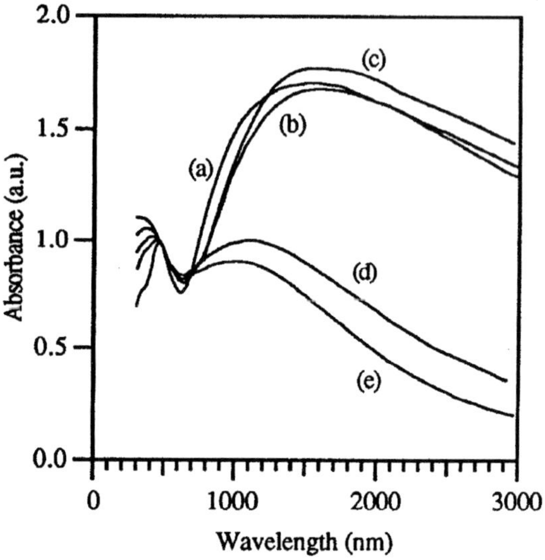

Figure 3. UV-VIS-NIR spectra of PPy film spin-cast from chloroform solution upon thermal heating at 200°C for (a) untreated, (b) 5, (c) 15, (d) 30 and (e) 60 minutes, respectively.

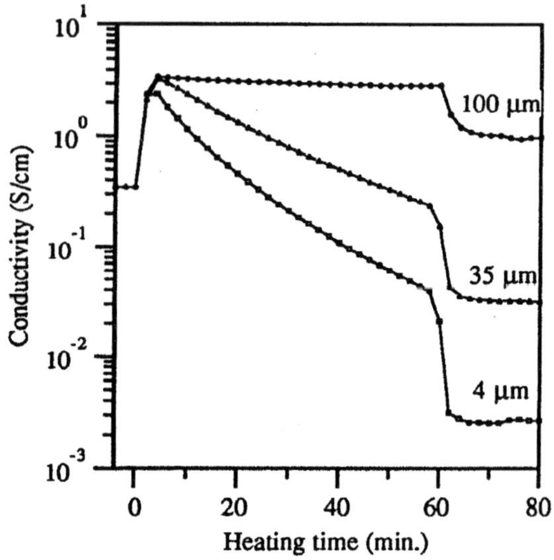

Figure 4. Electrical conductivity changes of PPy film with various thicknesses upon isothermal heating at 200°C.

It must be emphasized that the PPy film cast from the solution stands the heat treatment at up to 200°C for 60 minutes without losing the conductivity. The material may be used in polymer compounding processes as well as in electronic devices.

UV-VIS-NIR spectra of a PPy film with the thickness of 1 μm upon heating at 200°C for different periods of time are shown in Figure 3. The absorption at 1560 nm is considered as the electronic transition from the valence band to the bipolaron band[19] and the peak height may be correlated with electrical conductivity. The absorbance at 1560 nm changes little with the heating time of up to 15 minutes at 200°C. However, the absorbance decreases significantly after heating the polymer for 30 minutes, while the change in electrical conductivity of a 100 μm thick PPy film is not observed as shown in Figure 2.

Figure 4 shows the thicker a PPy film the more stable the conductivity on the thermal treatment at 200°C. The 35 and 4 μm thick PPy films gain the conductivities on heating at 200°C as much as that of the 100 μm thick PPy film does in the beginning. However, the conductivities of the thinner films decrease rather rapidly with time of the treatment while the conductivity of 100 μm thick film decreases little with the treatment time. The conductivities of the films always drop significantly when cooled to room temperature after heating and the drop becomes larger with a thinner PPy film.

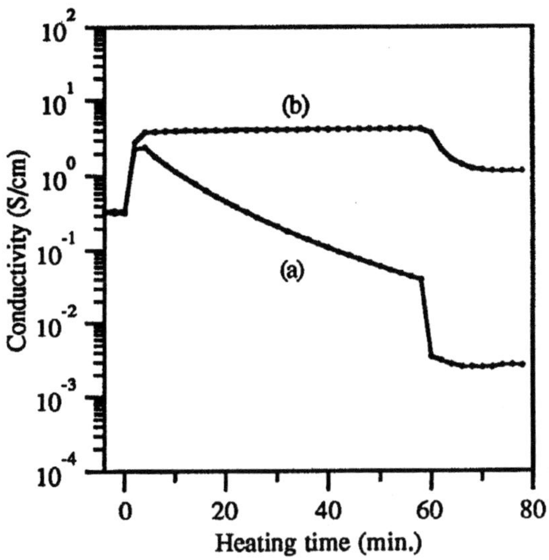

Figure 5. Changes in surface resistivity of 4 μm thick PPy cast film on heating at 200°C in (a) air, (b) nitrogen.

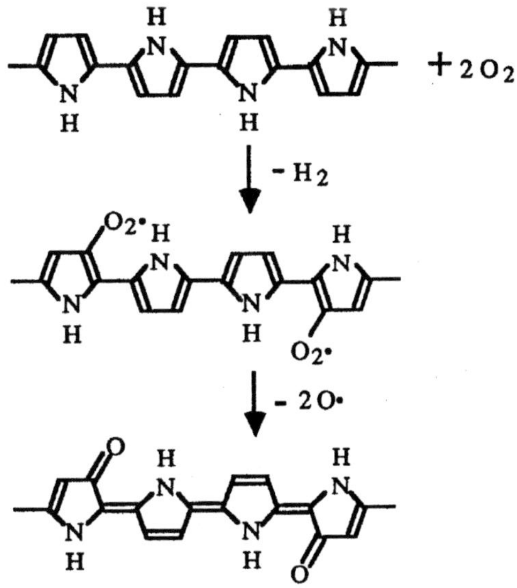

Figure 6. Chemical structural changes of PPy on heating in air.

The PPy film seems to be very sensitive to oxygen as shown in Figure 5. The 4 μm thick PPy film gains the conductivity on the initial heating at 200°C in nitrogen atmosphere as much as in air. The film heated at 200°C for 60 minutes in nitrogen shows higher conductivity at room temperature than that of the untreated film, while the conductivity of the film heated in air is lower than that of the untreated film by two decades. It is generally known that the carbon at the position of 3 or 4 in the pyrrole ring is vulnerable to oxygen and the hydrogen on the carbon is easily substituted by oxygen to cause the conjugation moiety to brake.[20-22]

The mechanism of the oxygen substitution is proposed as shown in Figure 6. The hydrogen bound to the carbon at the position 3 is extracted on heating to leave a free radical on the carbon. Oxygen is attached to the carbon with the free radical to make a peroxide group. Rearrangement of the peroxide group releasing an oxygen free radical leads to a carbonyl group. The conjugation in the polymer is then broken to show a lower electrical conductivity.

Even though the conductivity of a 100 μm thick PPy film changes a little on heating at 200°C in either air or nitrogen, the surface resistivity of the PPy film changes significantly when the film is treated in air at 200°C as shown in Figure 7. The surface resistivity drops while heating to 200°C either in nitrogen or air, indicating an increase of the conductivity. However, the resistivity increases faster on heating the specimen in air whereas the heating in

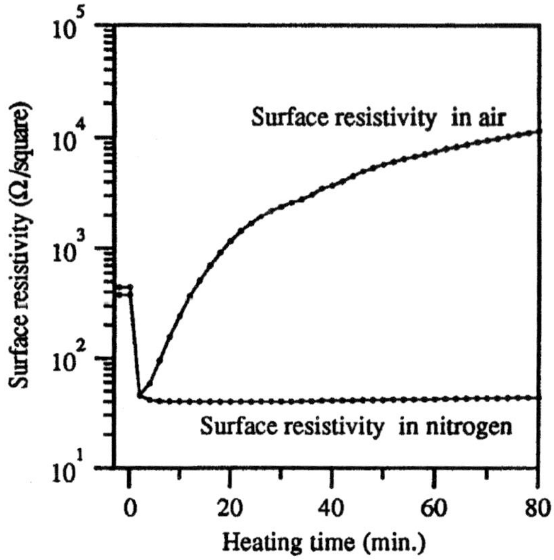

Figure 7. Changes in surface resistivity of 100 μm thick PPy cast film upon heating at 200°C in air and nitrogen.

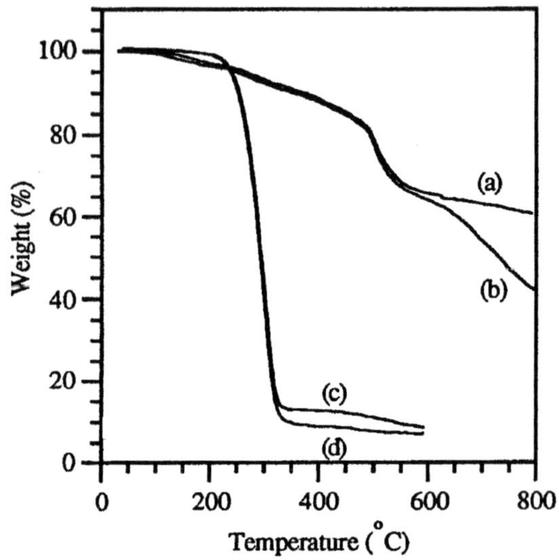

Figure 8. TGA curves of PPy cast film (a) in nitrogen and (b) in air and DBSA, (c) in air and (d) in nitrogen.

nitrogen causes no effect on the surface resistivity. The conductive polymer is a basically stable material but reactive to oxygen at elevated temperature.

The NIR spectra in Figure 3 indicated that there were changes in the chemical structure on heating the PPy film at 200°C. It is clear now that the thermal treatment in air damages the PPy film and the damage diffuses into the bulk from the film surface, while nitrogen causes no damage in the polymer on heating and the conductivity is preserved.

PPy doped with DBSA anion loses weight just a little until heated to 230°C while no change in weight of DBSA is detected on heating to the same temperature as shown in Figure 8. Since no weight change of the doping acid is observed on heating to 200°C, the slight weight loss in the polymer must be due to degradation of the polymer to release fragmented molecular segments. The decrease in the conductivity on heating in air, therefore, should not be correlated with decomposition of the dopant. The change in the conductivity on heating must be directly related to the conjugation length in PPy due to oxidation.

CONCLUSIONS

The concentration of the oxidant in polymerization controls molecular weight of PPy which has to exceed a critical point to show metal-like electrical conductivity. Heating the PPy film cast from the solution raises electrical conductivity and the higher the temperature up to 200°C the better the conductivity. The conductivity changes little with the heating time of 60 minutes at the isothermal temperature of up to 200°C. Thermal treatment of the polymer at 250°C raises the conductivity even higher at the early stage but lowers the conductivity gradually to that at 200°C in 60 minutes. Cooling the specimens to room temperature lowers the conductivity but still higher than the original one except the specimen treated at 250°C. The heating effect on the decrease of the conductivity is only observed when the polymer is heated in air and no change in the conductivity is observed when treated in nitrogen. The dopant is found to be very stable on heating to 200°C and shows no effect on the decrease in the conductivity on heating. It is proposed that the carbon at the position 3 in the pyrrole ring is vulnerable to oxygen and the attack by oxygen disturbs the conjugation in the polymer. The rate of the conductivity decrease seems to be controlled by diffusion of oxygen into the polymer.

REFERENCES

1 A.G. MacDiarmid, C.K. Chiang, C.R. Findher Jr., Y.W. Park, A.J. Heeger, H. Shirakawa, E.J. Louis, and S.C. Gau, *Phys. Rev. Lett.*, **39**, 1098(1997).
2 K. Kanazawa, A.F. Diaz, R.H. Geiss, W.D. Gill, J.F. Kwak, J.A. Logan, J.F. Robolt, and G.B. Steet, *J. Chem. Soc., Chem. Commun.*, 1979, 854(1979).
3 G. Tourillon and F. Ganier, J. *Electroanal. Chem. Interfacial Electrochem.*, **135**(1), 173(1982).
4 A.G. MacDiarmid, J.C. Chiang, A.F. Richter, and A.J. Epstein, *Synth. Met.*, **18**, 285(1987).
5 Y. Kudoh, S. Tsuchiya, T. Kojima, M. Fukuyama, and S. Yoshimura, *Synth. Met.*, **41**, 1133(1991).
6 L.G.M. Krings, E.E. Havinga, and J.J.T.M. Donkers, *Synth. Met.*, **54**, 453(1993).
7 T. Fuse, *Electronics*, 1987, 38(1987).
8 Incoblend Product Information Sheet (Zipperling-Kessler, Germany and Americhem, USA).
9 J.H. Burroughes and R.H. Friend, *Proc. of Mat., Res. Soc., Symp.*, **173**, 425(1990).
10 H. Masuda, S. Tanaka, and K. Kaeriyama, *J. Chem. Soc., Chem. Commun.*, **11**, 725(1989).
11 P. Audebert, P. Aldebert, N. Girault, and T.Kaneko, *Synth. Met.*, **53**, 251(1993).
12 M. Angelpoulos, A. Ray, A.G. MacDiarmid, and A.J. Epstein, *Synth. Met.*, **21**, 21(1987).
13 A. G. MacDiarmid and A.J. Epstein, **Science and Applications of Conducting Polymers**, edited by D.T. Clarke and E.J. Samuelson, *IOP Publishing Ltd.*, Bristol, 1990, pp. 117.
14 Y. Cao, G.M. Treacy, P. Smith, and A.J. Heeger, *Appl. Phys. Lett.*, **60**, 271(1992).
15 A.G. MacDiarmid and A.J. Epstein, *Synth. Met.*, **65**, 103(1994).
16 J.Y. Lee, D.Y. Kim, and C.Y. Kim, *Synth. Met.*, **74**, 103(1995).
17 J.Y. Lee, D.Y. Kim, K.T. Song, S.Y. Kim, and C.Y. Kim, *Mol. Cryst. Liq. Cryst., accepted.*
18 Y. Wang and M.F. Rubner, *Synth. Met.*, **41**, 1103(1991).
19 G.B. Street, **Handbook of Conducting Polymers**, Vol. 1., edited by T.A.Skotheim, *Marcel Dekker*, New York, Chap. 8, 1986, pp. 284.
20 L.A. Samuelson and M.A. Druy, *Macromolecules*, **19**, 824(1986).
21 W. Liang, J. Lei, and C.R. Martin, *Synth. Met.*, **52**, 227(1992).
22 H.H. Kuhn, A.D. Child, and W.C. Kimbrell, *Synth., Met.*, **71**, 2139(1995).

Creation of Electrically Conducting Plastics by Chaotic Mixing

Radu I. Danescu and David A. Zumbrunnen
Laboratory for Materials Processing and Industrial Mixing, Department of Mechanical Engineering, Clemson University, Clemson, SC, USA

INTRODUCTION

Electrically conducting plastics are commonly made by adding conductive fillers such as metallic powders or carbon black. When particulate additives are mechanically mixed with molten polymers, particle chains may arise by chance as a result of the random positions of individual particles. Such chains, or networks, provide conducting pathways. High shear rate flows, which characterize ordinary compounding methods, such as extrusion or batch mixing, lead to random and nearly uniform particle distributions. The likelihood for extended conducting pathways to form is accordingly low at small particle concentrations. If the particle concentration is higher than a certain value called the percolation threshold,[1] these associations among particles yield a network that spans the material and the solidified composite becomes electrically conductive.

Means to reduce the percolation threshold are of significant practical importance, as the addition of particles can degrade the desirable properties of plastic materials, introduce processing complexities, and fillers are in general costly. Creating a structured filler distribution is an effective way to reduce the percolation threshold since the particle density in some regions is higher than average. Structured distributions can be created by adding a conductive filler, such as carbon black, to a two-phase polymer blend where the filler becomes concentrated selectively in one phase or at interfaces.[2] Results were reported regarding polyethylene/polystyrene blends filled with carbon black.[3] A percolation threshold of 1.5 wt% was found where particles became concentrated in one phase and of only 0.5 wt% where the particles accumulated at interfaces. The processing conditions in the latter case are however very restrictive. Where particles concentrate in one phase, the conductivity depends appreciably on the blend morphology.

It has been shown that nearby fluid elements can be made to follow unique trajectories within cavities where fluid motion is laminar and induced solely by the simple, slow, periodic motion of bounding surfaces.[4] Deformable minor phase bodies become repeatedly stretched and folded and exhibit within physical space responses characteristic of chaotic behavior. For this reason, the collective motion of fluid elements in such conditions is generally referred to as chaotic mixing. Fibrous and lamellar microstructures of less than 1 micron diameter or thickness have been formed in this manner within polymer melts beginning with large minor phase bodies of about 4 mm diameter.[5,6] The microstructures have been associated with enhanced physical properties due to internal interconnections and morphological complexities.[7-9] Chaotic advection has been widely studied in an eccentric cylindrical cavity formed between two offset rotating cylinders.[10] Theoretical and experimental results are available regarding the processing conditions under which advection is chaotic.[11,12]

This paper reports results of an experimental investigation aimed to create electrically conductive polymer composites at low filler concentrations by chaotic mixing. Conductive carbon black was added to a polystyrene melt contained in an eccentric cylinder cavity and very fine-scale structures were created which were captured by rapid solidification. Micrographs of the composite structures and the electrical measurements are presented.

EXPERIMENTAL PROCEDURES

Conducting networks were formed by progressively stretching and folding by chaotic mixing initial minor phase bodies that were much larger than the ultimate characteristic dimension in the composites. These initial minor phase bodies were produced by blending in a batch mixer Rheomix 600 (Haake Inc., Saddle Brook, NJ) atactic polystyrene (Novacor Chemical, Inc.) and carbon black Printex XE-2 (Degussa Corporation, Akron, OH). This carbon black is characterized by a high porosity and a highly extended, bulky nature of aggregates, with considerable branching and chaining.[13] The carbon black content was 6 wt%, which was more than twice the percolation threshold for electrical conductivity (i.e., 2.8 wt%) for this carbon black in polystyrene.[14] After blending, the material was pelletized in a grinder, and pellets varying in size from 1 to 1.4 mm were retained using a series of standard sieves of decreasing hole size. In parallel, pure polystyrene, which was originally in a 2.5 x 4 mm granular form, was also ground. Pellets of 1 to 1.4 mm size were sorted using the same procedure. Filler-rich and filler-free pellets were combined at the desired overall carbon black content by shaking known masses of each in a plastic bag. Batches with a carbon black loading C varying from 0.4 to 2.5 wt% were produced. In this manner, normal mixing was relied upon to achieve compositional uniformity and chaotic mixing was used to develop structure *in situ*.

Two-dimensional chaotic mixing was achieved using an apparatus recently developed in conjunction with a prior study[15] and modified for the purposes of the present work. A 130 mm long cylindrical glass tube of inner diameter $D_1 = 36$ mm was housed in an electrically heated oven. The tube was attached to a lower shaft which was rotated by a servo motor. A cylindrical steel rod of diameter $D_2 = 13$ mm descended into the glass tube and was rotated by a separate servo motor via an upper shaft. The axes of the tube and the rod were parallel and offset by $e = 6.24$ mm. The servomotors were connected to the rod or tube via 100:1 speed reduction gear boxes and were controlled by a digital motion controller, with a servo-amplifier. The controller was programmable through a personal computer.

Prior to processing, the mixture of filler-rich and filler-free pellets was melted to remove voids. Additional pellet mixture was placed into the cavity in order to replace the void volume. In this manner, the mixing cavity was filled with a large, vertical, void-free span of pellets where two-dimensional mixing conditions could be induced. Prior to activating the servo motors, the mixture was melted by heating the oven to 210°C and maintaining this temperature for 15 minutes. In order to instill chaotic advection, each cylinder was rotated periodically and separately, for a specified angular displacement, θ, and then stopped. One complete set of rotations comprised one mixing period. Equal linear displacements at the surface of the two boundaries and equal times for rotation were imposed by setting the angular displacement and the angular velocity ratios equal to the diameter ratio: $\omega_1 / \omega_2 = \theta_1 / \theta_2 = D_2 / D_1$, with $\theta_1 = 0.75$ and $\omega_1 = 1$ rpm. The geometry and mixing protocol were selected according to the results of Swanson and Ottino,[12] such that chaotic mixing prevailed throughout the cavity. At the end of mixing, the sample was solidified by discharging an air jet array onto the glass tube.

After removal from the oven, the glass tube was broken and the steel rod was dislodged so as to free the sample for subsequent examination and testing. A 0.2 mm thin slice and 8 - 12 specimens were prepared from each sample. The slice was oriented perpendicular to the longitudinal axis of the sample and was located 30 mm from the bottom of the sample. This position was located in the portion of the sample where the mixing was strongly two-dimensional. The slice underwent optical analysis using a stereoscopic microscope equipped with a video camera and a high resolution monitor. Micrographs were digitized and the contrast of images was enhanced using image processors. Specimens from different locations within the cross section of the cavity were examined. Those located in the vicinity of the smaller gap between the two cylinders are referred to as group A specimens. Group B specimens were taken from the larger gap. Some specimens were fractured at room temperature in a plane normal to the longitudinal axis of the cavity and were examined by electron microscopy. The electrical resistivity of other specimens was measured. A conducting silver coating was applied on the opposite ends of these specimens to ensure good electrical contact between the specimens and the test leads. A precision digital multimeter was then used to determine the electrical resistivity, ρ_0.

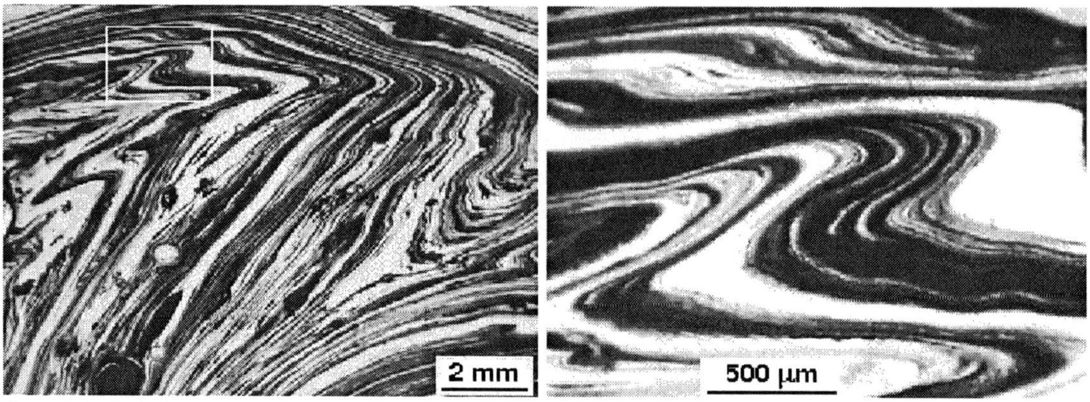

Figure 1. Optical micrographs at different magnifications showing thin, sinuous structures formed in the larger gap between the cylinders, for C = 0.4 wt%, D_1 = 36 mm, D_2 = 13 mm, e = 6.24 mm, θ_1 = 0.75, θ_2 = 2.08, N = 10, T = 210°C. (a - left, b - right).

In order to identify an optimal number of periods over which large, extended networks formed, separate samples were processed for $6 < N < 20$ for the same carbon black loading, C, of 1 wt%. Networks eventually broke down at longer processing times as particles became increasingly distributed by chaotic mixing. Additional samples were then processed with this optimal number of periods in a systematic study regarding the relationship between the filler content and the electrical resistivity.

RESULTS AND DISCUSSION

An example of the structures in group B specimens is shown in the optical micrograph of Figure 1a. Thin, sinuous striations were formed from individual carbon-rich pellets that were initially randomly dispersed. The structures include length scales much smaller than the physical size of the mixing cavity. An image at higher magnification of the boxed region in Figure 1a is shown in Figure 1b. In this micrograph, the separation distance between adjacent striations ranges from 10 to 20 μm. The similarity between patterns at different scales indicates that still smaller separation distances can be achieved by continuing the mixing process for a larger number of periods. If these distances are comparable to the distance for electron transfer across thin polymer gaps, the long and sinuous structures become interconnected, and conducting networks can arise.

In Figure 2, highly stretched and folded striations are connected to compact bodies of carbon black-rich material which appear as small convoluted regions. These compact bodies may serve as junctions to connect individual thin striations. This mechanism may arise when an initial carbon-rich region is subjected to both high and low stretching rates, such that portions of the region are formed into thin striations while other portions remain less deformed.

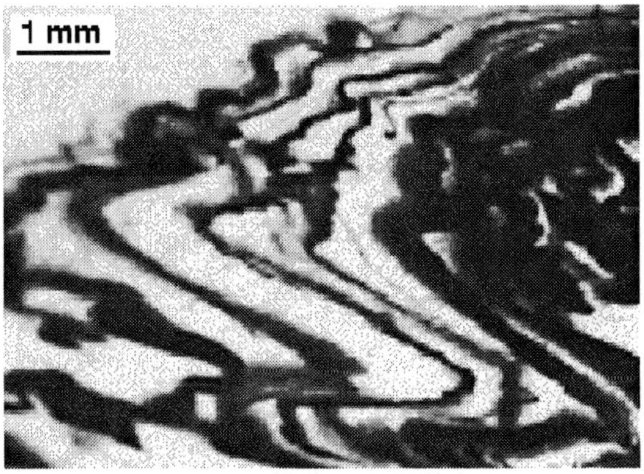

Figure 2. Carbon black structures in a polystyrene matrix revealing a mechanism for extended conducting network formation with C = 1%, D_1 = 36 mm, D_2 = 13 mm, e = 6.24 mm, θ_1 = 0.75, θ_2 = 2.08, N = 10, T = 210°C.

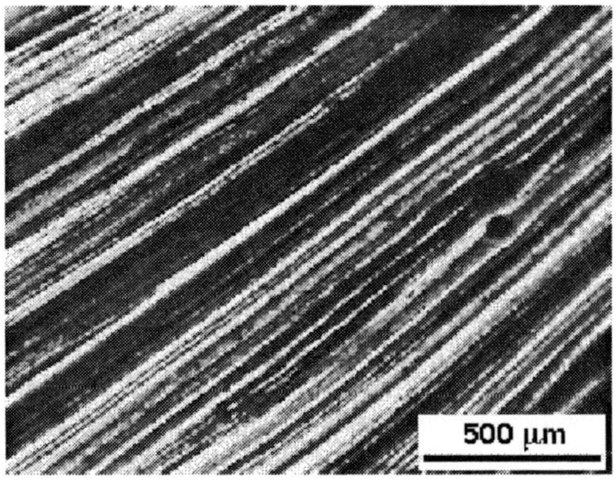

Figure 3. Thin, parallel, coherent microstructures for C = 1%, D_1 = 36 mm, D_2 = 13 mm, e = 6.24 mm, θ_1 = 0.75, θ_2 = 2.08, N = 10, T = 210°C.

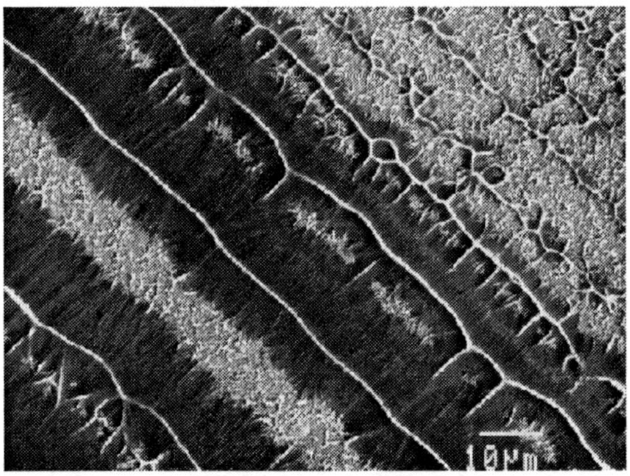

Figure 4. SEM micrograph displaying micro-size, parallel carbon black-rich structures in a pure polystyrene matrix for C = 1%, D_1 = 36 mm, D_2 = 13 mm, e = 6.24 mm, θ_1 = 0.75, θ_2 = 2.08, N = 10, T = 210°C.

Large differences in stretching rates characterize chaotic mixing, so that the presence of both compact bodies and filament structures is expected.

Microstructures found in the vicinity of the small gap between the cylinders are shown in Figure 3. In this region, chaotic mixing studies have shown that stretching rates are large in comparison to those in the large gap. Thin, parallel structures are formed as a result when the melt moves into this region. The close proximity of the striations provides opportunities for interconnections.

Several micron-size, parallel structures can be observed in the SEM micrograph in Figure 4 of a fracture surface. Black areas represent the pure polystyrene matrix, gray portions correspond to the carbon black-rich material, while the white lines are fracture lines in polystyrene. The distinctive boundary between the matrix and the conductive mixture indicates that the carbon black concentration inside the structures remained high. Some structures, such as the one starting at the bottom right corner, are so tenuous that they broke down into short segments, and electrical continuity was lost. However, solid continuous

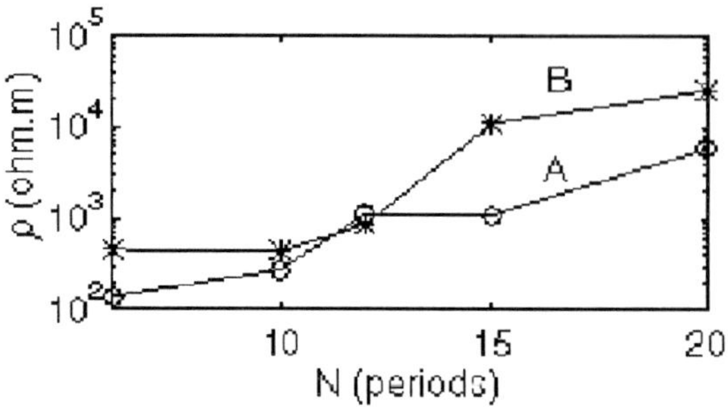

Figure 5. Dependence of the electrical resistivity on the mixing time for C = 1%, D_1 = 36 mm, D_2 = 13 mm, e = 6.24 mm, θ_1 = 0.75, θ_2 = 2.08, N = 10, T = 210°C.

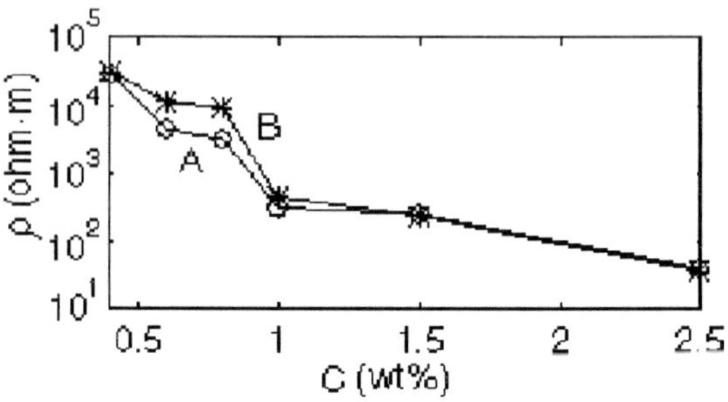

Figure 6. The effect of the carbon black loading on the electrical resistivity for C = 1%, D_1 = 36 mm, D_2 = 13 mm, e = 6.24 mm, θ_1 = 0.75, θ_2 = 2.08, N = 10, T = 210°C.

structures are present, such as the second striation in the lower left corner, which provide the desired connectivity. The distances between striations in the upper right corner are very small, so that opportunities exist for these structures to interconnect and form networks, as discussed in relation to Figure 3.

Electrical resistivities differed among individual specimens taken within each sample. This was a predictable outcome of chaotic mixing, where unique structures are produced at scales much larger than the filler particle size. The variability was greater for lower filler concentrations. The resistivity for both group A and group B specimens which were processed for different numbers of mixing periods is given in Figure 5 for 1 wt% carbon black. The resistivity was in general lower for group A than for group B specimens, owing to the better organized structures in the higher stretching rate regions. However, the differences are less than one order of magnitude. Resistivity for both groups increased as the carbon black became more uniformly distributed so that structures formed in the earlier stages of mixing were broken down. Optical examination and the resistivities in Figure 5 suggested that 10 periods was an optimal number of processing periods for subsequent tests.

The effect of the carbon black loading on the electrical resistivity of the composite is shown in Figure 6. A sudden and large reduction in resistivity occurred in the range 0.8 wt% < C < 1.0 wt%. Such reductions are indicative of a percolation threshold. It is notable that a percolation threshold of 0.8 wt% is 71% lower than the percolation threshold of carbon black in polystyrene attained by conventional compounding methods.[14] This value is also 47% lower than the percolation threshold of carbon black in two-phase polymer blends where the minor phase polymer promotes the formation of structure in order to reduce filler loading. The value is slightly higher than reported percolation thresholds where carbon black migrates to the interface between polymer phases.[3] For $C > 1$, the resistivity decreased gradually to about the resistivity of 15 ohm-m for the initial filler-rich minor phase bodies. Unlike in conventional mixing, the specimens remained conductive, although with significantly higher resistivities at carbon black concentrations below the percolation threshold. This result demonstrates that conducting structures can be formed at very low concentrations in contrast to current processes where thermoplastics become nearly perfect insulators as the concentration of added particles is reduced.

CONCLUSIONS

Chaotic mixing was induced in melts containing bodies rich with conducting particles to create percolating networks and produce conducting plastic materials. In contrast to common mixing processes, the development of structures among particles and the resulting percolation conditions did not rely on chance alone but were instead an outcome of stretching and folding events which characterize chaotic motions. Micrographs of carbon black conducting structures in polystyrene matrices showed coherent small scale structures. Measurements revealed a marked decrease in electrical resistivity that is characteristic of a percolation threshold at carbon black concentrations significantly lower than currently achieved by conventional methods. In contrast to mixtures produced by normal compounding methods, specimens also remained conducting at filler concentrations below the percolation threshold although with significantly higher resistivities.

ACKNOWLEDGMENT

Financial support from the National Science Foundation of the United States of America under Grant No. CMS-9253640 is gratefully acknowledged in conjunction with a Presidential Faculty Fellow Award to D. A. Zumbrunnen.

REFERENCES

1 Broadbent, S.R. and Hammersley, J.M., 1957, *Proc. Camb. Phil. Soc.*, **53**, 629-635.
2 Sumita, M., Sakata, M., Asai, S. And Nakagawa, H., 1991, *Polym. Bulletin.*, **25**, 265-271.
3 Gubbels, F., Blacher, S., Vanlathem, E., Jerome, R., Deltour, R., Brouers, F. and Teyssie, P., 1995, *Macromol.*, **28**, 1559-1566.
4 Aref, H., 1984, *J. Fluid Mech.*, **143**, 1-21.
5 Zumbrunnen, D.A., Miles, K. C. and Liu, Y.H., 1996, *Comp. A*, **27A**, 37-47.
6 Liu, Y. H. and Zumbrunnen, D.A., 1996, *Polym. Comp.*, **17**, 187-197.
7 Danescu, R.I. and Zumbrunnen, D.A., 1997, Proc. IMECE'97, Mat. Div., ASME, New York (in press).
8 Liu, Y.H. and Zumbrunnen, D.A., 1997, Proc. IMECE'97, Mat. Div., ASME, New York (in press); also: *J. Mat. Sci.* (in review).
9 Liu, Y.H. and Zumbrunnen, D.A., 1998, "Progressive Microstructure Development by Chaotic Mixing of Liquid Crystal Polymers and Thermoplastics and Corresponding Tensile Strengths," Submitted for Proc. ANTEC'98.
10 Ottino, J.M., Leong, C.W., Rising, H. and Swanson, P.D., 1988, *Nature*, **333**, 419-425.

11 Aref, H. and Balachandar, S., 1986, *Phys. Fluids*, **29**, 3515-3521.
12 Swanson, P.D. and Ottino, J.M., 1990, *J. Fluid Mech.*, **213**, 227-249.
13 Sichel. E.K., 1982, **Carbon Black Polymer Composites**, *Marcel Dekker, Inc.*
14 Probst, N., 1993, **Carbon Black, Science and Technology**, Donnet, J.B., Bansal, R.C. and Wang, M.J., Eds.; *Marcel Dekker, Inc.*
15 Miles, K.C., Nagarajan, B., and Zumbrunnen, D.A., 1995, *J. Fluid Eng.*, **117**, 582-588.

Production of Electrically Conducting Plastics at Reduced Carbon Black Concentrations by Three-Dimensional Chaotic Mixing

Radu I. Danescu and David A. Zumbrunnen
Center for Advanced Engineering Fibers and Films, Department of Mechanical Engineering, Clemson University, Clemson, SC 29631, U.S.A.

INTRODUCTION

Present and future applications, such as electrostatic dissipation, electromagnetic interference shielding, corrosion prevention or the production of chemical sensors, have created a demand for new electrically conducting plastics. The addition of minor phase materials is a popular method to induce conductivity in otherwise insulating plastics. However, the presence of conducting fillers may degrade other desirable properties of the polymer matrix, such as low weight and good mechanical toughness and strength. High filler loadings also increase costs since filler materials are in general more expensive than the polymeric materials themselves, and since processing can be made more difficult due to a significant increase in the apparent viscosity. Thus, means to reduce the minimum amount of filler necessary to achieve conductivity are of significant practical importance. Creating a structured filler distribution is an effective way to reduce the percolation threshold as the filler concentration in some regions of the material is higher than average. Hence, the probability for a network to form at a given filler loading is greater compared to the random, unstructured distribution which characterizes composites produced by common blending methods.

Chaotic mixing has been investigated in the last fifteen years as a model of mixing.[1,2] Its use to produce highly structured materials by repeated stretching and folding in the melt state of a minor phase within a major phase was proposed recently.[3,4] Chaotic motion in multi-phase polymer melts leads to coherent, fine-scale structures, like fibers and multilayer films, through recursive stretching and folding. Notably, these microstructures are formed from initially large minor phase bodies, as depicted in Figure 1. Such an "auto-processing"

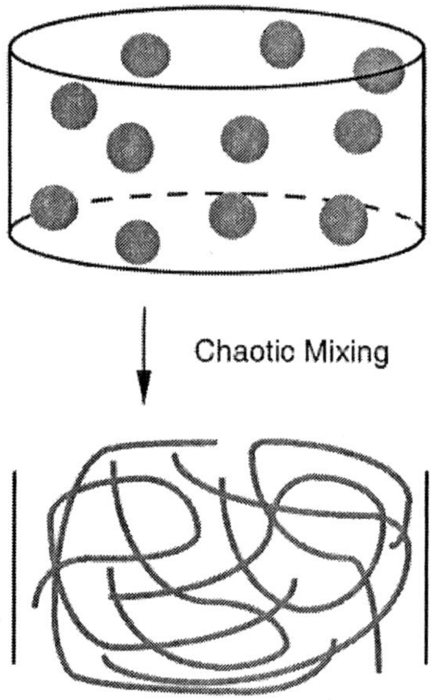

Figure 1. Formation of interconnected continuous structures in a melt from particle-rich molten minor phase bodies by three-dimensional chaotic mixing.

method has the potential of generating fibrous and lamellar blends directly in the mixing step. The structures can be subsequently captured by solidification.

The potential of chaotic mixing to form fine-scale structures offered the potential that conducting networks might be produced *in-situ* by combining in a similar manner plastics with conductive particulate fillers. With these observations, molten polymers were combined in a prior study with conducting carbon black to assess opportunities for the *in-situ* formation of extended particle networks through chaotic mixing.[5] Initially coarse filler-rich bodies, dispersed within a polymer matrix, were transformed into long, sinuous and intertwined structures. Electrical measurements indicated the existence of an optimum mixing time over which the filler particles were sufficiently distributed and extended networks were formed. Electrical conductivity was achieved at filler loadings approximately 70% lower than by ordinary mixing methods.

The results of the two-dimensional chaotic mixing study suggested still lower filler concentrations might be possible if greater interconnectivity between structures could be achieved. An efficient means to do this is to replace two-dimensional chaotic mixing with a three-dimensional process, in a geometry similar to the cylindrical cavity used in prior studies.[6,7] Unlike for the two-dimensional case, fine-scale structures can become entangled due to the motion of advected particles in three-dimensions so that more complex electrical networks can develop within the polymer matrix. This paper reports the results of a systematic study aimed to assess the effectiveness of three-dimensional chaotic mixing as a method to produce electrically conducting polymer composites at low filler loadings.

EXPERIMENTAL PROCEDURES

Three-dimensional chaotic mixing was instilled within a recently developed cylindrical cavity.[7] A 53 mm steel cylinder was housed in an electrically heated oven. Separate disks formed the upper and lower cavity surfaces. The upper disk was concentric with the cylinder, and the lower disk was offset by 32 mm. The cylinder was attached to the oven. The distance between the disks was set at 14 mm, so that the cavity aspect ratio was 0.26. Three-dimensional advection was induced by alternate rotation in opposite directions of the two disks. Each disk was driven by a separate servo-motor via speed reduction gearboxes and drive shafts.

Atactic polystyrene GPPS 555 (Novacor Chemicals, Inc., Leominster, MA) was used as the polymer matrix. Polystyrene was chosen since it is stable at temperatures above its softening point, does not adhere strongly after solidification to metallic surfaces, and morphology inspection is facilitated by its transparency. Highly conductive carbon black, Printex XE-2 (Degussa Corp., Akron, OH), was utilized as the conductive filler. Carbon blacks are popular additives and the electrical characteristics of carbon black-filled plastics are available in the literature.[8,9] High shear premixing was necessary to break up the carbon black pellets and agglomerates. This was achieved in a batch intensive mixer (model Rheomix 600, Haake, Paramus, NJ). A filler-rich polystyrene compound was prepared by mixing the two components for 15 minutes, at a rotational speed of 60 rpm, and at a temperature of 210±10°C. The carbon black content in this uniform blend was 6 wt%, which was more than twice the percolation threshold of 2.8 wt% reported for the same materials.[9] After blending, the material was pelletized in a grinder and pellets varying in size from 1 to 1.4 mm were produced. Pellets of the same size were also made from the polystyrene resin. Filler-rich and filler-free pellets were combined to give a specific overall carbon black content, C, by shaking known masses of each in a plastic bag. Samples with C varying from 0.3 to 3.4 wt% were prepared in this manner.

Prior to chaotic mixing, the cylindrical mixing cavity was filled with a mixture of filler-rich and filler-free pellets. The mixture was melted at 210°C. Chaotic mixing was then induced by rotating alternately the two disks, with the rotation of one disk composing one half period. The rotational displacement of each disk within one period, expressed as the fraction of a complete rotation, was called the perturbation strength and was denoted by μ. As aperiodic mixing protocols lead to more uniform fluid dispersion than periodic ones,[10,11] a 10-period recursive mixing protocol was used, where a sequence segment was always attached to its complement segment of the same length.[12] The geometric parameters that characterized the mixing cavity and those defining the mixing protocol were carefully selected based on experiments and computational modeling[13] to yield chaos throughout the cavity. After the mixing procedure was completed, the sample was maintained in the melted state for 10 to 12 hours at 175°C, then solidified by natural convection at room temperature, according to the procedure proposed elsewhere,[14] to increase linkage between carbon domains.

After solidification, the cylindrical sample was removed from the mixing cavity. Four 15x15x7.5 mm specimens for electrical measurements were prepared from different regions of the sample.[15] A conducting silver coating was applied on the opposite faces of each specimen to ensure good electrical contact between the specimens and the test leads, and the electrical resistance, R, was measured. The resistivity was calculated from R and the dimensions of the specimen. Similar measurements were made on specimens with various carbon black concentrations produced by normal blending in the batch intensive mixer for comparison purposes. Measurements were repeated over a period of 22 weeks after specimen production to assess the stability of electrical conductivity over time. For optical inspection, thin slices were taken from samples in the planes containing and perpendicular to the cavity axis. A stereoscopic microscope equipped with a video camera and a high resolution monitor was used. Samples were illuminated by reflected light from below. Micrographs were digitized using a video digitizer and image processing software.

RESULTS AND DISCUSSION

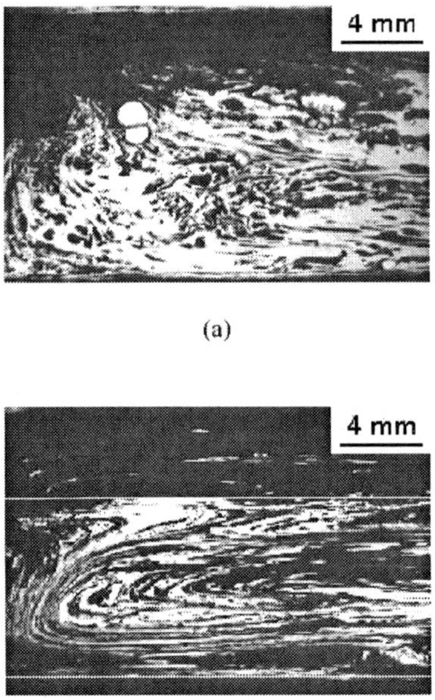

(a)

(b)

Figure 2. Optical micrographs showing structure development for different mixing times with C=1 wt%: (a) N=6, (b) N=10.

In order to understand the electrical resistivities, it is instructive to first consider an example of structure development. As the mixing time in terms of the number of mixing periods N was increased from 6 to 10 periods in Figure 2, stretching and folding of the minor phase bodies of Figure 1 occurred progressively and well-developed structures appeared, albeit at different formation rates. A 5 mm layer adjacent to the upper surface and a 1.5 layer adjacent to the bottom surface of the sample (i.e., beyond the added white lines in Figure 2b) were removed and were not used in electrical resistivity measurements in order to characterize properties for the bulk blend where mixing rates were more similar. Processing with N=10 periods was considered an optimum mixing time. For larger times, interparticle distances became too great to support electrical conduction in much of the sample.

Complex structures displaying the characteristics of chaos were observed at small scales within optical micrographs. For example, the morphology in Figure 3 resembles the complex manifolds of a hyperbolic point in a mathematical phase space.[2] The manifolds offer opportunities for electrical charge transport. The closed loops in the upper right corner suggest the existence of elliptic points where fluid particles locally follow orbit-like paths and remain unmixed in regions called islands. In three-dimensional chaotic mixing, however, such regions might become interconnected to the complex manifolds in distant regions and also serve as charge carriers.

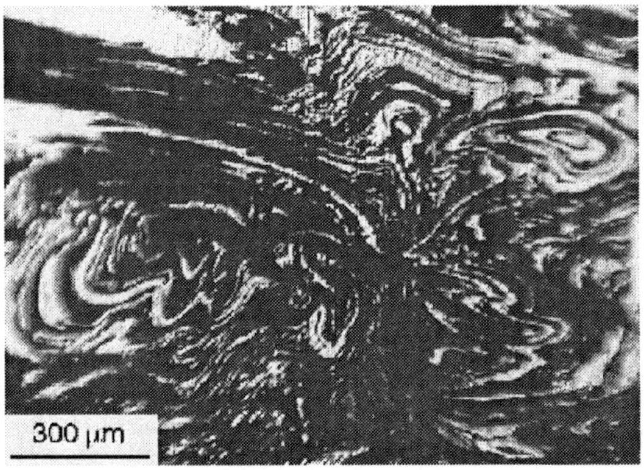

Figure 3. Structures displaying characteristics of chaos for C=1 wt% and N=10.

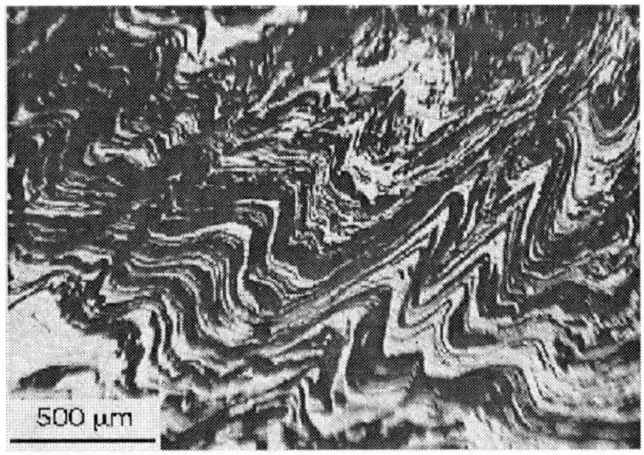

Figure 4. Fine-scale parallel striations developed from initially coarse bodies for C=1 wt% and N=10.

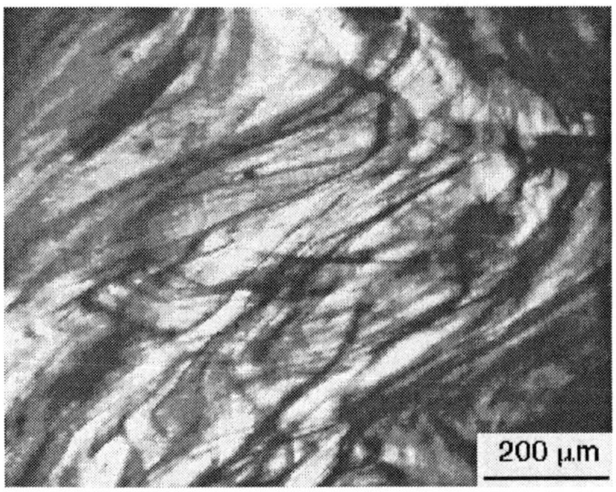

Figure 5. Optical micrograph revealing interconnections among structures for C=0.4 wt% and N=10.

Long, sinuous striations that developed from the much larger, initially coarse, carbon black-rich bodies are evident in the optical micrograph in Figure 4. The striation thickness in the central region of the micrograph is between 10 and 50 microns. Similar formations have been predicted theoretically.[5] Because of the chaotic motion of individual particles and resulting deviations of some particle positions from the striation, opportunities arise for electrical connections between adjacent striations and consequently for the formation of extended networks. The mechanisms for the formation of such networks have been disclosed computationally.[16]

Figure 5 displays a region from a 0.06 to 0.08 mm, thick slice parallel to the upper and lower surfaces of the mixing cavity. Unlike in similar materials formed by two-dimensional chaotic mixing where striations tend to orient about streamlines, striations can intersect and traverse in any direction. Any two striations that seem to intersect are separated in this case by a distance of the order of 10^{-2} mm or less. This close distance provides added opportunities for electrical interconnections and suggests that percolation thresholds might be smaller for three dimensional chaotic mixing than for two-dimensional chaotic mixing.

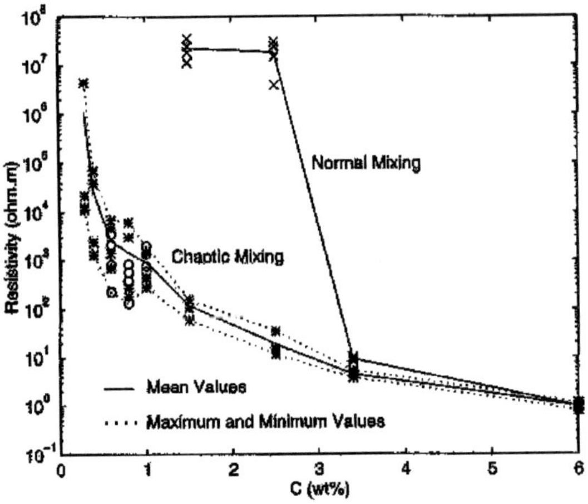

Figure 6. Influence of carbon black content on electrical resistivity of carbon black/polystyrene blends produced by normal mixing and by chaotic mixing for N=10.

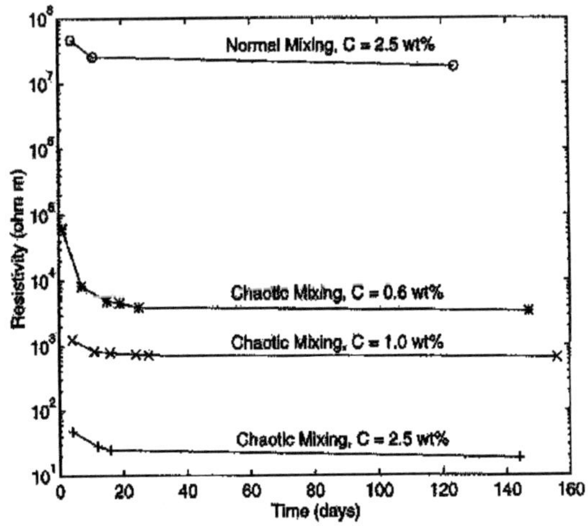

Figure 7. Stability of electrical resistivity over time for composites produced by normal and by chaotic mixing with N=10.

The electrical resistivity of specimens is shown as a function of the carbon black loading, C, in Figure 6. The asterisks and circles give values for various specimens taken from the same sample to indicate the variability in resistivities for the complex microstructures. The mean resistivity, calculated as the arithmetic average of all specimens with the same C, is given as a solid line. Dotted lines were drawn through the highest and lowest resistivities for clarity. Conducting specimens produced by chaotic mixing were produced even at very low filler concentrations of 0.3 wt%, where networks comprising a few thin structures were formed through mechanisms such as explained in relation to Figures 3, 4, and 5. A sharp transition from high to moderate resistivity is evident in Figure 6 for $0.3 < C < 0.6$ wt%. As C approached 6 wt%, the resistivity of chaotically mixed specimens equaled the uniformly mixed material from which the carbon black-rich pellets were prepared. Based on these data, the percolation threshold was estimated to fall between 0.4 and 0.6 wt%. The percolation threshold was between 2.5 and 3.4 wt% in the blends produced in this study by normal mixing with the same materials, and was consistent with the value of 2.8 wt% reported elsewhere.[9] A percolation threshold of approximately 0.5 wt% represents a decrease of over 80%. Three-dimensional chaotic mixing leads to better results than two-dimensional chaotic mixing, as the percolation threshold in the latter case was found to be between 0.8 and 1.0 for the same materials.[5]

The variation of electrical resistivity over time was found by repeating the electrical measurements during the first 22 weeks after the samples were produced. This variation is shown in Figure 7 for samples produced by chaotic mixing and also by normal mixing. A significant decrease in resistivity occurred in the first 7-10 days, but only a minor decrease was detected between the second and the fifth months. In order to eliminate the influence of this instability on the final results, resistivities measured 130 days or more after sample production were reported in Figure 6.

CONCLUSIONS

Three-dimensional chaotic mixing was used to produce conductive structures of carbon black in a polymer matrix. Due to the unique trajectories of individual carbon black particles contained within initially large minor phase bodies, continuous fine-scale structures among particles developed rapidly as processing progressed. Micrographs revealed connections along striations as well as between neighboring striations. Greater opportunities for the formation of electrically conducting networks were provided than when two-dimensional chaotic mixing is used. A percolation threshold occurred at a significantly lower concentration than is achievable by normal mixing since the particles form continuous structures in lieu of random dispersions.

ACKNOWLEDGMENT

Financial support from the National Science Foundation of the United States of America under Grant No. CMS-9253640 in conjunction with a Presidential Faculty Fellow Award to D.A. Zurnbrunnen and also from the Center for Advanced Engineering Fibers and Films, Grant No. ERC-9731680, is gratefully acknowledged.

REFERENCES

1 Aref, H., *J. Fluid Mech.*, **143** (1984), 1-21.
2 Ottino, J.M., **The Kinematics of Mixing**, *Cambridge University Press*, 1989.
3 Zumbrunnen, D.A., 1994, Proc. 1-st International Community for Composites Engineering, New Orleans, LA (1994), 601-602.
4 Liu, Y. H. and Zumbrunnen, D.A., *Polym. Comp.*, **17** (2) (1996),187-197.
5 Danescu, R.I. and Zumbrunnen, D.A., *J. Thermoplastic Comp. Mat.*, **11** (1998),299-320.
6 Zurnbrunnen, D.A., Miles, K. and Liu, Y.H., *Comp. A*, **27A** (1)(1996), 37-47.
7 Miles, K.C., Nagarajan, B. and Zurnbrunnen, D.A., *J. Fluids Eng.*, **117** (1995), 582-588.
8 Gubbels, F., Blacher, S., Vanlathem, E., Jerome, R., Deltour, R., Brouers, F. and Teyssie, P., *Macromolecules*, **28** (1995),1559-1566.
9 Probst, N., 1993, **Conducting Carbon Black, in Carbon Black, Science and Technology**, Donnet, J.B., Bansai, R.C. and Wang, M.J., (eds.), *Marcel Dekker,* 271-288.
10 Franjione, J.G. and Ottino, J.M., *Phil. Trans. R. Soc. Lond.* A (1992), 301-323.
11 Liu, M., Muzzio, F.J. and Peskin, R.L., *Chaos, Solitons and Fractals,* **4** (6) (1994), 869-893.

12 Liu, Y.H and Zumbrunnen, D. A., Toughness Enhancement in Polymer Blends Due to the In-Situ Formation by Chaotic Mixing of Fine-Scale Extended Structures, J. Mat. Sci., in press.
13 Nagarajan, B., M.S. thesis, Clemson University (1994).
14 Bouda, V., 1997, Proc. IMECE, 79 (1997), 281-298.
15 Danescu, R.I., Ph.D. dissertation, Clemson University (1998)
16 Danescu, R.I. and Zumbrunnen, D.A., Computational Simulation of the In-Situ Formation in Melts of Electrical Pathways among Particles by Two-Dimensional Chaotic Mixing, Proc. MIECE (1998), in press.

Preparation of Conducting Composites and Studies on Some Physical Properties

Jun-Seo Park
An Sung National University
Sung-Hun Ryu
Kyung Hee University
Ok-Hee Chung
Sun Chon National University

INTRODUCTION

The improvement of mechanical properties of conducting polymers has increased their potential for commercial applications. One of the immediate applications of the conducting polymers is in electrostatic protection and electromagnetic interference (EMI) shielding. In the last few decades, the synthesis of polyheterocyclic polymers, such as polypyrrole (PPy) and polythiophene (PPt), has received a great deal of attention due to their reasonably high environmental stability and electrical conductivity.[1,2] However, these tend to be insoluble and infusible. A useful approach to press these conducting polymers into useful and large articles is inclusion of conducting polymer in the matrix of a mechanically strong insulating polymers.[3,4] Although chemically prepared polyheterocyclic polymers are of poor quality and low electroconductivity compared with electrochemically prepared polyheterocyclic polymers, advantages of the chemical oxidative polymerization are simple preparation procedures, short reaction times, and mass production. The low penetration of conducting polymer into the host polymer limits the use of a non-porous host polymer. The problem of low penetration of conducting polymer in the matrix can be partially solved by employing a porous polymer as host polymer.[4,5] PPy impregnated conducting composites were prepared by employing filter paper or fabric as host polymer. However, there are limitations for this method regarding the practical applications due to the weakness of host polymer.

In this paper we report a new approach to prepare thick and large conducting composite objects. The composites, based on polyheterocyclic polymers in combination with a porous host polymer, has been prepared via chemical oxidative polymerization. The concentrated emulsion polymerization method is employed to obtain the porous host polymer and an imbibition technique is used to incorporate polyheterocyclic polymers in the host polymer.[5] The scope of this work is to study the effects of initial molar ratios of oxidant to monomer and the effect of environmental stability on the conductivity of composites. This study also includes the examination of the EMI shielding effectiveness of the conducting composites and the temperature dependence on the conductivity to study charge transport mechanism.

PREPARATION OF POROUS HOST POLYMER AND CONDUCTING COMPOSITES

A small amount of a mixture of styrene (6g, Aldrich) and divinyl benzene (1g, Aldrich) containing AIBN (0.01g, Aldrich) and sorbitane monooleate (1g, Span 80, Fluka) was placed in a flask (250ml capacity) equipped with a mechanical stirrer and addition funnel, in which distilled water (25g) was placed. The concentrated emulsion was prepared by dropwise addition of water to the stirred mixture of styrene and divinylbenzene. The polymerization was carried out at 50°C. Subsequently, water of the dispersed phase was removed by heating the polymer at elevated temperature.

The host polymer was first imbibed in monomer (pyrrole or 2,2'-biothiophene, Aldrich)-ether solution and then partially dried to evaporate ether. Subsequently, the partially dried host polymer was soaked again in excess oxidant($FeCl_3$, Aldrich) solution for 20 min. The conducting composite was rinsed with copious water and acetone.

INSTRUMENTATION

The standard four point probe method was employed to measure the conductivity at room temperature. The temperature dependence on the resistance at low temperature was conducted by employing 7T SQUID (Quantum Design, U.S.A.). EMI shielding effectiveness was measured by using the SET 19A Shielding Effectiveness Tester (Egal Instrumentation and Tester, Israel). This instrument is designed to measure the shielding effectiveness of planar materials under free space conditions by the coaxial transmission line method. The morphologies of the porous host polymer and conducting composites were measured by scanning electron microscopy (SEM, Amray, U.S.A.).

RESULTS AND DISCUSSION

Absorption test of the porous host polymer in various liquids indicates that the host polymer shows higher hydrophobicity and can absorb a large amount of organic liquids, i.e., 7.6 g chloroform/g host and 4.5 g toluene/g host.

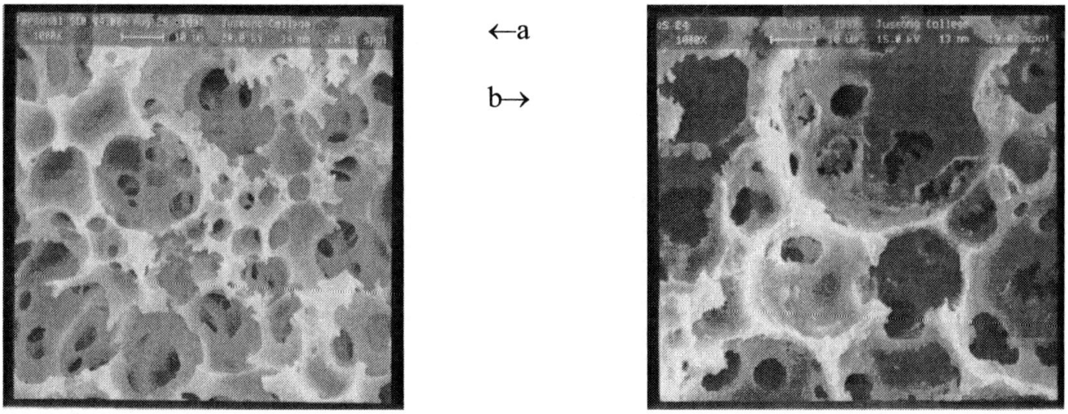

Figure 1. SEM micrographs: (a) the porous crosslinked polystyrene, (b) PPy impregnated composite.

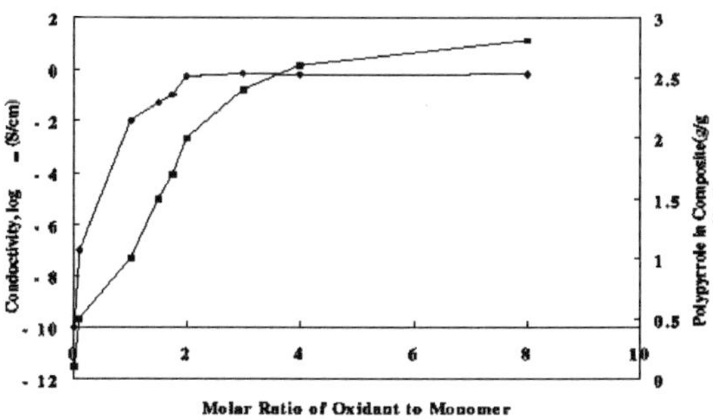

Figure 2. Conductivities and amount of PPy in composites vs. initial molar ratio of oxidant to monomer.

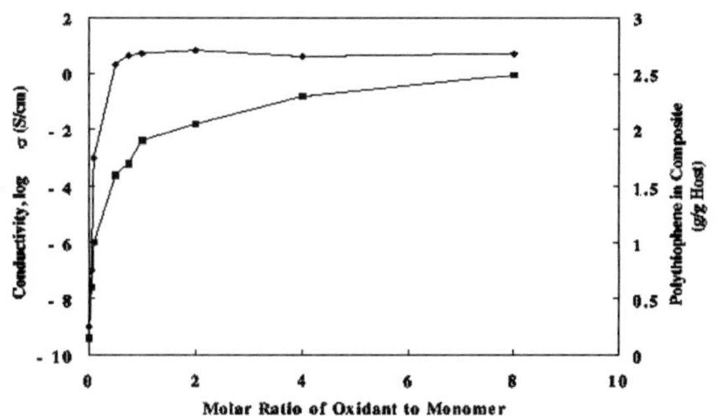

Figure 3. Conductivities and amount of PPt in composites vs. initial molar ratio of oxidant to monomer.

SEM micrographs of Figure 1 present the morphologies of highly porous host polymer and conducting composites. The similarity between host polymer and conducting composite indicates that the polyheterocyclic polymer coats uniformly the inner surface of the host polymer.

Figures 2 and 3 present the effect of the initial molar ratio of oxidant to monomer on the conductivity of the composite and on the amount of conducting polymers formed in the composites. An insulator-to-metal transition occurs at initial molar ratio of 1.5 and 0.75 for PPy impregnated composites and PPt impregnated composites, respectively. Larger amount of oxidant can function both as initiator and dopant. The leveling-off of the conductivity for high

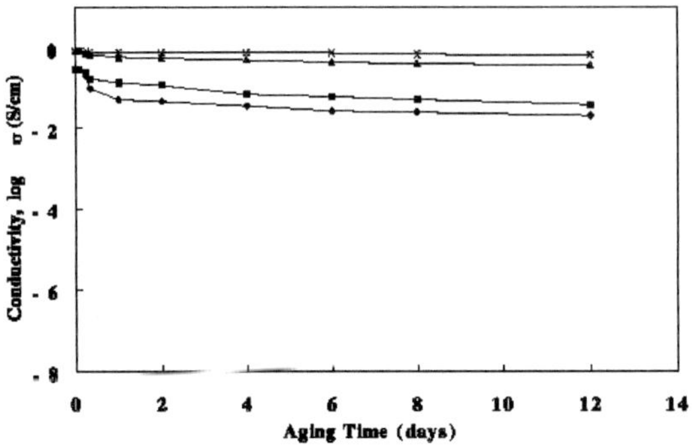

Figure 4. Conductivities of composites vs. aging time.

Table 1. Conductivity and EMI shielding. Effectiveness of conducting composites

	PPy composite	PPt composite
Conductivity, S/cm	0.82	4.20
Attenuation, dB		
1.0 GHz	7.5	10.9
1.5 GHz	14.5	23.0
2.0 GHz	15.2	22.5

molar ratio is due to saturation of the system in dopant. The conductivities of PPt impregnated composites are by an order of magnitude larger than those of PPy impregnated composites.

Figure 4 presents that the conductivity of PPy impregnated composites decreases slightly at room temperature, while no discernible decrease in the conductivity of PPt impregnated composites was observed during same period. Large decreases in conductivity of PPy impregnated composites were observed at 80°C. This thermal degradation of conducting polymer can be attributed to the result of a thermal driven undoping process in which conformational changes in the polymer backbone are activated by temperature.[6]

Table 1 presents the conductivities and EMI shielding effectiveness of conducting composites. The EMI shielding effectiveness of PPy and PPt impregnated composites reach 14.5dB at 1.5GHz and 23dB at 1.5GHz, respectively. The shielding effectiveness of the composites is attributed to the high conductivity of composites and uniform distribution of polyheterocyclic polymer in the composites.

The dependence of resistance on temperature for three dimensional hopping is proposed by the Mott equation as[7]

$$\rho \sim \exp\left[\left(T_o / T\right)^{1/4}\right]$$

The temperature dependence of the resistivity is plotted in Figure 5 as a function of $T^{-1/4}$ in the temperature range of 20-300K. This temperature dependence gives better fit with experimental data, indicative of three dimensional variable range hopping.

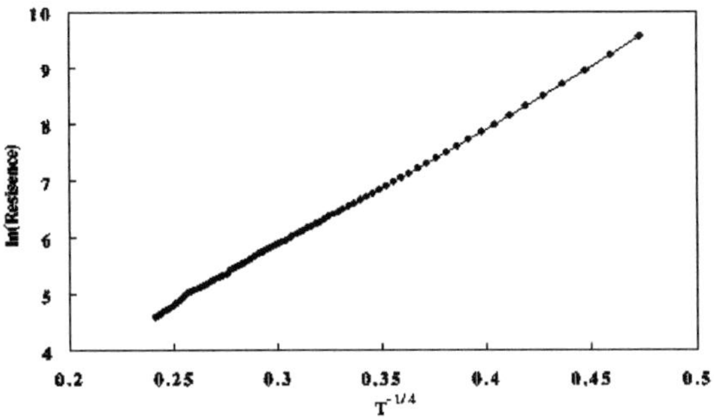

Figure 5. Resistance vs. temperature of PPy composites in the temperature range of 20-300K.

CONCLUSIONS

A method of preparing thick and large conducting objects is suggested. Morphology studies indicate that conducting polymer grows on the inner surface of the porous host polymer, which results in the formation of the electrical network in the composite. The initial molar ratio of oxidant to monomer is important for obtaining higher conducting composites. The temperature dependence of the resistivity indicates that conduction follows three dimensional variable range hopping model. With high EMI shielding effectiveness, the advantage of the conducting composites in light weight and versatility in fabrication can lead practical applications.

ACKNOWLEDGMENT

This work was supported by NON DIRECT RESEARCH FUND, Korea Research Foundation, 1996. J. S. Park wishes to thank Korea Research Foundation for financial support.

REFERENCES

1 R. E. Myers, *J. Electron. Mater.*, 1986, **15**, 21.
2 J. E. Osterholm, P. Passiniemi, H. Isotalo, *Synth. Met.*, 1987, **18**, 213.
3 C. Li and Z. Song, *Synth. Met.*, 1991, **40**, 23.
4 R. B. Bjorklund and I. Lundstrom, *J. Electron. Mat.*, 1984, **13**, 211.
5 E. Ruckenstein and J. S. Park, *Synth. Mat.* , 1991, **44**, 293.
6 Y. Wang and M. F. Rubner, *Synth. Met.*, 1990, **39**, 153.
7 D. S. Maddison and T. L. Tansley, *J. Appl. Phys.*, 1992, **72**, 4677.

Development of Electrohydrodynamic Flow Cells for the Synthesis of Conducting Polymers

P.C. Innis, V. Aboutanos, N. Barisci, S. Moulton and G.G. Wallace
Intelligent Polymer Research Institute, University of Wollongong, Northfields Avenue, Wollongong NSW, Australia, 2522.

INTRODUCTION

There are two approaches to electrochemical processing of conducting polymers that have been shown to be feasible on bench scale electro-hydrodynamic reactors.[1-4] Electroactive polymers can be produced in a water-soluble form by the incorporation of a polar substituent onto the monomer, prior to synthesis. Alternatively, dispersions of the conducting polymer can be formed by taking a suitable monomer and forming nanometer sized particles in the presence of a suitable steric stabilizer.

Early attempts at the above technologies have been achieved, on a limited scale, by bulk chemical polymerization. Chemical synthesis of a conducting polymer has little control over the level of polymer oxidation as well as little or no control over what anions can be inserted into the polymer. On the other hand, electrochemical synthesis of the conducting polymer permits fine control over the formation (and level of oxidation) of the conducting polymer. This technique also allows the use of a wider range of dopant molecules that can impart special properties to the final product such as corrosion inhibition, conductivity, enhanced solubility and bioactivity.

A major limitation in the use of conducting polymers is the insolubility and general lack of processability of the materials. Some solvent processability of the conducting polymer can be achieved by polymerizing with monomers which have been modified with either alkyl/phenyl[5-7] or carboxyl/sulfonate substituents[8-12] to give organic solvent or water solubility, respectively. Typically, the price of modification of the monomer by these substituents is the loss of conductivity in the final polymer with respect to a polymer produced with an unsubstituted monomer. In some specific cases, such as polyaniline, solubility can also be enhanced by the use of some specialized dopant anions, which can impart organic solvent solubility.[13,14] The major drawback of such systems is that organic solvents such as dichloromethane, *m*-cresol and N-methylpyrrolidone must be used, which are industrially undesirable.

Two potential product areas can be explored for large scale electrochemical production of conducting polymers, these being aqueous colloidal dispersions or water soluble polymers. A brief description of these polymer technologies is discussed below.

CONDUCTING POLYMER COLLOID PROCESSING

To overcome the water solubility limitations of conducting polymers, aqueous dispersions that are sterically stabilized[15-21] have been prepared. More recent approaches involve the use of a core/shell approach where an inner inert colloidal core, such as SiO_2 or TiO_2, is coated with the conducting polymer, in the presence of a steric stabilizer, in order to yield uniform colloid morphology and particle size distributions.[22-25] All of these approaches involve the production of the conducting polymer component *via* a chemical oxidative route.[26,27] This technique involves oxidizing the monomer using a chemical agent and has limitations as discussed above.

The electrochemical synthesis of conducting polymer colloids is based upon removing the polymer as it is formed above the electrode surface and stabilizing into colloidal particles with a polymeric surfactant (i.e., a steric stabilizer). In the early stages of electropolymerization, the process of oxidation and oligomerization is said to occur within the diffusion layer above the electrode surface.[28-31] As polymerization continues, the oligomer solubility (in the electrolyte solution) is exceeded and subsequently precipitates onto the electrode surface. If the electrolyte flows across the electrode surface fast enough, it is possible for the polymer to be swept away from the electrode before deposition occurs.[32-34] This process is further facilitated by the presence of the steric stabilizer.[35] The final product is a colloidal conducting polymer that is doped with the anion of the supporting electrolyte used during synthesis.

ELECTROCHEMICAL REACTOR DESIGNS

Two electrochemical cells designs were developed "in-house" at IPRI utilizing porous reticulated vitreous carbon (RVC – ERG Aerospace) foam electrodes. RVC is a three dimensional electrode material with a surface area to volume of 65.6 cm^2/cm^3 at a porosity of 100 pores per inch (PPI). RVC was chosen as a replacement for the traditional plate electrode in order to maximize the electrode surface area and minimize the cell volume. The two designs considered in this cell development study were either a facing anode/cathode design (Figure 1) or an sandwiched anode between two opposing cathodes configuration (Figure 2). Each cell compartment is separated by an anion exchange membrane (Neosepta™) and protected by insert-

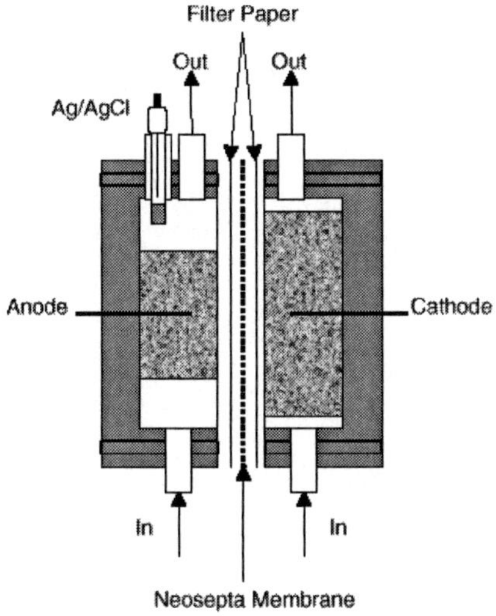

Figure 1. Parallel anode/cathode cell.

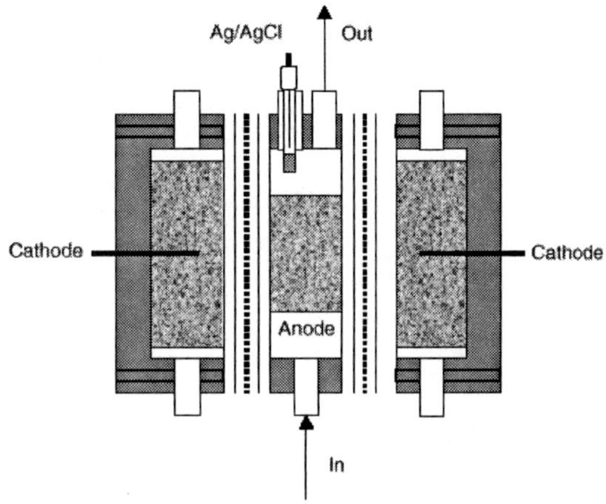

Figure 2. Sandwiched anode configuration.

ing filter paper between the membrane and electrode to minimize fouling. Electrolyte was flowed from separate catholyte and anolyte reservoirs.

EVALUATION OF ELECTROCHEMICAL REACTORS

In order to establish that the electrochemical cells were operating at optimal efficiency a series of standardized tests have been developed so that different cells can be compared to one another. The theoretical and experimental details are discussed below.

The cells were modelled as a Plug Flow Reactor type reactor. Equations exist for the design and evaluation of each type of reactor.[36] Parameters valid for continuous flow reactors are based upon the assumption of mass transport limited conditions. The essential equations for this evaluation process are:

Mass Transfer Coefficient (k_m)

$$k_m = i_L/zFAC \qquad [1]$$

i_L	limiting current
z	number of electrons of reaction
F	Faraday constant
A	electrode area

C concentration of reactant

In some cases it is preferable to use eq [1] as:

$$k_mA = i_L/zFC \quad [2]$$

In general, and particularly for cells with plate electrodes, it is possible to use:

$$k_mA_S = i_L/zFCV_R \quad [3]$$

$$A_S = A/V_R$$

V_R volume of reactor (containing electrode)

For cells with three-dimensional electrodes:

$$k_mA_e = i_L/zFCV_e \quad [4]$$

V_e electrode volume

Fractional Conversion (X_A)

$$X_A = (C_{(in)} - C_{(out)})/C_{(in)}$$

or

$$X_A = 1 - (C_{(out)}/C_{(in)}) \quad [5]$$

$C_{(in)}$ concentration of reactant IN

$C_{(out)}$ concentration of reactant OUT

For a PFR operated in single pass mode, and assuming first order mass controlled kinetics, the following relationship will hold:

$$C_{(out)} = C_{(in)} \exp[-(k_mA_e/Q_v)] \quad [6]$$

Q_v volumetric flow rate

$$X_A = 1-\exp[-(k_mA_e)/Q_v] \quad [7]$$

$$X_A = 1\text{-}exp[-\{k_m A_e / V_R\}\tau]$$

or

$$ln\ (C_{(out)}/C_{(in)}) = -\{k_m A_e / V_R\}\tau \qquad [8]$$

τ residence time (V_R/Q_V or V_e/Q_V for three dimensional porous electrode)
For a PFR operated in *recirculation* mode:

$$X_{A,t}^{PFR} = 1\text{-}exp[-(t\,X_A^{PFR}\ /\ \tau_T] \qquad [9]$$

t time
τ_T residence time in holding tank

FIGURES OF MERIT

Space-time Yield (ρ_{st})

$$\rho_{st} = m/V_R t \qquad [10]$$

m mass of product
In terms of k_m, at 100% current efficiency, ϕ

$$\rho_{st} = k_m A_e M C_{(in)} \qquad [11]$$

M molecular weight of reactant
Current Efficiency (ϕ)

$$\phi = Q_p/Q_T \qquad [12]$$

Q_p charge required to produce product
Q_T total charge used

EXPERIMENTAL.

CHEMICALS

Potassium hexacyanoferrate (II) and sodium nitrate were used as supplied from Ajax (Analar). Milli-Q grade water was used in all experiments.

EQUIPMENT

A PAR 363 potentiostat/galvanostat was used in all efficiency tests. All potentials were measured versus a Ag/AgCl reference electrode. A Shimadzu UV-1601 UV-Visible spectrophotometer was used for all concentration determinations. Data acquisition was made using a Maclab™ (AD Instruments) interface using Chart V3.3.5 software.

PROCEDURES

Each Cell design was tested by preparing a 1 L anolyte solution consisting of 2 mM $K_4Fe(CN)_6.3H_20_{(aq)}$ and 0.5 M $NaN0_{3(aq)}$. A 1 L catholyte solution of 0.5 M $NaNO_{3(aq)}$. The anolyte and catholyte solutions were then flowed through their respective compartments at 20, 40 and 60 mL/min by peristaltic pumps. Efficiency tests were carried out at a constant potential of +0. 8 V in single pass mode.

A chronoamperogram at the applied potential was recorded for both the duration of cell flushing and sample collection. Initially, 1 to 3 cell volumes were passed through the cell at +0.8 V, at the flow rate being investigated, and discarded to waste. With the potential still being applied, a minimum of 1 cell volume was then collected. The collection time (sec) and actual volume collected (mL) were also determined. The charge passed was then determined by integrating the area under the curve from the end of the data acquisition and back for the total collection time (t sec). The concentration of $Fe(CN)_6^{3-}$ produced from $Fe(CH)_6^{4-}$ test solution was determined by UV-Vis Spectroscopy.

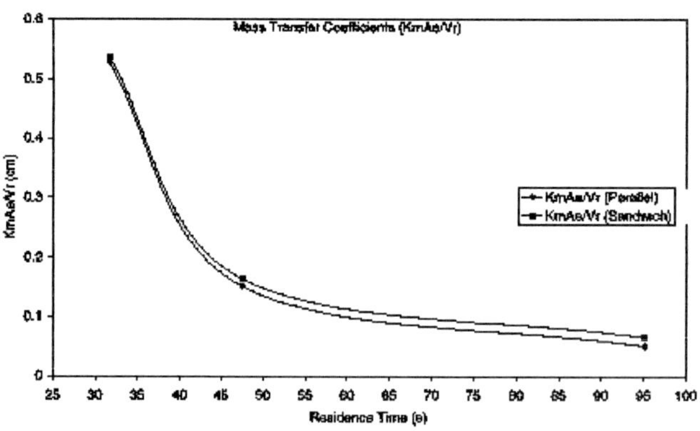

Figure 3. Effect of sandwich and parallel configuration on the mass transfer coefficient (K_mA_e/V_R).

RESULTS AND DISCUSSION

Analysis of the current efficiency parameters for both cell designs indicated that at + 0.8 V both cells designs had efficiencies of 99 % or better.

The dependence of the mass transfer coefficient is shown in Figure 3. At large residence times, or low flow rates, the mass transfer process in the sandwich configuration is slightly higher. At shorter residence times, higher flow rate, both cell designs have similar mass transfer characteristics due to greater turbulence induced with in

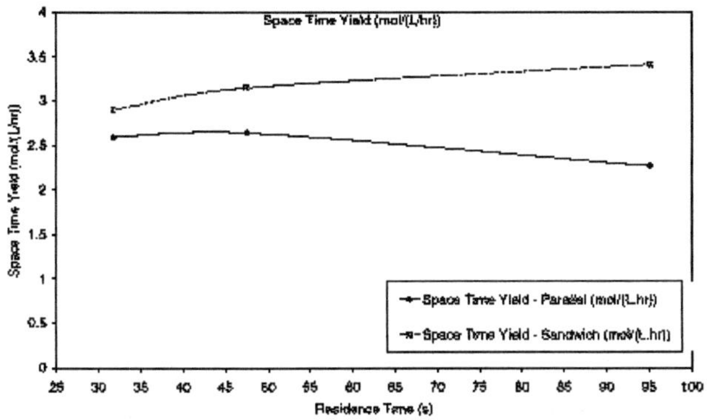

Figure 4. Effect of sandwich and parallel configuration on the space time yield, ρ_{st}.

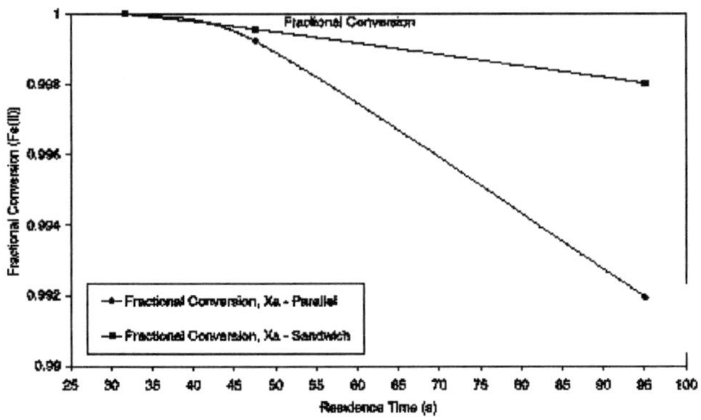

Figure 5. Effect of sandwich and parallel configuration on Fractional Conversion.

the RVC electrodes at higher flow rates. The effect of this turbulence is characterized by the rapid increase in mass transport at lower residence times (or higher flow rates).

The relationship of space time yield, ρ_{st}, to residence time, Figure 4, shows that the sandwich configuration is most efficient at all flow rates investigated. The sandwich configuration has higher ρ_{st} values due to minimizing the effect of electrode shielding and thereby utilizing more of the available anode surface. In the sandwich configuration, ρ_{st} decreases as the electrolyte residence time with in the cell is also reduced. This is indicative of insufficient electrolyte contact with the electrode at shorter residence times, but also shows that the cell was operating near its optimal at the lowest flow rate. Increasing the flow rate from 20 to 60 mL/min reduces the space time yield by 15%. In the case of the parallel configuration, the space time yield was at a maximum at shorter residence times. This result gives further evidence that this cell design was far less efficient than the sandwich configuration.

The fractional conversion, Figure 5, of $Fe(CH)_6^{4-}$ to $Fe(CN)_6^{3-}$ was better than 99% for both cell designs. The sandwich configuration was again a more efficient cell design. Interestingly, fractional conversion increased with decreasing residence time, or increasing flow rate, due to enhanced electrolyte turbulence.

EXAMPLES OF POLYMERS SYNTHESIZED USING THESE CELL DESIGNS

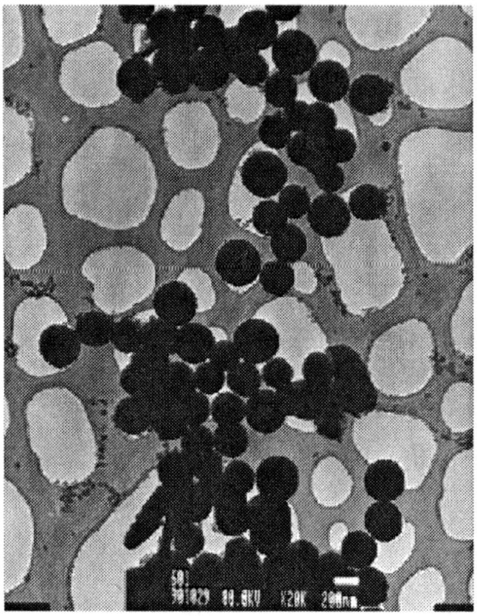

Figure 6. TEM of Polypyrrole-nitrate colloids.

A number of conducting polymer colloids have been synthesized using these cell designs. Colloids such as polypyrrole-nitrate,[2-4] polypyrrole-lactoferrin,[37] polyaniline-polystyrene sulfonate/camphorsulfonic acid[38] have been successfully synthesized. A TEM, Figure 6, of a typical polypyrrole nitrate colloid shows typical colloid morphologies achieved when using these cells. A major feature of this synthesis technique is that the colloids formed have a controllable and uniform size distributions.

CONCLUSIONS

In this paper we have presented a method of characterizing the efficiency and performance of electrochemical flow cells utilizing three-dimensional reticulated vitreous carbon foam electrodes. Cell design characterization is critical for the successful implementation and scale up of electrochemical cell, especially with respect to the scale up from laboratory to prototype and commercial conducting polymer synthesis.

REFERENCES

1 J.N. Barisci, P.C. Innis, L.A.P. Kane-Maguire, I.D. Norris and G.G. Wallace., *Synth. Met.*, 1997, **84**, 181-182.
2 J.N. Barisci, C.Y. Kim, D.Y. Kim, J.Y. Kim, J. Mansouri, G.M. Spinks and G.G. Wallace., *Colloids and Surfaces A.*, 1997, **126**, 129-135.
3 J. Barisci, J. Mansouri, G. Spinks, G. Wallace, D.Y. Kim and C.Y. Kim, *Synth. Met.*, 1997, **84**, 361-362.
4 V. Aboutanos, J.N. Barisci, P.C. Innis and G.G. Wallace., *Colloids and Surfaces* A., in press.
5 W-P. Hsu, K. Levon, K-S. Ho, A.S. Myerson and T.K. Kwei., *Macromolecules*, 1993, **26**, 318-1323.
6 R.M. McCullough, R.D. Lowe, M. Jayaraman and D.L. Anderson., *J. Org. Chem.*, 1993, **58**, 904-912.
7 Y. Wei and J. Tian., *Polymer,* 1992, **33**, 4872-4874.
8 H. Masuda and K. Kaeriyama., *Synth. Met.*, 1992, **13**, 461-465.
9 S. Rapi, V. Bocchi and G.P. Gardini., *Synth. Met.*, 1988, **24**, 217-221.
10 D. Delabouglise and F. Garnier., *New J. of Chem.*, 1991, **15**, 233-234.
11 S-A. Chem and G-W. Hwang., *J. Am. Chem. Soc.*, 1994, **116**, 7939-7940.
12 E.E Havinga, W. ten Hoeve, E.W. Meijer, and H. Wynberg., *Chem. Mater.*, 1989, **1**, 650-659.
13 Y. Cao, P. Smith and .J. Heeger., *Synth. Met.*, 1992, **48**, 91.
14 Y. Cao, P.Smith and A.J. Heeger., *App. Phys. Lett.*, 1992, **60**, 2711.
15 H. Eisazadeh, G. Spinks and G.G. Wallace., *Mater. Forum*, 1992, **16**, 341-344.
16 C. DeArmitt and S.P. Armes., *Langmuir,* **9**(1993)652-654.
17 B. Vincent., *Polym. Adv. Tech.*, 1995, **6**, 356-361.

18 M. Aldissi and S.P Armes., *Prog. Org. Coatings*, 1991, **19**, 21-58.
19 H. Eisazadeh, G.Spinks and G.G. Wallace., *Mater. Forum*, 1993, **17**, 241-245.
20 S.Y. Luk, W. Lineton, M. Keane, C. DeArmitt and S.P. Armes., J. *Chem. Soc., Faraday Trans.*, 1995, **91**, 905-910.
21 S.P. Armes and B. Vincent., *J. Chem. Soc., Faraday Trans.*, 1987, 228-290.
22 S.P. Armes, S. Gottesfeld, J.G. Beery, F. Garzon, *Agnew., Polym.*, 1991, **32**, 2325-2330.
23 R. Flitton , J. Johal, S. Maeda and S.P. Armes., *J. of Colloid and Interfacial Sci.*, 1995, **173**, 135- 142.
24 S. Maeda and S.P. Armes., *Synth. Met.*,1995, **69**, 499-500.
25 S. Maeda and S.P. Armes., *Chem. Mater.*, 1995, **7**, 171-178.
26 **German Patent P 37 29 566.7** Zipperling Kessler & Co.
27 **US Patent Application 823416, 823511 and 823512** Allied-Signal, Zipperling Kessler & Co. and Americhem Inc.
28 R. John and G.G. Wallace., *J. Electroanal. Chem.*, 1991, **306**, 157.
29 B.R. Scharifker and D.J. Fermin., *J. Electroanal. Chem.*, 1994, **365**, 35.
30 B.R. Scharifker, E. Garcia-Pastoriza and W. Marino., *J. Electroanal. Chem.,* 1991, **335**, 85.
31 D.E. Raymond and D.J. Harrison., *J. Electroanal. Chem.*, 1993, **355**, 115.
32 A.F. Diaz and B.J. Bargon in **Handbook of Conducting Polymers**, T.A. Skotheim Ed., Vol 1, *Marcel Dekker*, New York, 1986, 81.
33 C.K. Baker and J.R. Reynolds., *J. Electroanal.Chem.*, 1988, **251**, 307.
34 C. Lee, J. Kwak and A.J. Bard., *J. Electrochem. Soc.*, 1989, **136**, 3720.
35 H. Eisazadeh, G. Spinks and G.G. Wallace., *Mater. Forum*, 1992, **16**, 341.
36 F. Walsh., "A First Course in Electrochemical Engineering", The Electrochemical Consultancy, England, 1993.
37 V. Aboutanos, J.N. Barisci, G.R. Harper and G.G. Wallace, submitted ANTEC 98.
38 P.C. Innis, G.Spinks, G.G. Wallace., submitted ANTEC 98.

Hydroxyethyl Substituted Polyanilines: Chemistry and Applications as Resists

Maggie A.Z. Hupcey
Dept. of Materials Science and Engineering, Cornell University, Ithaca, NY
Marie Angelopoulos, and Jeffrey D. Gelorme
IBM T.J. Watson Research Center, Yorktown Heights, NY
Christopher K. Ober
Dept. of Materials Science and Engineering, Cornell University, Ithaca, NY

INTRODUCTION

In the field of conducting polymers, solubility of the polymer is a highly desirable quality. One method that has been widely used to enhance the solubility of conducting polymers is the incorporation of ring substituents on the polymer backbone.[1-3] However ring substituents have generally resulted in a decrease in the conductivity of the polymer. Steric constraints imposed by the substituents disrupt the coplanarity of the polymer chains as well as increase the interchain distance. Both factors reduce the mobility of the carriers and as a result lower conductivity is exhibited. In this work, we wish to report a new series of polyaniline copolymers possessing a hydroxyethyl substituent that improves the solubility of the conducting polymer and yet retains a high conductivity.

In the field of microlithography, conducting polymers have potentially many uses.[4-6] For example, it is thought that damage to the gate oxide and to the resist sidewalls during plasma etching is due to the charging of the insulating resist layer. In SEM metrology of devices and masks using a conventional insulating resist, charging of the resist yields errors in the measurement, a problem conducting resists have been shown to alleviate.[7] Conducting polymers have also been proven useful in electron beam (e-beam) lithography, where the exposing radiation is a beam of electrons. Excess electron charge builds up at the surface of an insulating resist causing severe pattern distortion; with a conducting resist the excess charge is dissipated.[8] With these new hydroxyethyl substituted polyanilines, a simple derivitization to make a crosslinkable conducting resist is possible.

EXPERIMENTAL

The polyaniline copolymers were synthesized following a well-known procedure.[9] Copolymer compositions were determined *via* integration on a 300 MHz Varian ^{1}H NMR. Molecular weights were measured *via* GPC eluted with NMP with 0.5% LiCl added to each sample. For UV-Visible spectra and conductivity measurements, films were spun coat from 5% (wt/wt) NMP solutions filtered through 5 μm filters onto polished quartz discs, then baking the samples in an 85°C oven for 15min. Conductivities were measured *via* four point probe. Doping with camphorsulfonic acid (CSA) and acrylamidomethyl-propanesulfonic acid (Aampsa) was done by adding the dopant in the ratio of 2 mole acid per mole repeat unit to a 5% (wt/wt) NMP solution of the base polymer, and processed as for the undoped. For HCl doping, the undoped spun coat films were immersed in 1M HCl overnight and dried under light vacuum.

The methacrylate functionalization was achieved by dissolving the poly(o-hydroxyethyl)aniline homopolymer in cyclohexanone at 3 wt%, adding isocyanoethylmethacrylate (IEM) and stirring for 48 hours until complete disappearance of the isocyanate peak in the IR spectrum.

RESULTS AND DISCUSSION

COPOLYMERIZATION

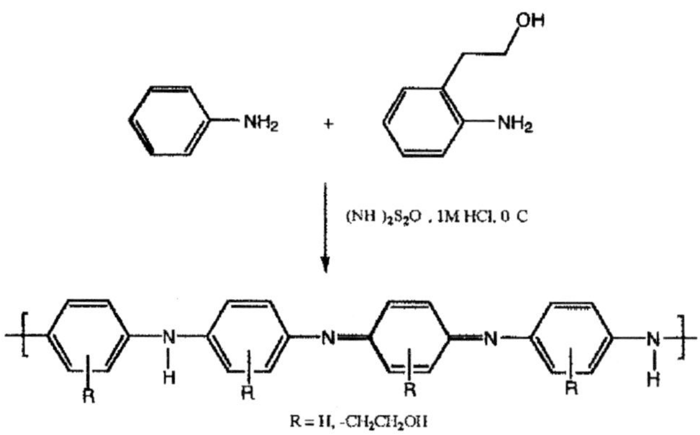

Figure 1. Poly(o-hydroxyethyl)aniline copolymer synthesis.

Seven copolymers were synthesized by varying the molar amount of o-hydroxyethylaniline monomer in the polymerization (Figure 1). It was found that the incorporated functionality always exceeded the feed (Table 1). The GPC results show that the molecular weight decreased from a high of M_n=22K and leveled off at around M_n=12K for feeds larger than 30%.

ELECTRONIC PROPERTIES

In the undoped or base form, the hydroxyethyl substituted polymers exhibit a red shift in the exciton absorption in the UV-Visible spectrum as compared to the unsubstituted emeraldine base. A λ_{max} of 609 nm is observed for the thin film of the unsubstituted polyaniline base whereas a λ_{max} of 629 nm is observed for the fully substituted poly(o-hydroxyethyl)aniline. The fully substituted

Table 1. Poly(o-hydroxyethyl)aniline copolymer synthesis data

o-hydroxyethyl aniline molar feed, %	Actual molar fraction incorporated, %	M_w	M_n	PDI	Conversion
1	2.1	42500	22000	1.93	19.3
10	23	22600	14300	1.59	18.5
20	35	20700	13000	1.59	19.0
30	44	18700	12200	1.54	19.1
50	66	17700	11700	1.51	17.7
100	100	19500	12600	1.55	27.0

poly(o-ethoxy)aniline synthesized and processed in the same way exhibits a dramatic blue shift to a λ_{max} of 580 nm as compared to emeraldine base and poly(o-hydroxyethyl) aniline. The ethoxy functionality and the hydroxyethyl substituent basically pose similar geometric constraints on the polymer. However, the hydroxy functionality provides a means for hydrogen bonding between chains thereby reducing the interchain distance. In addition the hydroxy group provides a means of hydrogen bonding with the processing solvent which may allow for an expanded chain conformation to be adopted. Previously we have shown a significant dependence of the exciton absorption on the morphology of the polymer.[10-11]

When the copolymers are doped in solution with CSA or Aampsa, the conductivity of the spun-coat thin films are ~10^{-2} Scm^{-1}. The hydroxyethyl homopolymer exhibits a conductivity ~10^{0} Scm^{-1}, which is of the same order of magnitude as the conductivity of the unsubstituted polyaniline. The high conductivity exhibited by the poly(o-hydroxyethyl) aniline is due to the altered interchain hydrogen bonding as well as improved polymer/solvent interactions resulting in a more expanded coil conformation. In agreement with this is the highly delocalized free carrier tail extending into the near-infrared that is observed for the poly(o-hydroxyethyl)aniline.

LITHOGRAPHY

The presence of the hydroxy substituent also provides a convenient site for further functionalization of the polymer to provide radiation sensitivity. We have developed a simple isocyanate-alcohol reaction (Figure 2) in order to add methacrylate functionality without affecting the oxidation state of the backbone. This development is significant because a traditional Schotten-Baumen esterification with methacrylic acid yields an insoluble mixture of the desired product along with the HCl doped product. The isocyanate reaction on the other hand is a reaction that produces no side products. The substituted polymer is produced in solution from which it can be easily spin coated into thin films.

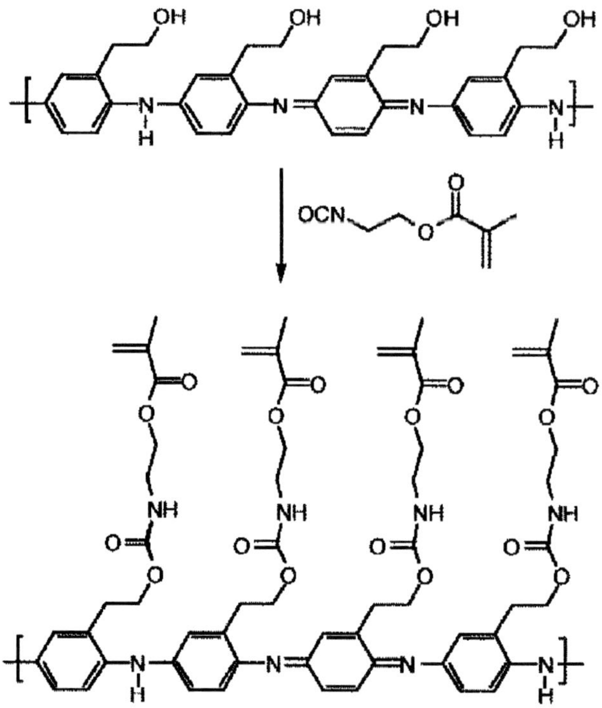

Figure 2. PaniIEM synthesis.

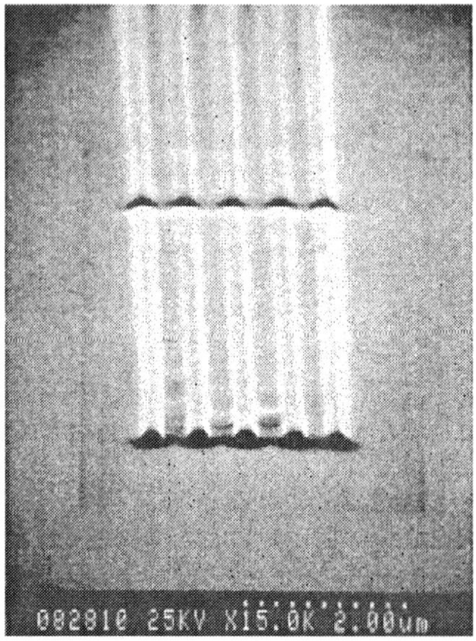

Figure 3. PaniIEM e-beam exposure: 0.075 μm equal lines and spaces. Sample: PaniEtOH 100% methacrylated, CSA doped. Exposure: 250 $\mu C/cm^2$ at 25 keV. Develop: 1 min ethyl lactate.

This new substituted polyaniline, PaniIEM, is useful as a new negative-tone resist. Upon e-beam and broad band UV light exposure the methacrylate crosslinks and becomes insoluble in the developer. Because of the improved solubility of the polymer due to the presence of the substituent, the developer can be a simple polar solvent, in this case ethyl lactate. Figure 3 shows 75 nm lines patterned *via* $250\mu C/cm^2$ of e-beam radiation at 25KeV accelerating voltage and developed for 1 minute in ethyl lactate. These patterns are the smallest produced thus far in a single-layer negative tone conducting e-beam resist. Work is in progress to improve the sensitivity and contrast in order to reduce the backscatter exposure between the patterned lines.

PaniIEM is also sensitive to broadband UV-Visible light exposure. Further studies to evaluate the ultimate speed and resolution are underway.

CONCLUSIONS

Poly(o-hydroxyethyl)aniline copolymers have been shown to possess interesting electronic properties, namely a high conductivity upon doping and a red shift in the UV-visible spectrum indicating a high degree of hydrogen bonding between the chains. These new copolymers show considerable promise as resists as they can be derivatized to the radiation sensitive methacrylate substituted polymer.

ACKNOWLEDGMENTS

The authors would like to thank Jim Bucchignano for assistance with the e-beam exposures and Nancy LaBianca for helpful discussions. M.H. would also like to thank the SRC Microlithography program.

REFERENCES

1 Roy, B.C.; Gupta, M.D.; Ray, J.K., *Macromolecules*, 1995, **28**, 1727.
2 Leclerc, M.; D'Arano, G.; Zotti, G., *Synth. Met.*, 1993, **55-57**, 1527.
3 Leclerc, M.; Guay, J.; Dao, L.H., *Macromolecules*, 1989, **22**, 649.
4 Lowe, J.; Holdcroft, S., *Macromolecules*, 1995, **28**, 4608.
5 Abdou, M.S.A.; Diaz-Guijada, G.A.; Arroyo, M.I.; Holdcroft, S., *Chem. Mater.,* 1991, **3**, 1003.
6 Persson, S.H.M.; Dyreklev, P.; Inganas, O., *Adv. Mater.*, 1996, **8**, 405.
7 Angelopoulos, M.; Shaw, J.M.; Lecorre, M.-A.; Tissier, M., *Microelectronic Eng.*, 1991, **13**, 515.
8 Angelopoulos, M.; Patel, N.; Shaw, J.M.; LaBianca, N.C.; Rishton, S.A., *J. Vac. Sci. Tech. B*, 1993, **11**, 2794
9 MacDiarmid, A.G.; Chiang, J.C.; Richter, A.F.; Somasiri, N.L.D.; Epstein, A.J., **Conducting Polymers**; *Alcacer Reidel Publications*: Dordrecht, 1987; pp 105- 120.
10 Seery, T.A.P., Angelopoulos, M., Levon, K., Seghal, A., *Synth.Met.,* 1997, **84**, 79.
11 Angelopoulos, M., Dipietro, R., Zheng, W.G., MacDiarmid, A.G., Epstein, A.J., *Synth.Met.*, 1997, **84**, 35.

Electroformation of Polymer Devices and Structures

G. G. Wallace, J. N. Barisci, A. Lawal, D. Ongarato, and A. Partridge
Intelligent Polymer Research Laboratories, University of Wollongong, Australia

INTRODUCTION

Table 1. Inherently conductive polymers

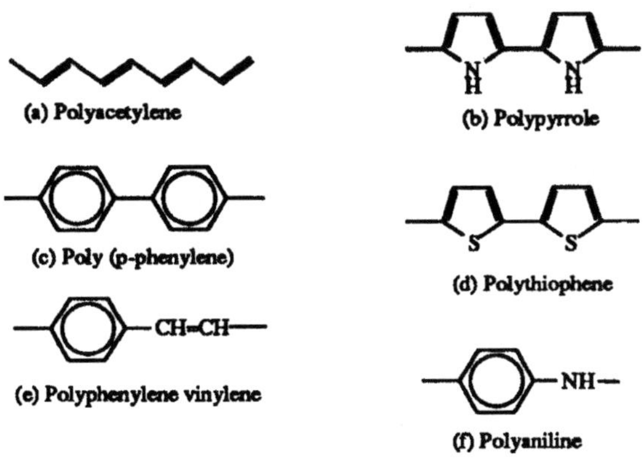

The discovery of inherently conducting polymers (see Table 1) just 17 years ago has led to the widespread use of polymers in devices used for sensing, chemical separations, electromechanical actuators, electronic components and electrochromics, amongst many others.[1-4]

Not long after the discovery of conducting polymers it was revealed that simple materials such as polypyrroles, polythiophenes and polyanilines can be produced using electrolytic methods, akin to metal deposition. For example, the electropolymerization of polypyrrole can be depicted (rather simplistically) as,

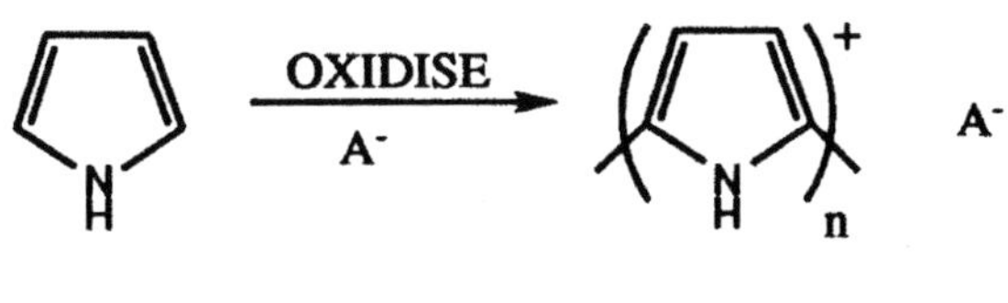

This reaction can also be induced using a chemical oxidant such as $FeCl_3$. Electroformation, however has several advantages.

- A wide range of counterions (A^-) can be used.

- The oxidizing strength of the solution can be accurately controlled as a function of polymerization time. This can be used to control the rate of reaction and to avoid unwanted side reactions that result in deterioration of the polymer properties.
- Electroformation can be used to obtain spatial control in the assembly of polymer structures. Patterned electrodes can be addressed with different electroformation potentials and/or in different feed solutions.
- Electroformation can be used to produce layered structures.
- Electroformation can be used to coat very small surfaces of micro dimensions.

ELECTROFORMATION AT CONVENTIONAL (LARGE) ELECTRODES

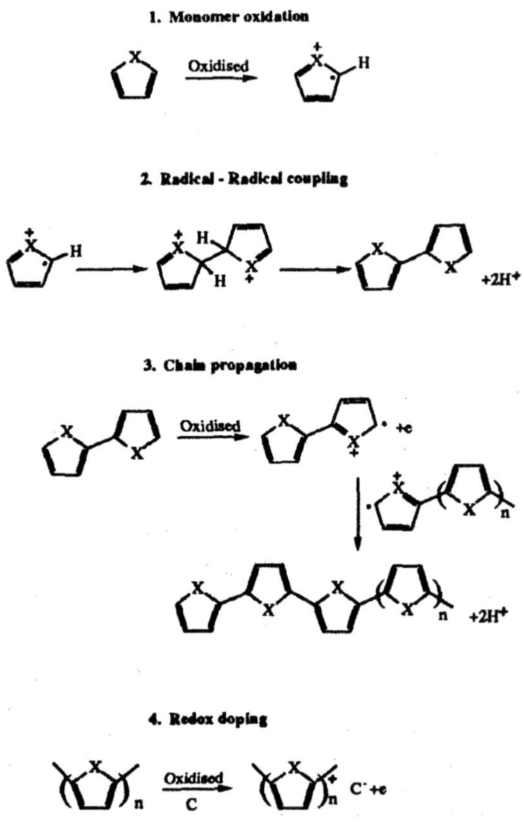

Scheme 1: Electroformation of polypyrrole.

As alluded to above the processes involved in electroformation of conducting polymers are often oversimplified. Closer examination[5] reveals that several steps are involved (Scheme 1).

Once initiated, the reaction proceeds in solution until sufficient polymer (of high enough molecular weight) is produced and exceeds the solubility limit. The polymer material then precipitates on to the electrode provided the chemical nature of the surface is compatible.

Subsequent polymerization is controlled by a heterogeneous surface (some bare/some polymer coated) until it is completely covered. Polymer continues to be deposited and/or add to reactive oligomers attached to the electrode surface. Polymers deposited in this fashion can be removed from the electrode as mechanically stable stand alone films. They are dense structures with the chemical and physical properties varying across the depth of the material. This asymmetric effect can be enhanced by use of appropriate structure ordering counterions such as surfactants. For example, the resultant materials can be forced to have hydrophobic properties on one side and hydrophylic properties on the other.

The use of the electrodeposition process is restricted in that materials of limited dimensions and porosity can be produced. This in turn restricts the use of electroformation to produce polymer based devices. In addition, in many cases the incorporation of more exotic counterions to provide molecular functionality often results in a deterioration of mechanical properties.

ELECTROFORMATION IN HOST POLYMERS

In order to overcome the above limitations, conducting polymers can be grown electrochemically inside other host structures. This is most readily achieved by simply precoating the electrode substrate with a suitable host polymer (e.g., polyvinyl alcohol). The host material must be porous enough to allow ingress of solvent, monomer and supporting electrolyte so that the electropolymerization can proceed. In some instances the host polymer can in fact be used to provide charge balance, e.g., nafion coatings cast from ethanol can be employed.

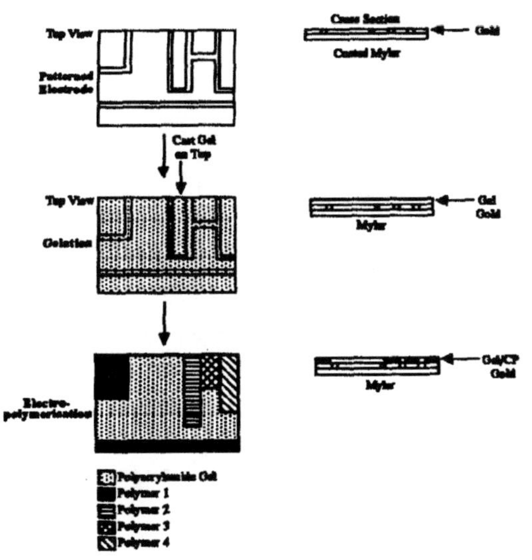

Figure 1: Schematic representation showing electroformation of patterned hydrogel-conducting composite structures.

A particularly interesting class of host polymers are hydrogels. For example, we have utilized crosslinked polyacrylamide with water contents greater than 90%. Electrodes may be cast inside those materials at the time of gelation. The open porous nature of these materials makes them ideal for electropolymerization. It has been found that the gel networks provides a template for polymer growth the resulting polymer composite still retaining high (>90%) water content, but also having electronic conductivity and electroactivity. These conducting polymer containing materials can be grown to large dimensions with the final shape determined by that of the gel at the time of gelation.

Spatial distribution of conducting polymers throughout the hydrogel networks is easily achieved either by dispersing addressable electrodes throughout the gel or by using patterned electrode surfaces (Figure 1).

ELECTROFORMATION OF MICROSTRUCTURES

Electrochemical processes behave differently at micron sized electrodes than at conventional (large) electrodes. Of particular importance, as far as electropolymerization is concerned, is the enhanced rate of mass transport of both reactants and products to and from the electrode, respectively. The former should prevent depletion and hence avoid polymer overoxidation; the latter, however, prevents deposition.

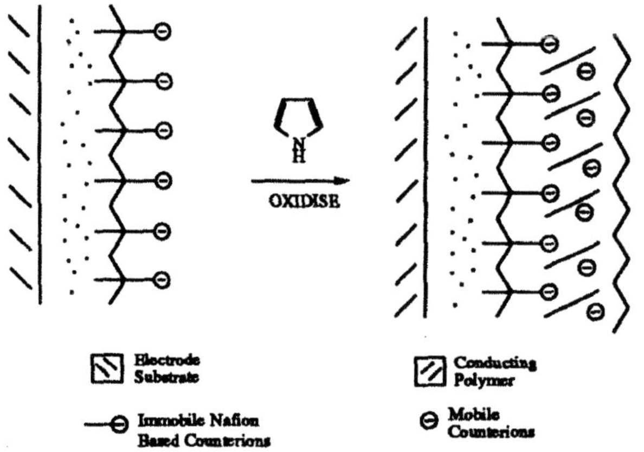

Figure 2: Schematic representation showing use of Nafion precoatings to facilitate electroformation on micro structures.

This means that deposition of polymers on microelectrodes becomes more difficult. It is interesting to note that deposition is not so difficult on line microelectrodes (10 μm x 1mm). Presumably the effective concentration at the longer electrodes increases sufficiently to exceed the solubility limit before the products are transported from the electrode.

Nafion precoatings can be used to advantage in cases where deposition is difficult in that the ionic nature of the predeposited material helps attract and trap the conducting polymer on the micro surface. The Nafion is precoated by simple evaporation from an ethanol solution (Figure 2).

The effect of the presence of Nafion on the deposition of polypyrroles containing a range of counterions is shown in Table 2.

Other interesting aspect of micro electrofabrication is that very low currents are encountered and so polymerization in resistive media can be used. This enables an increased range of counterions to be incorporated. In fact we have recently shown that this enables electropolymerization in the gas phase. The cell used is shown in Figure 3.

The micro electrode is precoated with Nafion and pyrrole vapor is present in the enclosed atmosphere. Imposition of appropriate current densities results in polymerization/deposition on the micro surface. Cyclic voltammograms recorded in solution after growth revealed the presence of a conductive electroactive polymer material.

Table 2. Electroformation on 10 μm electrodes using galvanometric growth

Polymer	Current density mA/cm^2	Growth potential V	Comment on polymer coating
PP/NDSA	2.0	0.49	poor
PP/NDSA/naf	2.0	0.56	good
PP/EBSA	2.0	0.54	extremely poor
PP/EBSA/naf	2.0	0.43	good
PP/BSA	2.0	0.80	extremely poor
PP/BSA/naf	2.0	0.75	good
PP/HBSA	0.2	1.60	extremely poor
PP/HBSA/naf	0.2	0.84	good
PP/4SBA	2.0	0.64	extremely poor
PP/4SBA/naf	2.0	0.60	good

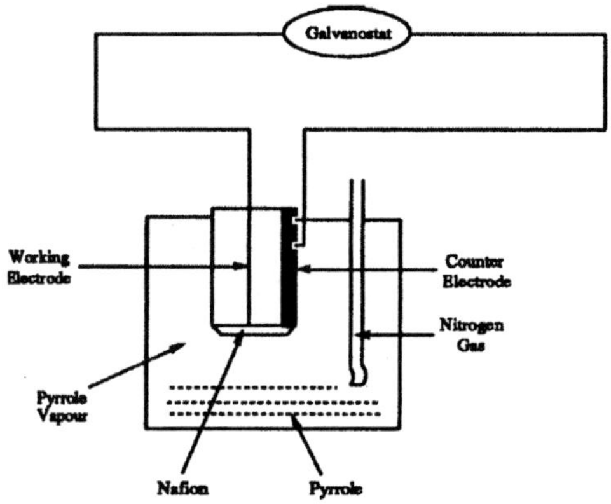

Figure 3: Schematic diagram of gas polymerization cell.

CONCLUSIONS

Advances in the use of electropolymerization enable the use of this approach to produce materials of varying size and dimension. This in turn will enable the production of a wider range of polymer based devices and structures.

REFERENCES

1 Adeloju, S.B.; Wallace, G.G., *Analyst*, 1996, **121**, 699-703.
2 Bakhshi, A.K., *Bull. Mater. Sci.*, 1995, **18**, 469-495.
3 Gardner, J.W.; Bartlett, P.N., *Nanotechnology*, 1991, **2**, 19-32.
4 Baughman, R.H., 1991, 51, 193-215.
5 John, R.; Wallace, G.G., 1991, 306, 157.

APPENDIX 1

This immobilization of PE also facilitates growth between conducting polymers on insulating substrates as required for gas sensing systems. In this case polymer must grow laterally between the four gold tracks to make electrical connection. The resistance of the polymer in the presence of target volatiles is then monitored by passing a current between the outer tracks and measuring the potential of the inner tracks. Precoating of polyelectrolyte (Nafion) between the tracks facilitates this lateral growth.

Microelectronic Encapsulation and Related Technologies: an Overview

Stephen L. Buchwalter
IBM Corporation, Thomas J. Watson Research Center, Yorktown Heights, New York

BACKGROUND

Encapsulation is the term commonly used for device and interconnection protection because of its long history in electronics, even predating semiconductor devices; and indeed encapsulation by molding plastic around the silicon and a metal leadframe is the predominant form of device protection in the microelectronic industry on the basis of the sheer volume of packages manufactured. Encapsulation is a misnomer, however, when it is used as a general term to refer to all forms of device protection, many of which do not entail total enclosure of the device in plastic or any other single material. In this overview, device protection is used as an abbreviation for device and interconnection protection and is meant to be an inclusive term to emphasize the common functions of a variety of technologies including, but not limited to encapsulation.

These common functions fall into two categories—those that are intrinsic to device protection and those which are extrinsic but closely coupled to device protection. In the first category there are the important functions of mechanical protection and protection against corrosion. The microcircuits on silicon devices and connections between the silicon and the next level of packaging are delicate and, if unprotected, can be damaged by incidental contact during assembly or actual use. Similarly, the microcircuits and their interconnections must be protected from environmental effects which can cause rapid failure of the device from corrosion of the metallic conductors in the circuitry.

The actual connection of the device to the next level of packaging falls into the extrinsic category of functions for device protection. This relationship can readily be seen by considering the molded plastic package as it exists in a variety of forms designed for surface mount assembly to a printed circuit card. If, for example, because of high stress or high moisture content, the plastic causes the wirebonds in the package to fail during solder reflow, the material has failed its function of providing a package suitable for surface mount assembly. This function can be seen to directly parallel the function of conductive adhesives as used, for example, to attach driver chips to flat panel, active matrix displays. A second extrinsic function of device protection is heat dissipation. It is readily apparent, for example, that the pathways offered for heat dissipation by a plastic molded package will be different from those offered by flipchip on a ceramic module. Recent trends in chip integration have increased requirements for both the density of interconnections and heat dissipation. In addition, the proliferation of semiconductors in mobile computer and communication devices has added size and weight limitations. These trends towards increased integration and miniaturization show no signs of abating; and thus, it is clear that the requirements of interconnection and heat dissipation should be even more tightly integrated into the device protection technology in order to achieve functional, reliable, compact and cost-effective semiconductor packaging for the future. For the remainder of this paper, a cursory evaluation of the main device protection technologies will be given in terms of how well they integrate interconnection and heat dissipation with device protection and what potential they offer for further improvement in this regard. Two other inter-related packaging considerations directly impacted by the choice of device protection are chip test and burn-in and reworkability of the assemblies. These aspects will be touched on in this overview as well, including a brief description of one approach to achieving reworkability without using molded plastic packaging

PLASTIC PACKAGING

Molded plastic encapsulation[1-2] has been the dominant form of device protection, starting with the dual inline package from the early days of microelectronics. Because of its dominance in packaging the huge volume of memory chips, plastic packaging has had a solid technology base from which the requirements of more specialized applications have been met by incremental improvements. Plastic packaging has continued to thrive by increasing the number of interconnections (I/Os) that can be handled and by reducing the thickness of the package. The latter has helped overcome some of the thermal limitations of encasing the chip in epoxy, a poor thermal conductor, as well as meeting requirements for miniaturization. A continuing attractive feature of plastic packaging is the fact that the packaged chips are easily handled individually both for test and burn-in to eliminate the chips most likely to fail and for rework of defective assemblies. An important limitation for the future, however, is the fact that all plastic packages use wirebonding between pads on the periphery of the chip and a leadframe. Although the demise of wirebonding has been prematurely predicted before, it does seem that momentum has shifted to area array interconnection (flipchip attach) because of its intrinsic advantage for interconnecting high I/O chips. Flipchip is not compatible with molded plastic encapsulation.

GLOBTOPS/DIE ATTACH ADHESIVES

Implicit in the choice of globtops for device protection[3] is the fact that if the top of the chip is protected with the globtop, the bottom or opposite face of the chip is in contact with some other material. This other material varies with the specific application—examples include ceramic substrates, printed circuit cards, flex substrates, and thermal substrates—but in many cases the globtop is used because the requirements of interconnection or heat dissipation prevent the use of molded plastic encapsulation. The choice of the globtop is then coupled with the choice of some other material, usually a thermally conductive die attach adhesive, to bond the other face of the chip to the substrate, and reliability of the package is a function of the properties of both these materials and any interaction between the two. In short, globtop packaging does increase the options for interconnection and heat dissipation, but globtops are not suitable for flipchip attach, the mode of interconnection needed for the highest I/O chips. Also, the chip test/burn-in and reworkability features of plastic packages are lost with globtop packaging, although some reworkability may be achievable with appropriate materials (see below).

UNDERFILL FOR FLIPCHIP ATTACH

Chip interconnection by use of an array of solder balls bonded to the chip surface significantly increases the number of I/Os that can be handled in comparison to peripheral attach *via* wirebonding.[4] Thermal mismatch between silicon and the substrate in many cases requires that a reinforcing material be applied in the space between the chip and the substrate, completely surrounding the solder connections.[5-8] Like globtop, this packaging option makes the opposite face of the chip available for heat dissipation, if the array of solder joints do not provide sufficient thermal conductivity to dissipate the heat generated by the device. Depending on the application, even if no cap/heat sink is needed for thermal reasons, some mechanical protection for the exposed surface of the chip may be needed such as a globtop or metal cap. In terms of the criteria of this overview, flipchip with underfill is similar to globtop in providing more options for interconnection and heat dissipation, with the added advantage of being designed to handle the high I/O chips. For chip test/burn-in and reworkability, plastic packages maintain their advantage, although efforts to provide known good chips and to enable rework of flipchip with underfill may reduce this advantage somewhat.

CONDUCTIVE ADHESIVES

Including electrically conductive adhesives in the category of device protection may seem to be a stretch, but one can view a conductive adhesive as a packaging option in which the wirebonds or solder balls have been replaced by conductive particles in the adhesive. With the anisotropic conductive adhesives,[9] i.e., those which are conductive in the direction perpendicular to the plane of the adhesive film and insulating in the plane, assembly of a chip to a substrate simply involves aligning the pads on the two surfaces with the adhesive film in-between and applying heat and pressure to activate the adhesive. To date, these materials have largely been limited to applications in which the joint conductance and interconnection density which they can provide are adequate, such as in attaching driver chips to active matrix flat panel displays. Improvements in this technology, however, would make conductive adhesives an attractive low-cost option for smaller, thinner, lighter packaging of semiconductor devices. The epoxy adhesives normally used in these materials are not reworkable, which would be one disadvantage of this option unless reworkable materials can be developed.

REWORKABLE EPOXY

Stand-alone plastic packages are a convenient, inexpensive form of packaging especially with respect to chip test and burn-in and rework of microelectronic assemblies. All the other options discussed in this short review sacrifice this convenience in order to achieve advantages in terms of I/O density or heat dissipation. To at least partially mitigate these disadvantages, there has been an effort in IBM to develop an inherently reworkable epoxy.[10-12] Conventional epoxy materials, as formulated for all of the packaging options discussed in this overview, are not reworkable because they are thermosets, i.e., crosslinked, insoluble and infusible plastics. The cleavable epoxy materials developed at IBM Research are also thermosets, much like those used in conventional liquid epoxy formulations, but they include special chemistry in the crosslinks to allow the network to be broken down and washed away for rework. The specific application which has been targeted first is for flipchip underfill on ceramic modules,[13] but formulations suitable for globtop and conductive adhesives are also envisioned.

CONCLUSIONS

Perhaps the ideal chip package for the smaller, thinner, lighter microelectronics of the future would combine: a) the stand-alone convenience of a plastic package; b) the capabilities for dense arrays of I/Os and efficient heat dissipation of flipchip with underfill; and c) the low-cost, simple assembly of anisotropic conductive adhesives. Such a combination does not seem likely to be available in the short-term, but making reworkability possible for all the packaging options seems to be an appropriate step towards this ultimate goal.

REFERENCES

1 Manzione, L.T., **Plastic Packaging of Microelectronic Devices**, *Van Nostrand Reinhold*, New York, 1990.

2 Kinjo, N., Ogata, M. , Nishi, K., Kaneda, A., Epoxy Molding Compounds as Encapsulation Materials for Microelectronic Devices, in *Adv. in Polym. Sci.*, **88**, K. Dusek, ed., Springer-Verlag, Berlin, 1989, 1-48.

3 Burkhart, A., *Int. SAMPE Electr. Conf.*, **6**, 1992, 243-255.

4 Koopman, N.G., Reiley, T.C., and Totta, P.A., **Microelectronics Packaging Handbook**, *Van Nostrand Reinhold,* New York, 1989, 361-453.

5 Nakano, F., Soga, T., Amagi, S., *ISHM Proc.*, 1987, 536-541.

6 Suryanarayana, D., Hsiao, R., Gall, T.P., McCreary, J.M., *IEEE Trans. Comp. Hybrids Manuf. Technol.*, **14**, 1991, 218-233.

7 Wang, D.W., Papathomas, K.I., *IEEE Trans. Comp. Hybrids Manuf. Technol.,* **16**, 863-867.

8 Tsukada, Y., Mashimoto, Y., Nishio, T., Mii, N., Proc. 1st ASME/JSME Adv. Elect. Packaging Conf., 827-835.

9 Chang, D.D., Crawford, P.A., Fulton, J.A., McBride, R., Schmidt, M.B., Sinitski, R.E., Wong, C.P., *IEEE Trans. Comp. Hybr. Manuf. Technol.*, **16**(8), 1993, 828-835.

10 Buchwalter, S.L., Kosbar, L.L., Gelorme, J.D., *Polym. Mat. Sci. Eng.*, **72**, 1995, 450-451.

11 Buchwalter, S.L., Kosbar, L.L., *J. Polym. Sci. Polym. Chem. Ed., in press.*

12 Buchwalter, S.L., Kosbar, L.L., Gelorme, J.D., Afzali-Ardakani, A., Pompeo, F.L.., Newman, B., *U.S. Patents, pending.*

13 Pompeo, F.L., Call, A.J., Coffin, J.T., Buchwalter, S.L., *Adv. in Electr. Packaging ASME, EEP*, **10-2**, 1995, 781-787.

Fabrication and Characterization of Conductive Polyaniline Fiber

Hsing-Lin Wang, Benjamin R. Mattes
Chemical Science and Technology Division, Los Alamos National Laboratory, Los Alamos NM, 87545
Yuntian Zhu, James A. Valdez
Material Science and Technology Division, Los Alamos National Laboratory, Los Alamos NM, 87545

We previously reported the concept of "gel-inhibitor" assisted processing of ultra-high molecular weight emeraldine base (EB) into wet-spun fiber. This method uses small amounts of secondary amine additives, e.g., 2-methyl aziridine (2MA), to form thermodynamically stable, particle-free, and highly concentrated (20% w/w) EB solutions. 2MA is a relatively toxic compound. Here we report that wet-spun fibers with similar physical characteristics may be obtained by utilizing non-toxic heptamethyleneimine (HPMI) as the gel inhibitor. As-spun EB fiber was prepared and then subsequently immersed in a variety of different Bronsted acids. Room temperature DC conductivity values for the doped fibers ranged from 3 to 10^{-5} S/cm depending on the acid dopant. The as-spun fibers were of low density and they contained closed-cell porous microstructures riddled with macro-voids due to residual solvent entrained during coagulation. Each fibers diameter was observed to either contract or expand depending upon which acid was used for doping the fiber segment. We also report the observed differences in fiber density, mechanical strength and conductivity as a function of the acid type selected for doping studies. Optical spectroscopy of the solutions used to prepare fiber with HPMI showed no evidence for polymer degradation. The thermal and mechanical properties of the as-spun and doped EB fibers are presented.

INTRODUCTION

Polyaniline has emerged as one of the most promising conducting polymers for industrial applications due to its combination of low cost and high environmental stability. The main engineering limitation of this polymer for fiber production is that it is only sparingly soluble in just a few organic solvents. Polar aprotic solvents such as N-methylpyrolidinone, dimethylpropylene urea, and dimethylsulfoxide have been used to process emeraldine base (EB) powder from solution into solid-state film and fiber. However, processing polyaniline into textile fiber is constrained by the propensity of this material to irreversibly gel in short periods of time. It is known that poor solubility and rapid gelation are both correlated with strong intra- and inter-chain hydrogen bonding between secondary amine and tertiary imine groups found in the polymer repeat units.

We previously reported that certain amine additives, e.g., 2-methyl aziridine, serve as "gel-inhibitors" (GI) for emeraldine base in solution.[1-3] We call these amine additives by this name since small quantities, on the order of one to two molecules per polymer repeat unit in solution, tend to dramatically reduce solution viscosity and prolong time to gelation after the solution is formed.[3] We believe this is due to a reversible hydrogen-bond complexing mechanism caused by physical interactions between imine (or amine) sites along the EB repeat unit and the electron lone pair (or proton) associated with the nitrogen atom of the gel inhibitory agent. These additives are completely removable from the film or fiber by extraction with water or by thermal evaporation. The GI-EB complex in solution serves to inhibit the reformation of EB inter-chain hydrogen bonds which, in the absence of GIs, leads to the rapid development of a gel network, and importantly, this occurs for periods of time sufficient to wet spin fiber. Highly concentrated, stable solutions may be prepared from high molecular weight forms of emeraldine base with the assistance of this class of gel inhibitory agents. In this paper, we present the results obtained for wet-spun EB fiber prepared when non-toxic heptamethyleneimine (HPMI) is used in place of 2-methylaziridine.

EXPERIMENTAL

MATERIALS AND EQUIPMENT

Emeraldine base was purchased and used as received from Neste Oy (Helsinki, Finland). N-methyl-2-pyrrolidinone (NMP) and heptamethylene imine (HPMI) was used as received. Differential scanning calorimetry and thermal gravimetric analysis measurements were made with a Perkin-Elmer 7 Series thermal analysis system at a heating rate of 5°C/min. A RVDV-III Brookfield Cone and Plate Viscometer was used at a constant shear rate of 0.8 s^{-1} to obtain viscosities of 1% solutions prepared with HMPI of Neste EB and of EB synthesized at -40°C of known M_W ($M_w = 6x10^5$). Vis-UV spectra of polymer solution and film were obtained using the Perkin-Elmer UV-Vis-NIR spectrometer. Standard 4-probe conductivity measurements were made with a Hewlett-Packard Model 3478A Digital Multimeter to measure the DC conductivity of fibers. Four copper wires were glued to the doped EB fiber by way of using conductive paste (DuPont Conductor Composition) as electrode leads. Alligator clips were then clipped to the ends of copper wires leads and the other end of the alligator clips were connected to the multimeter.

PREPARATION OF CONCENTRATED POLYMER SOLUTIONS WITH HPMI

82.8 g of N-methyl-2-pyrrolidinone (NMP) was mixed with 7.47 g (6.61×10^{-2} moles) of heptamethyleneimine (98% Acros). This mixture was placed inside a 500 ml resin kettle equipped with a mechanical stirrer and wrapped with heat tape for temperature control. 22.5 g (6.22×10^{-2} moles) of EB powder was added to this solution over a 5 minute period. The temperature of the polymer solution was maintained at 32°C. The mixture became homogeneous and very flowable after vigorous stirring for 60 minutes. The EB solution had a GI/EB molar ratio of 1.06 and the mass content of EB in this solution was 20% w/w. We have reported the details on the fiber spinning conditions and mechanical measurements previously.[1]

FIBER DOPING

Three inch lengths of the as-spun fiber were immersed in 500 ml of their respective aqueous acid solutions for 48 hours. They were removed from the doping solution, and then dried under dynamic vacuum ($\sim 10^{-3}$ torr) for another 48 hours. The acid solutions used for doping the fibers were: 1.0 M HCl, 4.0 M acetic acid (HOAc), 1.0 M trifluoroacetic acid (TFA), and an aqueous solution of benzene phosphinic acid [BPA (pH=-0.37)].

RESULTS AND DISCUSSION

OPTICAL SPECTRA

We have observed that deleterious concurrent reductive substitution reactions take place when EB is mixed together with strongly basic secondary amines, e.g., pyrrolidine, and solvent at GI/EB mole ratios >3. This reduction reaction leads to severe degradation of the mechanical properties of the EB film cast from solution, and therefore it is very essential to maintain the original oxidation state of the polymer by using near stoichiometric amounts of base additives. Han *et al.*[4] recently reported that pyrrolidine (pKb = 2.36) by itself will dissolve the emeraldine base form of polyaniline; however, he also observed that it serves to reduce the polymer to a lower oxidation state as revealed by a significantly altered solid-state ^{13}C NMR spectra. He attributed this change to concurrent reduction by nucleophilic substitution, at the ortho- or meta- positions of the semiquinone ring of EB, by the strongly basic pyrrolidine molecule. We were concerned that this sort of substitution reaction might occur in our system which contained HMPI, albeit at low concentrations, since it is also a strong base (pKb = 3.06).

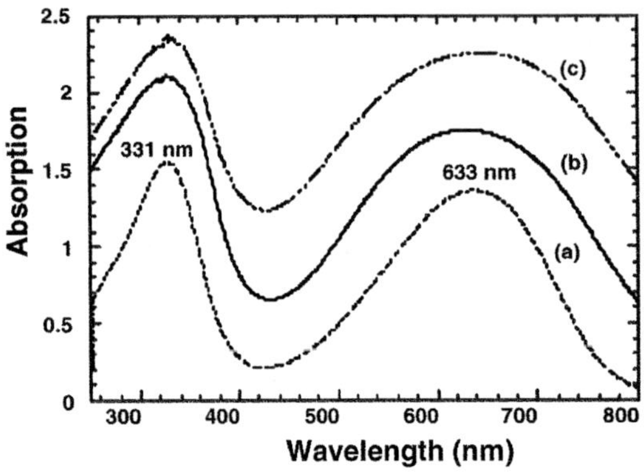

Figure 1. UV-VIS spectra of EB solution and thin film.

The UV-Vis spectra of the EB solution used to spin fiber and an ultra-thin film prepared from this solution are shown in Figure 1. Figure 1b shows the spectrum of the concentrated EB/NMP/HPMI solution (20 wt%).

There are two absorption peaks at 331 nm ($\pi - \pi$ *) and 633 nm (exciton peak). This is consistent with the solution UV-Vis spectrum of a 1 wt% EB/NMP solution without the addition of HMPI to the solution as shown in Figure 1(a). A thin EB film was obtained by spin casting the concentrated EB solution with HMPI on top of a quartz plate, and subsequently immersing it in H_2O for 1 hour and in CH_3OH for 30 minutes in order to remove the residual NMP and HPMI from the film. The UV-Vis spectra of this thin transparent film is shown in Figure 1(c). Again, the absorption spectra of the thin film and the EB solution prepared without HMPI are identical. There are two extreme oxidation states of polyaniline, the fully reduced leucoemeraldine base (LEB) and the fully oxidized pernigraniline base (PNB) forms. EB has an oxidation potential in between these two extremes. The UV-Vis spectrum of LEB has only one absorption peak at 330 nm. The UV-Vis spectrum of PNB has two absorption maxima: one at 330 nm and the other at 535 nm. These spectral results (Figure 1a-1c) show that the UV-Vis spectra of diluted EB/NMP solution, the concentrated EB/NMP/HPMI solution used to prepare fiber, and the solid-state thin film prepared from the fiber spinning solution are all identical; and moreover, they show no features in common with either LEB or PNB spectra. It is, therefore our conclusion that the oxidation state of the EB polymer is not altered by HMPI. FTIR and solid state ^{13}C NMR spectra obtained for the thin film and fiber prepared respectively from the spinning solution showed no indications of ring substitution.[7]

THERMAL ANALYSIS OF EB POWDER AND FIBER

Figure 2 shows the differential scanning calorimetry (DSC) scans of the as-spun EB fiber. The first scan has an exotherm peak at 220°C which is presumably due to a crosslinking reaction between the polyaniline chains. After reaching 300°C the sample was cooled to 50°C and then scanned for a second time. The second scan shows no further reactions to 300°C. Similar result have been reported for EB powder crosslinking reactions at 220°C. We found no evidence for a glass transition with this fiber in the temperature range tested.

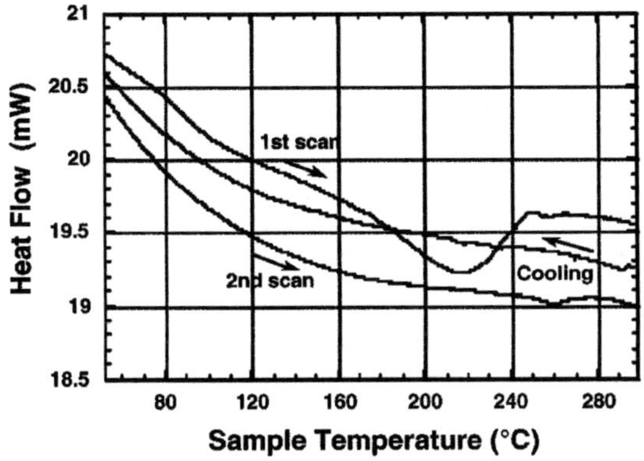

Figure 2. DSC scan of EB fiber.

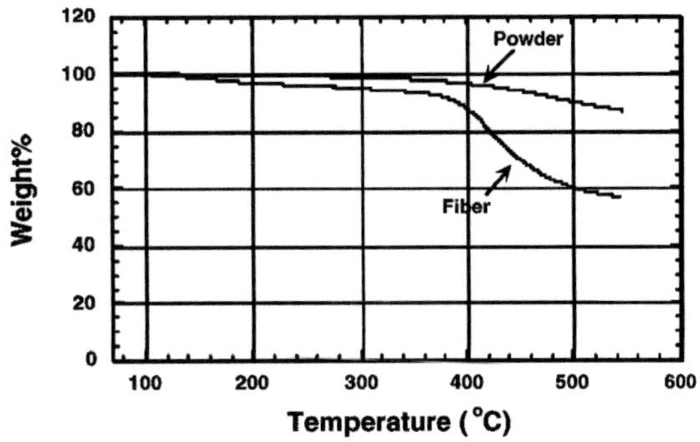

Figure 3. TGA scans of EB fiber and powder.

Table 1. Fiber diameter, conductivity, and density values for polyaniline

Dopant	Diameter, μm	σ, S/cm	Density, g/cm^3
As cast	209	$<10^{-10}$	0.26
HCl	180	3.3	0.47
TFA	157	0.23	0.48
HOAc	221	$5x10^{-5}$	0.19
BPA	241	2.4	0.60

Figure 3 shows the results from thermogravimetric analysis of the EB powder compared to the as-spun EB fiber. The fiber had been immersed in water for 3 days and subsequently dried in dynamic vacuum for 48 hours. EB powder is thermally stable to almost 390°C under nitrogen after which the polymer decomposes. We define the decomposition temperature as the corresponding temperature of the intersection of the two tangent lines derived from the TGA curve. The EB fiber is also thermally stable to up to 375°C. There is a 2.7% weight loss for the EB fiber when the temperature reaches 210°C. The EB powder loses only 0.7% of its weight at this temperature which is attributed to residual adsorbed water. This result suggests that there is still ~2.0 wt% of the NMP or HPMI residue remaining in the fiber. This residue can be removed by immersing the fiber in CH_3OH for 24 hours. Thermal stability of the EB fiber is fairly consistent with the EB powder itself which suggests that no processing induced degradation of the polymer occurred.

CONDUCTIVITY, FIBER DENSITY, AND FIBER DIAMETER AFTER DOPING

The results obtained for the acid doped EB fibers with respect to fiber diameter, fiber density, and DC conductivity are presented in Table 1. We were surprised to observe that the fiber diameter changes, either increasing or decreasing, depending on the type of acid used for doping the as-spun fiber. HCl and trifluoroacetic (TFA) acid doped fiber both show volume contraction while the acetic acid (HOAc) and benzenephosphinic acid (BPA) doped fiber expand the volume of the fiber. Acid doping leads to increases of the fiber density. The HCl, TFA and BPA doped EB fiber all have significantly higher densities as compared to the as-spun EB fiber. However, the acetic acid doped EB fiber has lower density as compared to the as-spun EB fiber. We attribute this anomalous behavior to the fact that acetic acid initially fully dopes the imine nitrogens of the polymer, thus increasing internal stress and reorganizing the polymer free-volume. However, acetic acid is a very weak acid which is removed from the porous fiber structure when it is pumped under dynamic vacuum during the drying process. This dedoping phenomenon left the dopant sites created by acetate counter ion empty, thus creating more free volume on the mesoscopic level inside the fiber. This is further evidenced by the low conductivity value of the HOAc doped fiber in comparison with the other 3 acids used in this study.

FIBER MECHANICAL PROPERTIES

Table 2 shows the results obtained for mechanical testing of each sample. The Young's modulus for the HCl (1.75 GPa) and the TFA (1.38 GPa) doped fibers were greater than the as-spun fiber (1.26 GPa), while the BPA doped fiber (0.90 GPa) was smaller. The Young's modulus of the HOAc doped fiber (1.75 GPa) was higher that of the as-spun EB fiber. Calculation of specific Young's modulus (gram per denier) takes into account the fiber density which is simply the bulk modulus divided by the fiber density. 4.0 M acetic acid doped EB fiber had the highest specific Young's modulus of 103.6 g/d.

The failure strength of the EB fibers range from 14.0 MPa (HCl doped fiber) to 20 MPa (trifluoroacetic acid doped fiber). These fibers break at strain from 1.72% (BPA doped fiber) to 3.35% (trifluoroacetic acid doped fiber). This result indicates that these fibers are hard and brittle. We previously reported the mechanical properties of EB fiber (450 μm diameter) spun from solution prepared from high molecular weight EB utilizing 2-methylaziridine as gel-inhibitor in the solution.[1] The failure strain of those as-spun fibers was 9%, and we were

Table 2. Mechanical properties of polyaniline fibers

Dopant	Young's modulus, GPa	Tenacity, g/d	Specific Young's modulus, g/d	Failure strain, %	Failure strength, MPa
As cast	1.26	1.47	54.6	2.77	15.0
HCl	1.75	1.13	41.9	1.85	14.0
TFA	1.38	0.87	32.4	3.35	20.2
HOAc	1.75	2.79	104	1.80	17.2
BPA	0.90	0.45	16.9	1.72	14.2

able to tie a knot with this fiber without breaking it, i.e., the fiber was hard and strong. The Young's modulus of our current fiber (1.26 GPa) is stronger as compared to the previously reported fiber (0.52 GPa), however, the failure strain is significantly smaller in the present case. The viscosity values measured for a 1% (w/w) EB solution prepared with polymer obtained from Neste and used in the present study was 3.6 kPa s, whereas the ultra-high MW EB used in our previous study[1] had a solution viscosity of 14.1 kPa s, both solutions prepared at 20% w/w. We believe that the present difference in the mechanical properties are likely due to the difference in molecular weights used for the two studies.

Our current as-spun fiber exhibit low density (0.26 g/cm^3) with microporous features similar to fiber reported by Gregory *et al.* which was spun from concentrated EB/DMPU solution.[6] These investigators were able to increase their fiber density and, as a result, the conductivity of the fiber by increasing the take-up speed during production. All the results that we have discussed so far have been focused on the as-spun fiber without heat treatment or stretch orientation which are known to improve the mechanical properties and conductivity of the fiber.[1,7] The tenacity of our as-spun fiber is as high as (1.47 g/d) before stretch orientation. It is expected that higher mechanical strength and conductivity values can be achieved by applying these processing methods to the HMPI processed as-spun fiber.

CONCLUSIONS

We have fabricated and characterized polyaniline emeraldine base (EB) fiber from highly concentrated (20% w/w) EB/NMP/HPMI solution. HMPI can be used in place of 2-methyl aziridine as the gel-inhibitor, and this is desirable from an environmental point of view. UV-Vis data shows that the oxidation state of EB polymer in solution and in the fiber is unaltered by the presence of HMPI. Acid doping of as-spun EB fibers leads to conductivity values ranging from 10^{-5} to 3.3 S/cm. The volatile weak acetic acid is removed by mechanical pumping in vacuum which explains the reduced conductivity value compared to the other samples. Acid doping of the EB fiber results in changes in fiber volume (expansion or contraction), and thus, fiber density. Differences in failure strength and strain of the present fibers compared to previously reported fibers are likely due to differences in the molecular weight of the EB utilized for each study.

ACKNOWLEDGMENTS

We wish to thank Dali Yang and Robert Romero for their assistance in fiber spinning. This work was conducted under the auspices of the US Department of Energy through the Office of Industrial Technology (AIM), and supported (in part) by funds provided by the University of California for the conduct of discretionary research by Los Alamos National Laboratory.

REFERENCES

1 B. R. Matttes, H. L. Wang, D. Yang, Y. T. Zhu, W. R. Blumenthal, M. Hundley *Synthetic Metals*, 1997, **84**, 45.
2 B. R. Matttes and H. L. Wang Stable, Concentrated Solutions of High Molecular Weight Polyaniline and Article Therefrom, *U.S. Patent Application, June, 1996.*
3 B. R. Matttes, H. L. Wang and D. Yang, Proceedings of the SPE 55th Annual Conference, ANTEC 1997, 1463.
4 C. C. Han and R. C. Jeng, *Chem. Commun.*, 1997, 553.
5 X. H. Wang, Y. H. Geng, L.X. Wang, X. B. Jing and F.S. Wang, *Synthetic Metals*, 1995, **69**, 263.
6 A. P. Chaco, S. S. Hardaker, B. Huang, R. V. Gregory Proceeding of Material Research Society, Boston, Fall 1995
7 H. L. Wang, R. Romero, Y. T. Zhu, B. R. Matttes, *to be published.*

Electrically Conductive Polyaniline Fibers Prepared by Dry-Wet Spinning Techniques

Benjamin R. Mattes, Hsing-Lin Wang, and Dali Yang
Chemical Science and Technology Division, Los Alamos National Laboratory, Los Alamos, NM 87545, USA

INTRODUCTION

Dopable π-conjugated polymers (alternating double and single bonds along the polymer main chain repeat units), such as those found in the family of polymers known as polyaniline, show potential for a variety of commercial applications such as chemical separations, electromagnetic interference shielding, protection of metals from corrosive environments, and anti-static coatings and current carrying fibers. Polyaniline is a commercially attractive polymer since, unlike many other dopable π-conjugated polymers, it is both environmentally stable and can be made electrically conducting by acid treatment.

Polyaniline is a promising candidate for commercial fiber applications. However, introduction of commercial products based on polyaniline fiber has been slow to develop because it is difficult to process it by traditional fiber production techniques, i.e., melt extrusion or wet spinning methods. There are several critical reasons for this which include: (a) poor polymer solubility in organic solvents; (b) rapid polymer gelation times at low (<5% w/w) total solids content; and, (c) the inability to utilize high molecular weight polyanilines at concentrations exceeding 20% w/w.

We have discovered[1] a class of chemical agents that selectively complex the imine nitrogens of the emeraldine base repeat unit through a hydrogen bond formation, and thus disrupt inter- and intrachain hydrogen bonding responsible for rapid gelation times and low solubility in organic solvents. We call this class of compounds Gel-Inhibitors (GI). Figure 1 schematically illustrates the likely mechanism responsible for the above mentioned problems with the emeraldine base form of polyaniline when dissolved in solutions not containing gel-inhibition agents. Rapid gelation at low polymer concentrations is due to the very strong hydrogen bonding forces that develop between the secondary amine hydrogens on one poly-

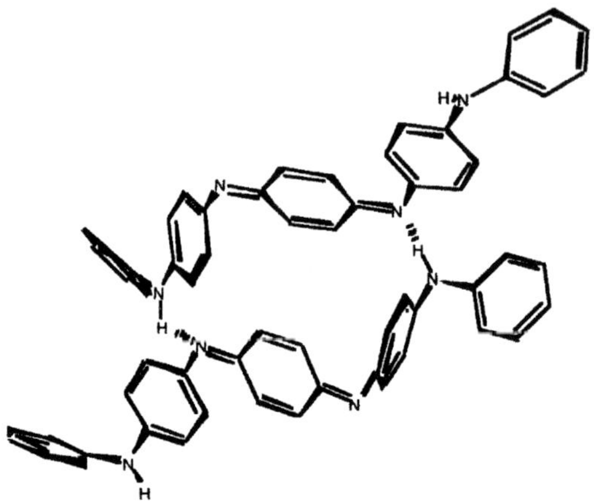

Figure 1. Schematic representation of hydrogen bonding between imine nitrogens of one polymer chain and the secondary amine hydrogens on the next nearest neighboring polymer repeat unit.

mer chain and the imine nitrogen atoms on the next nearest neighboring chain. We reasoned that it would be possible to form thermodynamically stable, highly concentrated EB solutions if we could inhibit this tendency for the polymer to develop interchain hydrogen bonds. Presently we discuss one such approach which utilizes the heterocyclic secondary amine, 2-methyl-aziridine, which forms intermediate EB-2MA hydrogen bonded complexes in >20% (w/w) solutions of high molecular weight polyaniline suspended in N-methyl-pyrrolidinone (NMP).

EXPERIMENTAL

POLYMER SYNTHESIS

High molecular weight polyaniline was synthesized by dissolving 100g (1.074 mole) of aniline in 1500 ml of 1 M HCl together with enough LiCl to make a 5 M salt solution. This solution was transferred to a 4 L resin kettle, and subsequently immersed in a cyclohexanone/CO_2 ice bath, where it was mechanically stirred throughout the course of the polymerization reaction. After 1 hour the reaction temperature of the aniline/LiCl solution reached a temperature of -45°C. Ammonium persulphate [131 g (0.574 mole)] was dissolved in a separate flask which contained 1200 ml of 1 M HCl and 5 M LiCl. This oxidant solution (25°C) was added to the aniline solution at a rate of 8 mL/minute by means of a metered syringe pump. The reaction mixture was maintained at -45°C for 48 hours. The emeraldine hydrochloride powder was collected by vacuum filtration and, subsequently washed with 2 L increments of 1 M HCl until the filtrate become colorless. The polymer was then washed with 2 L of water and then transferred to a 4 L beaker containing 2.5 L of 0.1 N NH_4OH, stirred for 1 hour, and subsequently vacuum filtered to collect the deprotonated emeraldine base powder. The powder was further reacted with another 2.5 L of 0.1 N NH_4OH aqueous solution for another hour, and subsequently vacuum filtered to recover the EB powder. The polymer was dried under dynamic vacuum at 10^{-2} torr for 72 hours. The yield was 45%.

RHEOLOGICAL MEASUREMENTS

Viscosity measurements were performed with a RVDV-III Brookfield Cone and Plate Viscometer. Experiments performed under conditions of constant shear were set at 0.8 sec^{-1} under isothermal conditions at a chosen temperature.

SOLID FIBER SPINNING

A solution for spinning EB solid fibers was prepared as follows: 31.32 g of N-methyl-2-pyrrolidinone (NMP) was mixed with 4.879 g ($7.9x10^{-2}$ mole) of 2-methylaziridine [90%, 2-MA, Aldrich]. This mixture was placed in a 60 ml glass jar with a teflon lined screw cap at 60°C for one hour, after which 9.109 g ($2.5x10^{-2}$ mole) of EB was quickly added to this NMP/2-MA mixture (GI/EB =3.1), and vigorously stirred for a few minutes to wet the polymer powder. The glass jar was tightly sealed and returned to the oven set at 100°C for about 30 minutes. During this time, the EB/NMP/2-MA mixture was removed every 10 minutes and vigorously stirred. After this time, a flowable homogeneous liquid solution free from gel particles formed. The concentration of EB in this solvent system was 20.1 wt%.

This EB solution was transferred to a hydraulic stainless steel cylinder and cooled to room temperature. A gear pump motor, fed by a nitrogen gas at 100 psi, was used to drive the EB fluid through 3/8" stainless steel tubing, and through a spinnerette (500 μm O.D.), at a pressure of 250 to 1,000 psi. The polymer solution was extruded through a 1 inch air-gap directly into a water coagulation bath (5°C) where the solvent and GI where removed from the nascent polyaniline fiber by de-mixing and solvent/non-solvent exchange in the bath. The take-up speed was varied between 3 to 10 feet per minute. The nascent fiber was continuously wound on a series of two water bath godets maintained at 15°C, and collected on a bobbin by means of a Leesona Winder. The fibers were placed in water extraction baths for 48 hours to remove residual solvent, and dried under dynamic vacuum.

FILM AND FIBER DOPING

Six inch segments of the stretched and unstretched EB fiber and film were immersed in 400 ml of their respective aqueous acid solutions for 48 hours. They were removed from the doping solution, dried under dynamic vacuum for another 48 hours, and their conductivity was measured by a 4-probe method. The acid solutions used for doping the solid fibers and films were: 1.5 N HCl, 1 N acetic acid, and an aqueous solution of benzene phosphinic acid [BPA (pH= -0.37)].

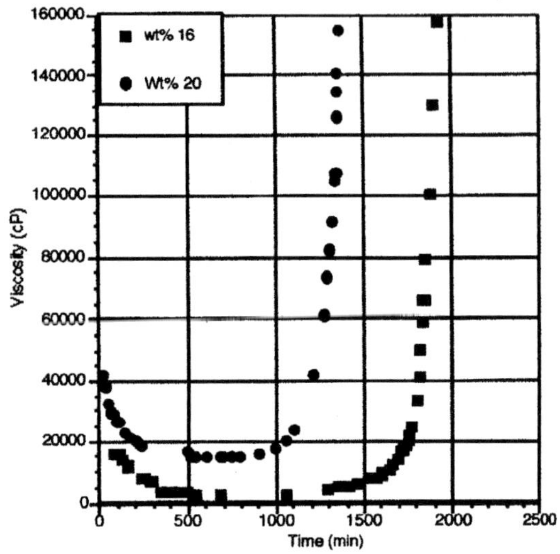

Figure 2. Viscosity as a function of time for 2MA/EB ratio = 2.5 prepared at 16% and 20% (w/w) total solids at 25°C.

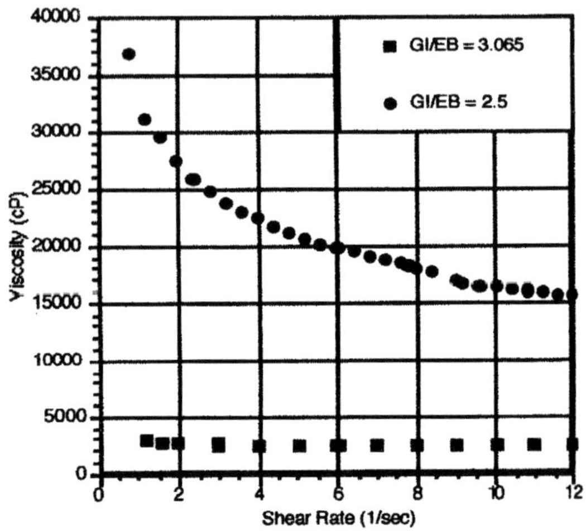

Figure 3. Viscosity as a function of shear rate at two different 2MA/EB ratios both prepared at 20% (w/w) total solids at 25°C.

TEMPERATURE DEPENDENT CONDUCTIVITY MEASUREMENTS

The temperature-dependent resistivity was measured with a four-probe DC apparatus that employed Keithley model 182/181 voltmeters and a model 220 constant current source. A colloidal graphite suspension was used to make high-quality ohmic contacts to the polyaniline specimens.

RESULTS AND DISCUSSION

SOLUTION RHEOLOGY

GPC analysis confirmed that our samples were indeed of high molecular weight (M_n=6.7x10^4 and M_w=6.8x10^5). EB prepared under the synthetic conditions described above exhibits a high degree of polydispersity. None-the-less, it is possible to form thermodynamically stable EB solutions above the 20% weight total solids level by adjusting the molar ratio of 2MA to EB repeat unit in the ranges lying between 0.5 to 4.0. The addition of 2MA at molar ratios greater than 4.0, i.e., more than one 2MA molecule for each imine nitrogen atom in the EB repeat unit, results in greatly reduced mechanical and physical properties for the EB fiber or film, e.g., moduli and conductivity. Figure 2 illustrates the viscosity behavior for a dope solution prepared at the 2MA/EB molar ratio of 2.5 as a function of time at 25°C at constant shear

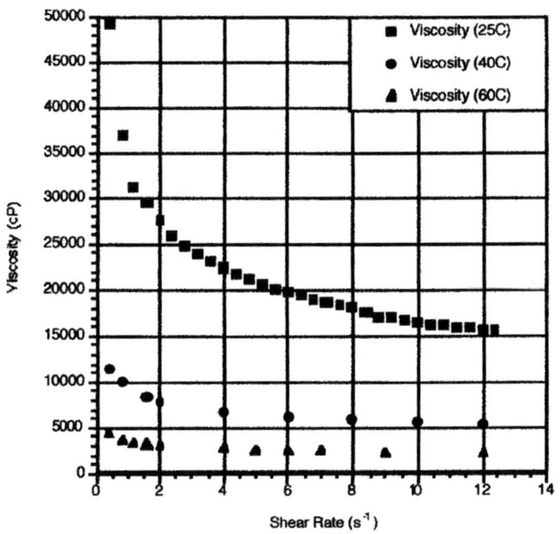

Figure 4. Viscosity as a function of shear rate at three different temperatures for 2MA/EB ratios = 2.5, all prepared at 20% (w/w) total solids.

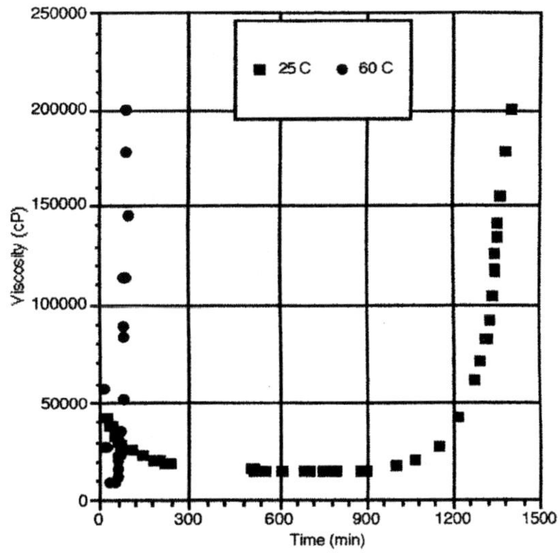

Figure 5. Viscosity as function of time at two different temperatures: 2MA/EB ratios = 2.6 with 20% (w/w) total solids.

rate. There is gradual decrease in the solutions viscosity during the first 3 hours of testing until equilibrium mixing is achieved. The measured viscosity then remains relatively constant for periods of 20 and 11 hours for the 16% and the 20% (w/w) solutions respectively. These times present the "window of opportunity" for fiber spinning. It is clear that gel-inhibition agents, such as 2-methyl aziridine, serve to simultaneously reduce viscosity as well as increase time to gelation. Figure 3 shows the relationship between viscosity and shear rate for two solutions of EB in NMP prepared at the 2MA/EB ratio of 3.1 and 2.5 respectively, while maintaining a constant total solids level of 20% (w/w). This example demonstrates the sensitivity of the 20% (w/w) 2MA/EB/NMP solutions to the number of GI molecules coordinated to imine nitrogens in the repeat unit of the polymer under conditions of increasing shear rates. In general, increasing the 2MA/EB molar ratio leads to reduced solution viscosity at constant polymer mass. Constant viscosity with respect to increasing shear ratio is observed for solutions prepared with more than 1.5 2-methyl-aziridine molecules associated per polymer repeat unit.

Figure 4 gives a plot of viscosity vs. shear rate for three solutions prepared at a 2MA/EB molar ratio of 2.5, each solution having 20% (w/w) total solids, respectively measured at 25°C, 40°C, and 60°C. The sample measured at room temperature exhibits shear thinning behavior. It is clear that increases in temperature from room temperature to 40°C and 60°C,

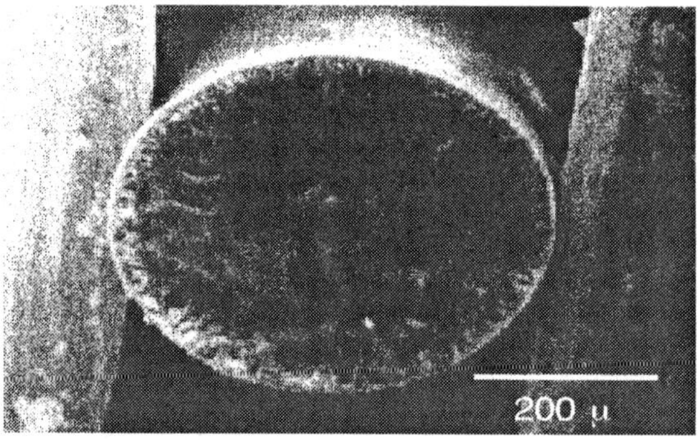

Figure 6. Cross-sectional SEM photograph of a representative as-spun emeraldine base fiber.

respectively, leads to drastic reductions in solution viscosity at constant 2MA/EB ratios for solutions prepared at the same concentration. The rheological properties of these solutions exhibit Newtonian flow patterns under conditions of low shear rate. Moreover, the higher temperatures result in relatively constant viscosity with increasing shcar rates. It is important to note that elevated temperatures are required to form the 2MA/EB intermediate complex, but as illustrated in Figure 5, that prolonged exposure to such temperature increases results in rapid gelation. It is necessary initially to provide thermal energy to form the intermediate hydrogen bonded polymer/gel-inhibitor complexes; however, prolonged gel inhibition times require cooling the solutions back to (or lower than) room temperature. Once the thermodynamically stable particle-free solution is prepared, it may be stored indefinitely at 0°C.

FIBER PROPERTIES

Figure 6 shows the cross-sectional SEM photograph from one of the representative as-spun fibers. The relative absence of very large macrovoids in this fiber relates to the fact that the high total solids content of the polymer solution instantaneously precipitates in the coagulation bath at the moment of extrusion. The solvent and GI are removed from the nascent polyaniline fiber by: 1) rapid de-mixing and solvent/non-solvent exchange in the coagulation bath; 2) length of residence time on the godets; and 3) final extraction procedures. The kinetics which govern this precipitation process and, ultimately, the fiber's morphology on the mesoscopic scale, may be controlled by changing the polarity of the coagulation bath and/or the bath's temperature. Higher SEM magnification imaging revealed a non-interconnected pore structure with average pore radii in the 0.1-0.2 μm range. The density of these as-spun fibers was measured at 0.52 g/cm^3. A 4x stretch ratio increases fiber density to 0.92 g/cm^3, which is still significantly lower than EB powder which was measured at 1.315 g/cm^3.[7] The maximum draw ratio depends on the amount of residual plasticizing solvent and the temperature of the hot tip. Overdrying the fiber may reduce the draw ratio due to the lower NMP content. Residual NMP acts as a plasticizer which increases interchain mobility and, also depresses the glass transition temperature of EB.

CONCLUSIONS

We have introduced the concept of gel-inhibitory agents which improve the solution processing parameters of emeraldine base. These agents are quite sensitive to the concentration range described in this report, i.e., GI/EB molar ratio between 1-4. The preparation of highly concentrated EB solutions utilizing high molecular weight polyaniline is easily achieved with this processing advantage. The solutions are stable for periods of time sufficient to dry-wet spin fiber. The fiber moduli, strength, and DC conductivity are improved by stretch alignment. Mineral acids tend to weaken the fibers following doping protocols, while organic acids such as BPA preserve good mechanical and conductive properties. All of the doped samples showed temperature dependant DC conductivity. This data can be fit to disorder models such as the quasi-one dimensional VRH model. To values indicate that disorder increases in the following order: stretched film > stretched fiber> unstretched film and fiber. Acetic acid is an anomalously poor dopant for polyaniline. BPA is a good dopant.

ACKNOWLEDGMENTS

We wish to thank J. Thompson and D. McBranch at LANL for helpful technical conversations, and R. Romero for his assistance with fiber spinning. We also thank M. Winokur for analyzing samples with XRD. This work was sponsored by Los Alamos National Laboratory Directed Research and Development program through the Industrial Partnership Office.

REFERENCES

1 Mattes, B. R. and Wang, H.L. *Manuscript in preparation.*
2 Tzou, K. T., R. V. Gregory, *Synthetic Metals*, **69**, 109-112, 1995.
3 Mattes, B. R. and Wang, H.L. Stable, Concentrated Solutions of High Molecular Weight Polyaniline and Articles Therefrom, *US Patent Application, June, 1996.*
4 Winokur, M. *Personal communication.*
5 Hsu, C.-H., J.D. Cohen, & R.F. Tietz, *Synthetic Metals*, **59**, 37 (1993).
6 MacDiarmid, A.G. et.al., **Conducting Polymers**, Alcacer, L., ed., *Riedel Pub.*, 1986, p.105.
7 Pellegrino, J. P., Radebaugh, R., and Mattes, B. R., *Macromolecules*, 1996, **29**, 14, 4985-4991.
8 Blades, H., **US Patent 3,869,430**, 1975.
9 Wang, Z. H., Scherr, E. M., MacDiarmid, A, G., and Epstein, A., *J. Phys. Rev. B.*, 1992, **45**, 4190-4202.

Conductive Thermoplastic Compounds for EMI/RFI Applications

Larry Rupprecht
RTP Company, 580 E. Front Street, Winona, MN

INTRODUCTION

The rapid growth of electronic devices has increased the demand for injection moldable thermoplastics for housings and structural components. Many of these electronic devices must also be protected against electromagnetic interference (EMI/RFI). Unfortunately, the common thermoplastics used in electronic housings and structural members are transparent to EMI/RFI. Shielding previously meant employing metal housings and components or, more recently, post-mold applied coatings to thermoplastic parts. Today the use of conductive modifiers in thermoplastics has brought to the electronic industry the design freedom of thermoplastics with intrinsic EMI/RFI shielding.

While conductive modifiers for thermoplastic resins have been available for many years, their use in EMI/RFI shielding applications has only recently experienced growth. This is attributed to advancements in compounding and processing techniques and improvements in the quality of conductive modifiers. Such improvements have provided enhanced performance and reliability in conductive thermoplastics for shielding.

Electrically conductive thermoplastics combine a matrix resin and a conductive modifier. The matrix resin includes a thermoplastic resin with reinforcement, modifiers, or additives to impart particular physical properties to the composite. The conductive modifier is chosen to achieve specific conductive/shielding properties and be compatible with the matrix resin for minimal effect on the composite's other properties.

EMI/RFI SHIELDING

EMI CONCEPTS

Electromagnetic interference (EMI) is radiation with adverse effects on performance of electronic devices. While EMI exists across the entire electromagnetic spectrum, from dc electricity at less than 1 Hz to gamma rays above 10^{20} Hz, the great majority of EMI problems are limited to that part of the spectrum between 25 kHz and 10 GHz. This portion is known as the radio frequency interference (RFI) area and covers radio and audio frequencies. The acronym EMI is generally used to represent both EMI and RFI.

EMI PROBLEMS

Electronic devices are both sources and receptors of EMI. This fact creates a two-fold problem for manufacturers since both operational integrity of and emissions from products must be dealt with. In addition, electromagnetic waves have both magnetic and electric components. The magnetic and electrical wave components each have unique EMI effects depending upon frequency and distance from the radiation source. Resolution of EMI problems includes consideration of all these factors.

EMI SOLUTIONS

Securing operational integrity and reducing emissions to meet applicable standards requires some modification to commonly used thermoplastic enclosure materials. Thermoplastics are electrical insulators, transparent to electrical energy, and, therefore, EMI waves pass through them without loss. A range of material options for providing EMI shielding to thermoplastic enclosures is presented below.

EMI/RFI TESTING/QUALIFICATION

The objective of a test for shielding effectiveness in materials should be determined so that appropriate value can be obtained. Objectives are usually classified into one of three categories:

- Compliance Testing
- Engineering Evaluation
- Audit/Screening

COMPLIANCE TESTING

Compliance testing is a precise and absolute evaluation of a fully assembled and functioning device. The open field test technique is generally required for compliance under various agency standards prior to commercialization of most wireless electronic devices. Free space measurements for both radiated and conducted emissions are taken in an open field. Such testing requires sophisticated equipment and must be performed by accredited test labs.

ENGINEERING EVALUATION

The objective of engineering evaluation is to determine suitability of materials to perform shielding. Less precise than compliance testing, engineering evaluation still requires a controlled environment for accuracy. Generally, an open field test site is used for greater accuracy. Equipment is less sophisticated than required for agency compliance. Common techniques used include dual chamber shielded box as used in ASTM ES-7-83 (discontinued), transverse electromagnetic cell (TEM-cell), and coaxial transmission line as in ASTM D 4935-89.

AUDIT/SCREENING

This testing methodology is primarily a screening process, providing a quick look at shielding characteristics of a conductive thermoplastic composite. Volume and surface resistivity, microwave reflectance, and attenuation tests on a spectrum analyzer are screening methods that can be used. Results obtained are useful for ranking and screening of candidate shielding composites. Further validation of test results is required through engineering evaluation and compliance levels.

VOLUME AND SURFACE RESISTIVITY

Volume resistivity is electrical resistance through a unit mass of material and is expressed as "ohm-cm." There is a good correlation between volume resistivity and shielding effectiveness of conductive thermoplastic composites. Volume resistivity of 10^0 to 10^{-3} ohm-cm is generally thought to be acceptable for effective shielding composites.

Surface resistivity is electrical resistance over a unit area of material's surface and is expressed as "ohms/square." Surface resistivity and shielding effectiveness can be correlated where electrical conductivity is a surface property as in applied conductive films or coatings. In some applications, coating thickness is a variable in shielding effectiveness. However, in the instance of incorporated conductive modifiers, surface resistivity has little value in correlating shielding effectiveness. The value of low surface resistance of plastics in shielding is in providing electrical continuity between components of an assembly.

MICROWAVE REFLECTANCE

A reflectometer measures microwave energy reflectivity of planar material surfaces. The instrument focuses energy at a frequency of 10.525 GHz through a rectangular waveguide aperture onto a target surface. The relative return loss of a sample is compared to a metal plate reference. The relationship of this reflection coefficient to shielding effectiveness of the material can be determined through higher level testing as described under "engineering" or "compliance" levels.

SPECTRUM ANALYZER

A spectrum analyzer with tracking generator and transmitting/receiving probes can be used to screen attenuation values of conductive thermoplastic composites. A molded specimen is passed through an electromagnetic field and the reduction in field intensity is measured. Data obtained is useful for ranking and comparison of candidate materials. Determination of the relationship of tested attenuation to shielding effectiveness can be made through higher level testing as described under "engineering" or "compliance" levels.

The key features of testing just described are summarized in Figure 1.

Test Method	Field	Specimen	Comments
Compliance testing			
1. Open field	far	device	Used for various agency approvals; sophisticated equipment; high cost; excellent reproducibility.
Engineering evaluation			
1. ASTM D4935-89	far	doughnut	Modest cost; good reproducibility and correlation; suitable for ranking.
2. TEM cell	far	flat plaque enclosure	Modest cost; good reproducibility and correlation; suitable for ranking tests enclosures.
3. Dual chamber	near	flat plaque	Low cost; poor correlation.
Audit/screening			
1. Volume resistivity	N/A	small parts	Low cost; limited correlation.
2. Surface resistivity	N/A	small parts	Low cost; limited correlation.
3. Microwave reflectance	far	any	Uncertain correlation; in-process tool
4. Spectrum analyzer	near	any	Uncertain correlation; in-process tool

Figure 1. EMI test methods.

SHIELDING MATERIALS

SHIELDING ADDITIVES

Metallic substances including stainless steel, copper, nickel, and silver in fiber, flake, or particulate form, and metal-coated substrates including glass or carbon fiber, minerals, or glass beads are typical additives found to provide shielding characteristics to thermoplastics. Primary form of metallic additive for shielding composites is a fiber. The high aspect ratio of fiber relative to other material forms enables the formation of a conductive pathway through the resin matrix at low concentration.

METALLIC SUBSTANCES

Stainless steel fibers are the most common metal fibers but copper, nickel, silver, and aluminum are also used. Available products include stainless steel fibers of 7-8 micron diameter and copper fibers of 50-micron diameter, while nickel and silver are of various diameters. Stainless steel fibers are the primary commercial product in this group and are supplied typically as chopped bundles containing 12,000 filaments. Each bundle is wetted with an appropriate sizing or thermoplastic resin and is chopped to a 4-7 mm length. Advancements in stainless steel fiber manufacture have led to improved composite properties, including aesthetics and processability.

Metal flakes, powders, and particulates are also utilized as shielding additives but, with reduced aspect ratio compared to fiber forms, these materials find little economic value.

METAL-COATED SUBSTRATES

Metal-coated substrates are composites of metal, typically nickel or copper, and various substrates including glass and carbon fibers, glass beads, and minerals like mica and titanium dioxide. The primary commercial product in this group is nickel-coated carbon fiber (NCCF) supplied typically in a form similar to stainless steel fibers. Much development work has been performed in NCCF fiber, bringing improved economics and enhancement of composite properties and processability.

Nickel and silver-coated minerals and glass beads have found usage as shielding additives, particularly where reinforced strength features are not desired. Applications include seals and gaskets in elastomeric thermoplastics where elongation and elastic memory are desired but high modulus is not.

SHIELDING COMPOSITES

An injection moldable shielding composite can be either of two physical forms – an extruded compound of shielding additive and thermoplastic resin or a physical blend of these components. These are referred to as compounded blend and cube blend, respectively.

COMPOUNDED BLEND

A compounded blend is a molding compound produced through extrusion compounding of shielding additive, thermoplastic resin, and possibly other additives such as flame retardant, wear additive, and pigment. This compounded blend is fed directly to the injection-molding machine. Compounded blend's advantage is in uniformity of molded parts as dispersion of shielding additive is generally complete. Extrusion compounding does cause fiber attrition and loss of conductive and shielding properties, remedied by increased metal-coated carbon fiber content.

CUBE BLEND

A cube blend is a molding compound that is fed directly to the injection-molding machine. It consists of two or more physically blended components - a shielding additive concentrate, a resin matrix consisting of thermoplastic resin and possibly other additives such as flame retardant, wear additive, and pigment, and sometimes any of these additives as a concentrate. The thermoplastic resin with its additives is extrusion compounded prior to blending with the shielding additive concentrate. A cube blend's advantage is primarily in economics, as less shielding additive is typically needed to achieve a given conductivity and shielding effectiveness. In a cube blend, the shielding additive is not subjected to shear of extrusion compounding, resulting in reduced fiber attrition. Additional concentrate can be added during molding if conductivity and/or shielding fall below specification. Cube blends demand a higher level of technical skill for complete dispersion of the shielding additive in molded parts.

MATERIAL DATA

Nickel-coated carbon fiber (NCCF) and stainless steel fiber were evaluated as EMI shielding additives in polycarbonate thermoplastic resin. No other modifiers or reinforcements were included in this study. Physical and electrical properties were determined for unfilled polycarbonate and NCCF and stainless steel fiber composites. Shielding investigations for this paper were performed at the audit/screening level by electrical conductivity (volume and surface resistivity) and microwave reflectance and at the engineering level by ASTM D4935-89. Effect of shielding additive content was studied at varying weight percents. Both physical forms - cube blend and compounded blend - were included.

Tables 1 through 4 detail physical, electrical conductivity, and microwave reflectance properties of stainless steel fiber and NCCF at five, ten, fifteen, and twenty weight percent in polycarbonate. Figure 2 through 5 detail shielding effectiveness (SE) testing per ASTM D4935-89 of stainless steel fiber and NCCF at varying weight percents in polycarbonate. Specimens are injection molded "doughnut" shapes of five-inch diameter by 0.125 inch thick.

CONCLUSIONS

Note the physical changes, including increased specific gravity, changes in mold shrink rate, and increases in strength and stiffness of the NCCF products over neat polycarbonate. Also note the minimal physical effect imparted by stainless steel fiber to polycarbonate. Conductivity features are similar between the two fiber types in cube blends. NCCF retains more conductivity (as measured by volume resistivity) in compounded blends than does stainless steel.

Table 1. NCCF in polycarbonate - cube blend

Nickel-coated carbon fiber, wt%	0	5	10	15	20
Specific gravity	1.20	1.24	1.28	1.32	1.36
Mold shrinkage, 0.125" thick, %	0.60	0.28	0.25	0.22	0.19
Tensile strength, Kpsi	9.5	11.5	13.0	14.5	17.2
Tensile modulus, Mpsi	0.35	0.8	1.0	1.2	1.6
Tensile elongation, %	100	4.0	2.9	2.1	1.5
Flexural strength, Kpsi	13.5	19.3	21.8	24.2	27.3
Flexural modulus, Mpsi	0.34	0.7	0.9	1.2	1.6
Izod impact strength, ft-lb/in					
Notched	3.0	2.2	1.9	1.5	1.4
Unnotched	>40.0	20.5	15.0	7.6	6.9
Deflection temperature, 264 psi, °F	270	283	284	285	285
Volume resistivity, ohm-cm	$>10^{16}$	240	0.50	0.15	0.09
Surface resistivity. ohms/square	$>10^{16}$	10^5	10^3	10^2	10^1
Microwave reflectivity, %	30	85	90	95	97

Table 2. NCCF in polycarbonate - compounded blend

Nickel-coated carbon fiber, wt%	0	5	10	15	20
Specific gravity	1.20	1.24	1.28	1.32	1.36
Mold shrinkage, 0.125" thick, %	0.60	0.35	0.33	0.28	0.25
Tensile strength, Kpsi	9.5	10.1	10.8	12.1	13.5
Tensile modulus, Mpsi	0.35	0.7	0.9	1.1	1.3
Tensile elongation, %	100	6.0	4.2	3.2	2.5
Flexural strength, Kpsi	13.5	16.2	17.4	18.8	19.7
Flexural modulus, Mpsi	0.34	0.6	0.8	1.1	1.4
Izod impact strength, ft-lb/in					
Notched	3.0	1.8	1.6	1.5	1.5
Unnotched	>40	25.8	18.1	9.3	8.6
Deflection temperature, 264 psi, °F	270	283	284	285	285
Volume resistivity, ohm-cm	$>10^{16}$	10^5	260	3.05	0.95
Surface resistivity. ohms/square	$>10^{16}$	$>10^{13}$	10^8	10^5	10^5
Microwave reflectivity, %	30	73	82	87	91

Table 3. Stainless steel in polycarbonate - cube blend

Nickel-coated carbon fiber, wt%	0	5	10	15	20
Specific gravity	1.20	1.24	1.30	1.36	1.42
Mold shrinkage, 0.125" thick, %	0.60	0.70	0.65	0.60	0.55
Tensile strength, Kpsi	9.5	9.2	9.5	9.7	9.8
Tensile modulus, Mpsi	0.35	0.35	0.40	0.43	0.44
Tensile elongation, %	100	2.0	1.5	1.2	1.1
Flexural strength, Kpsi	13.5	13.1	13.6	13.9	14.1
Flexural modulus, Mpsi	0.34	0.36	0.38	0.40	0.41
Izod impact strength, ft-lb/in					
Notched	3.0	2.0	1.8	1.5	1.3
Unnotched	>40.0	37	31	26	24
Deflection temperature, 264 psi, °F	270	270	270	270	270
Volume resistivity, ohm-cm	$>10^{16}$	1.7	0.67	0.21	0.08
Surface resistivity. ohms/square	$>10^{16}$	10^5	10^4	10^4	10^3
Microwave reflectivity, %	30	74	93	95	95

Table 4. Stainless steel in polycarbonate - compounded blend

Nickel-coated carbon fiber, wt%	0	5	10	15	20
Specific gravity	1.20	1.24	1.30	1.36	1.41
Mold shrinkage, 0.125" thick, %	0.60	0.68	0.66	0.62	0.59
Tensile strength, Kpsi	9.5	9.3	9.4	9.4	9.5
Tensile modulus, Mpsi	0.35	0.32	0.46	0.39	0.41
Tensile elongation, %	100	2.4	2.1	1.9	1.8
Flexural strength, Kpsi	13.5	13.0	13.2	13.3	13.5
Flexural modulus, Mpsi	0.34	0.34	0.34	0.36	0.37
Izod impact strength, ft-lb/in					
Notched	3.0	2.1	1.8	1.7	1.5
Unnotched	>40.0	40	32	21	18
Deflection temperature, 264 psi, °F	270	270	270	270	270
Volume resistivity, ohm-cm	$>10^{16}$	$>10^{13}$	$>10^{13}$	10^5	10^3
Surface resistivity. ohms/square	$>10^{16}$	$>10^{13}$	$>10^{13}$	$>10^{13}$	10^5
Microwave reflectivity, %	30	35	55	65	88

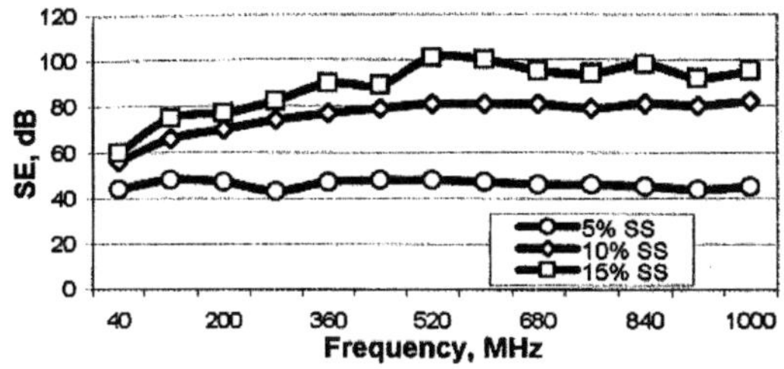

Figure 2. Shielding effectiveness per ASTM D4935-89. Stainless steel fiber in polycarbonate - cube blend.

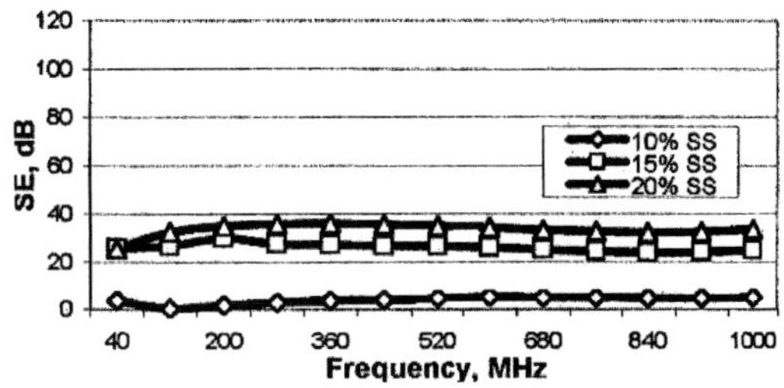

Figure 3. Shielding effectiveness per ASTM D4935-89. Stainless steel fiber in polycarbonate - compounded blend.

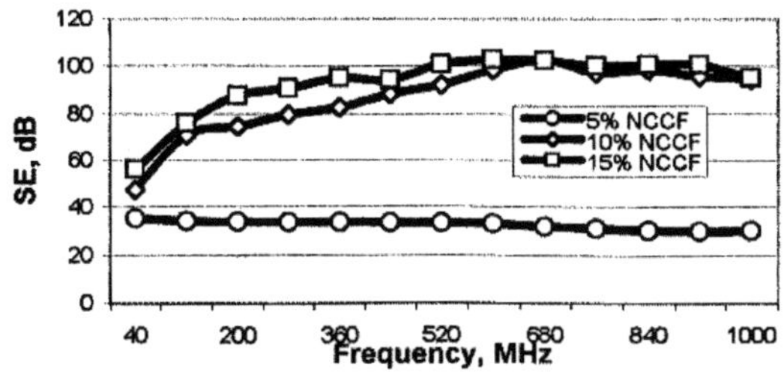

Figure 4. Shielding effectiveness per ASTM D4935-89. Nickel-coated carbon fiber in polycarbonate - cube blend.

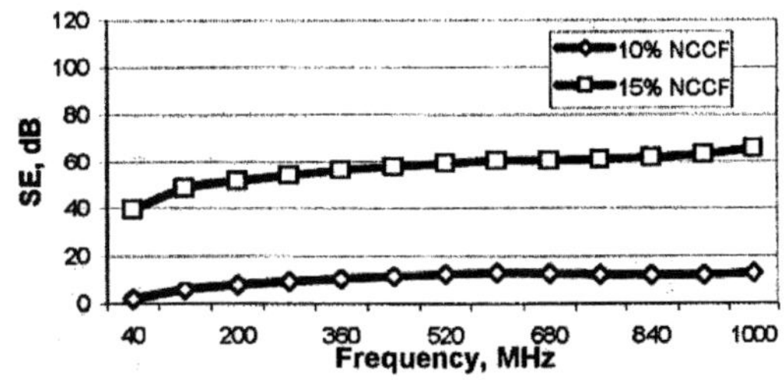

Figure 5. Shielding effectiveness per ASTM D4935-89. Nickel-coated carbon fiber in polycarbonate - compounded blend.

ASTM D4935-89 testing of shielding effectiveness shows differences between additive types and product blends. Shielding effectiveness at given concentrations of either EMI additive in cube blends is higher than equivalent compounded blends. The attainment of significant shielding effectiveness appears to be between five and ten weight percent in cube blends and is not yet maximized at twenty weight percent in compounded blends. Compounded blends of NCCF retain more shielding effectiveness than compounded blends of stainless steel fiber.

The development process for candidate materials in specific EMI shielding applications would utilize such information as presented here. Evaluation of physical, conductive, and shielding properties of thermoplastic EMI composites leads to optimization of content, form, and processing method. Understanding test methods and objectives is important in qualifying candidates for EMI/RFI protection.

Stainless steel fiber and NCCF in thermoplastic materials are shown to provide strong electrical conductivity and shielding through audit/screening and engineering level evaluations. Both stainless steel fiber and NCCF composites provide desirable features to composites for shielding applications. Numerous commercial citings of both EMI additives in injection molded applications are found in published literature, including trade journals and promotional releases from various suppliers.

Crystallization Kinetics in Low Density Polyethylene Composites

Brian P. Grady and W. B. Genetti
University of Oklahoma

INTRODUCTION

Electrically conductive polymer composites consist of an electrically insulating polymer matrix filled with an electrically conductive filler, which is often a metal particle. Although metal-filled systems have been studied extensively for property changes such as enhanced thermal and electrical conductivity, rheological and mechanical properties, and density, little work has been done on the effect metal has on the crystallization kinetics of semicrystalline thermoplastic composites. Maiti *et. al.*[1] studied the crystallization kinetics of polypropylene (PP) in nickel-PP composites, but this work was limited to volume fractions below where a continuous network of nickel particles had formed; i.e. below the percolation threshold. We are aware of no studies on metal-thermoplastic systems to determine the effect of a metal on the crystallization kinetics at concentrations equal to or above the percolation threshold. Recently, we observed that the fractional crystallinity of nickel-filled low-density polyethylene (LDPE) increased with increasing filler loading during calendering. This work prompted us to examine the effect of the filler on the crystallization kinetics of nickel-filled LDPE composites.

THEORY

In isothermal crystallization experiments, the transformation from an amorphous melt to a semicrystalline solid begins after some specific time period, the nucleation time, during which no measurable crystallization occurs. During nucleation, growth centers form, and these growth centers provide templates for crystallization. After a growth center has formed, the polymer begins to crystallize through lamella formation into a spherulitic structure. Crystal growth continues until a spherulite impinges upon another growing spherulite, another phase, or the polymer no longer has enough chain mobility for continued growth.

A theoretically derived equation for first order, heterogeneous nucleation with three-dimensional crystal growth is presented in Eq. [1].

$$\ln\left[\frac{1}{1-\chi_r}\right] = \frac{4}{3}\pi\frac{\rho_s}{\rho_m}N_0G^3t^3 \qquad [1]$$

where,

$$\chi_r(t) = \frac{\chi(t)}{\chi(\infty)} \qquad [2]$$

N_0 is the total number of heterogeneous particles added to the system, G is the constant linear crystal growth rate, t is the time, and ρ_s and ρ_m are the crystalline and melt densities. For systems with one-dimensional and two-dimensional growth, the power of t and G are 1 and 2, respectively. In real polymer systems, the assumptions used in deriving Eq. [1] are seldom accurate, but the derivation does give insight into the variables that affect crystallization.

Avrami proposed a semi-empirical equation where the nucleation and linear growth rates are embodied in the crystallization rate constant, K, and the dimensionality is characterized by the Avrami exponent, n, as shown in Eq. [3].[2]

$$\chi_r(t) = 1 - e^{-Kt^n} \qquad [3]$$

The Avrami exponent is rarely a whole number and is a function of both the nucleation mechanism and the dimensionality of growth, so dimensionality cannot be fully determined from kinetic data alone. Some commonly accepted guidelines state that n varies from 1 to 4 and is related to the dimensionality since the regions from 1 to 2, 2 to 3, and 3 to 4, generally indicate 1, 2, and 3 dimensional growth, respectively.

The Avrami equation only applies during the rapid crystalline growth clearly visible in an isothermal crystallization experiment. After rapid crystalline growth ceases, a pseudo-equilibrium level of crystallization is obtained. However, if the polymer remains at the isothermal crystallization temperature indefinitely, secondary crystallization will occur over long times at an extremely slow rate. The experiments described in this work only measured crystallization during this rapid growth period.

EXPERIMENTAL

Films of randomly cut nickel flake (Alfa Aesar), with a maximum diameter of 44 μm and an average thickness of 0.37 μm, supported by LDPE were produced with varying concentrations of filler. The crystallization rate, fractional crystallinity, electrical conductivity, and thermal conductivity ratios were measured as a function of nickel content. Composites were produced using a Killion KT-100, 3 zone extruder with a mixing section. To ensure uniform mixing, each concentration was processed three times, the first two using a pelletizing die and the third using a film die and calendaring system built at the University of Oklahoma.

The conductivity of the composite films was obtained using a Keithley Model 6105 Resistivity Chamber and a Keithley Model 610C Electrometer. The reported conductivity at each volume fraction is an average of several samples.

Isothermal crystallization experiments at varying temperatures between 95 and 104°C were conducted using a Perkin-Elmer DSC II differential scanning calorimeter. The sample was heated to 117°C and held at that temperature for 5 minutes to ensure complete melting. The sample was then quickly cooled at a nominal rate of 320°C/minute to a specific isothermal crystallization temperature. In order to isolate the heat evolution of crystallization from that of sample cooling, for each isothermal crystallization temperature a companion cooling curve was subtracted from the isothermal crystallization curve. Crystallization data was fit to the Avrami equation, Eq. [3].

The ratio of the composite thermal conductivity, k_c, to the polymer thermal conductivity, k_p, was determined from the cooling curves according to the following procedure. If the resistance to heat transfer at the interface between the sample and the DSC cell was much smaller than the resistance due to thermal conductivity of the sample, then the rate of heat transfer for the DSC sample is dependent on the following dimensionless parameter:

$$\frac{\alpha_c t_c}{b_c^2} = \frac{\alpha_p t_p}{b_p^2} \qquad [4]$$

where t is time, b is sample thickness, and α is the thermal diffusivity. Measured values of the thermal conductivity ratios were compared to the values predicted by the model developed by Nielsen.[3]

RESULTS AND DISCUSSION

Figure 1 shows the percolation diagram for the nickel-filled LDPE composites and indicates that the percolation threshold is between 5 and 7.5 percent nickel by volume. This critical vol-

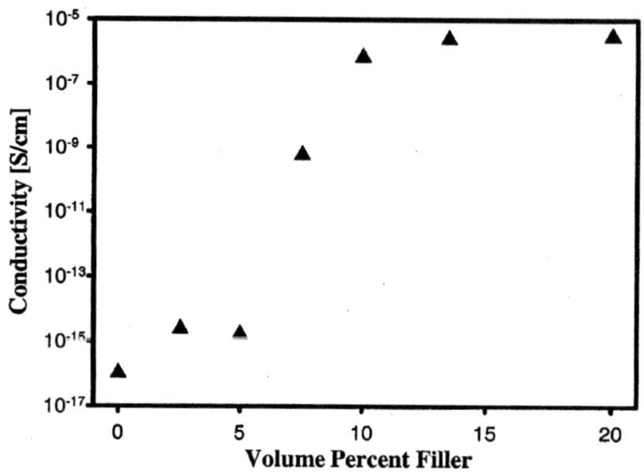

Figure 1. Percolation diagram for nickel-filled low-density polyethylene composites.

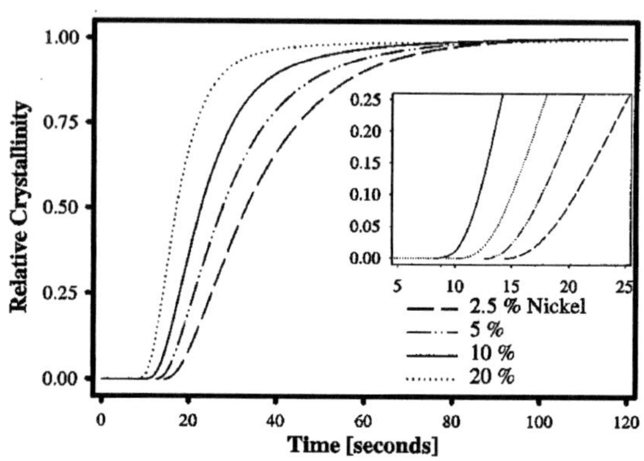

Figure 2. Relative crystallinity of low-density polyethylene filled with varying percent of nickel determined from isothermal crystallization experiments at 100°C.

ume fraction is lower than in most metal-filled systems because the flakes are randomly cut and anisotropic; therefore the flakes have a higher number of contacts per particle. The plateau region, where a continuous network of conductive filler particles has formed, began at approximately 10 percent by volume.

Isothermal crystallization studies were limited to temperatures between 95 and 104°C because at higher undercoolings the DSC scans indicated significant crystallization before the isothermal crystallization temperature was reached. Figure 2 shows the crystallinity as a function of time for 2.5, 5, 10, and 20 percent nickel by volume at 100°C. As Figure 2 indicates, the nucleation time decreases as the concentration of filler increases. The higher slope between 0 and 90 percent relative crystallinity for materials with more nickel indicates a faster crystallization rate. Finally, at higher nickel contents, the curves terminate more abruptly. The first two observations are explained by the increase in thermal conductivity of the composite, while the abrupt termination is probably due to impingement by nickel particles.

Crystal growth is an exothermic process, and conditions that increase the heat transfer rate would be expected to increase both the rate of nucleation and the linear growth rate, G. However, as evidenced by the invariance of the Avrami exponent with nickel content as shown in Figure 3, no increase in the rate of nucleation seemed to be present. Hence, the increased nucleation time shown in Figure 2 does not seem to be caused by an increase in nucleation rate. In fact, this decrease with nucleation time scales roughly with the increase in

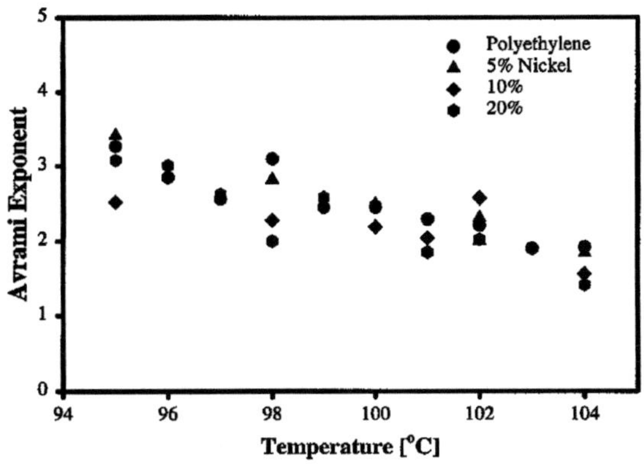

Figure 3. Avrami exponent, n, as a function of temperature and percent nickel filler.

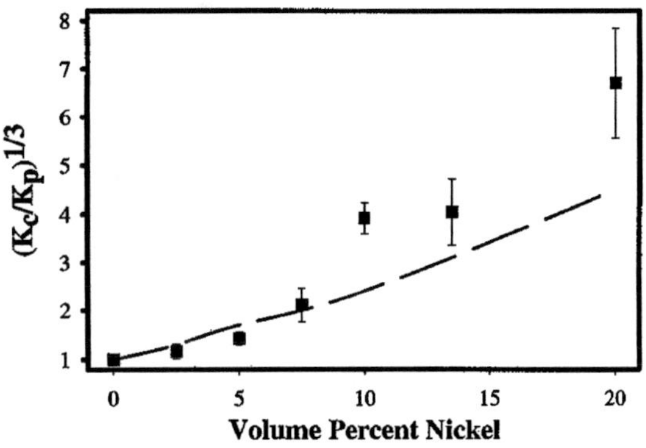

Figure 4. Linear growth rate ratio of polyethylene crystallites. Line represents the thermal conductivity ratio.

cooling rate in the cooling curves, hence this nucleation time is almost certainly caused by the increase in thermal conductivity caused by the addition of nickel filler.

The Avrami rate constant, K, increased with temperature, but this change was much smaller than that due to the increase in filler volume fraction. Figure 4 shows the cube root of the ratio of composite growth rates to polymer growth $(K_c/K_p)^{1/3}$ which should be proportional to the ratio of linear growth rates for the composite and polymer, G_c/G_p. The line in Figure 4 represents the prediction of the thermal conductivity from the Nielsen model. Within experimental error the Nielsen model predicted exactly the experimentally measured thermal conductivity ratios.

At low volume fractions, the growth rate ratio follows the thermal conductivity, but G_c/G_p increases much more rapidly than the thermal conductivity above the critical region. In fact, complete network formation appears to cause a very steep jump in growth rate. As far as we know, this study represents the first time this phenomena has been observed. We believe the continuous network of nickel particles at high concentrations may cause increased local heat transfer as the energy is more efficiently dissipated. In other words, as the concentration of filler is increased, local temperature gradients caused by the heat of fusion of the polymer chains will be reduced.We believe this phenomenon probably occurs for any high thermal conductivity filler.

CONCLUSIONS

We have shown that the network of nickel particles described by percolation statistics, used to characterize the electrical conductivity, is also important in determining crystallization kinetics. A large jump in electrical conductivity over a very short range of concentration occurs when the first continuous pathways of nickel particles form; complete network formation corresponds to the plateau region in electrical conductivity. In crystallization, complete network formation causes a sharp increase in linear crystal growth rate, much larger than the bulk thermal conductivity would predict. We believe that complete network formation causes the energy released during crystallization to be more effectively removed in a microscopic sense. The continuous network of nickel particles dissipates the heat of crystallization, thus allowing crystallization to occur more rapidly.

ACKNOWLEDGMENTS

The authors would like to thank A. Lowe for designing the die used in processing the composite films and P. Hunt for assistance in sample preparation. We also thank PFS Thermoplastics for supplying the LDPE used in this work. Funding was provided by NSF EPSCoR (Cooperative Agreement No OSR-9550478).

REFERENCES

1 S . N. Maiti, and P. K. Mahapatro, *J. Appl. Polym. Sci.*, **37**, 1889, (1989).
2 A. Kumar and R. K. Gupta, **Fundamentals of Polymers**, *McGraw Hill*, New York, 349,1998.
3 L. E. Nielsen, *Ind Eng. Chem., Fundam.*, **13**, 17, (1974).

Development of Conductive Elastomer Foams by *in Situ* Copolymerization of Pyrrole and N-Methylpyrrole

R. A. Weiss and Yueping Fu
University of Connecticut
Poh Poh Gan and Michael D. Bessette
Rogers Corporation

INTRODUCTION

Among the intrinsically conductive organic polymers, polypyrrole (PPy) is especially attractive for commercial applications, because of its relatively good environmental stability and facile synthesis. It can be synthesized by either an oxidative chemical or electrochemical polymerization of pyrrole, though conductive PPy is insoluble and infusible. Polymerization of β- or N-substituted pyrrole with alkyl chains having more than six carbons yields polymers with improved solubility in organic solvents.[1-31] However, β-alkyl substituted PPy's exhibit conductivities 1-2 orders of magnitude lower than PPy[2] and N-alkyl substituted PPy's have conductivities about 5-6 orders of magnitude lower than PPy.[3-5]

Electrochemical copolymerization of pyrrole and N-substituted pyrroles have been used to a limited extent to control the electronic properties of conductive polymer films.[6-9] The monomer oxidation potentials of pyrrole (1.15 V vs. SCE) and N-methylpyrrole (1.19 V vs. SCE) are very close, and the monomers have very similar polymerization reactivity. The polymer redox potential for PPy is ca. 0.5 V less than poly(N-methylpyrrole) (PMPy), however, which indicates that PPy is more oxidatively stable than PMPy.[5,10] The conductivity of PPy/PMPy copolymers depends upon the composition and is intermediate between that of PPy (10-100 S/cm) and PMPy (10^{-4} - 10^{-7} S/cm).

The mechanical properties and processability of conductive polymers may be improved by preparing polymer blends or composites by either directly dispersing conducting polymer particles into an insulating polymer matrix or by an *in situ* polymerization of the conducting polymer within a polymer host.[11] The *in situ* polymerization of pyrrole may be accomplished by a diffusing pyrrole into a polymer matrix containing a suitable oxidant, and this approach has been used to prepare conductive blends based on a variety of different polymer matrices, including poly(vinyl chloride), poly(vinyl alcohol), cotton, poly(phenylene terephthalamide), and polyurethane.[11]

This paper describes the preparation of conductive polyurethane (PU) foams by using a vapor phase *in situ* polymerization to incorporate PPy or pyrrole/N-methylpyrrole copolymers. The conductivity of the resulting composite may be controlled by varying the copolymer composition and the amount of conductive polymer. Compared with dense polymers, foams have an advantage for vapor phase *in situ* polymerization in that the monomer may penetrate the porous structure much more easily. However, one problem with an *in situ* polymerization within a foam is that the polymerization may occur within the cells of the foam, which allows the conducting polymer to be easily removed by abrasion or handling. Loss of the conducting polymer by mechanical handling of the foam not only decreases the conductivity of the composites, but it also may result in undesirable marking of the foam on a surface with which it comes into contact. Therefore, an important objective for preparing conductive foams is to restrict the conducting polymer to the polymeric walls, or struts, of the foam.

EXPERIMENTAL DETAILS

Pyrrole and N-methylpyrrole (Aldrich) were distilled and stored in a refrigerator. Ferric chloride hexahydrate, $FeCl_3$-$6H_2O$, (Aldrich, 98%) was used without any further purification, and ferric chloride solutions with different concentrations were prepared in methanol. Polyurethane foams with mass densities of 0.24 g/cm^3, 0.30 g/cm^3, and 0.35 g/cm^3 were cut to a size of 15x10x5 mm. The PU foam samples were first immersed in a $FeCl_3$/methanol solution for ca. 4 hours to swell the PU foam and allow the $FeCl_3$ to diffuse into the foam. After incorporation of the oxidant, the foams were dried for 3-4 hours and were then exposed to a pyrrole/N-methylpyrrole vapor in a desiccator under a pressure of ca. 0.5 torr. The vapor composition was estimated from the composition of the liquid fed to the desiccator and Raoult's law. Following the vapor-phase polymerization, the composite foams were washed with methanol several times to remove unreacted oxidant and byproducts, (e.g. $FeCl_3$), dried in air for 1-2 hours and finally dried under vacuum at room temperature for 24 hours. The polymer concentration was estimated from the change in mass of the foam before and after polymerization.

Conductivity measurements were made using a 4-point probe method. A home-made testing fixture consisting of four parallel copper wires separated by 4 mm was pressed onto the foam samples. A constant current supplied by a Keithley 224 Programmable Current

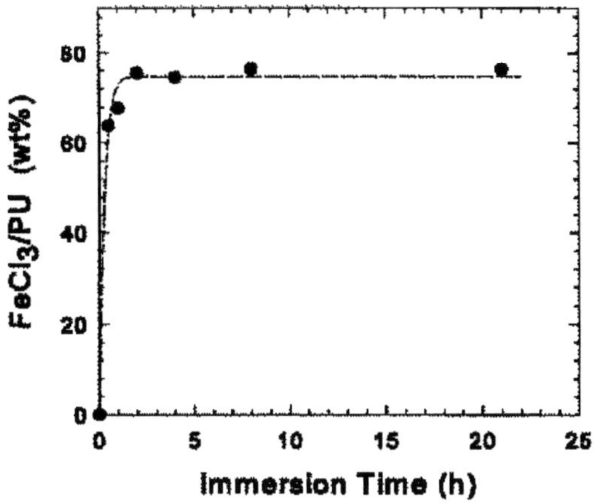

Figure 1. Sorption of $FeCl_3$ vs. time by a 0.24 g/cm^3 PU foam immersed in a 0.5 M $FeCl_3$/MeOH solution.

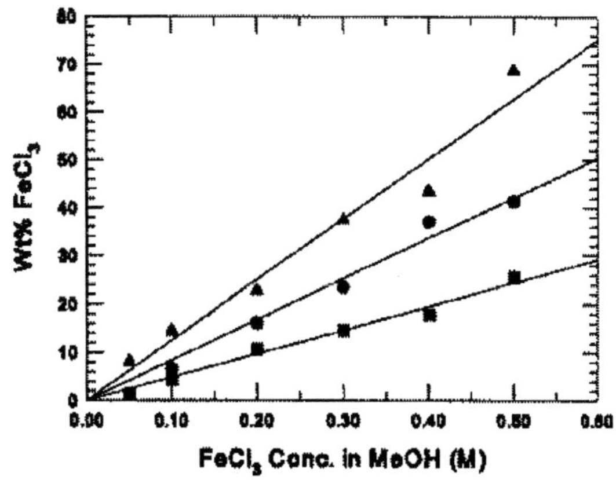

Figure 2. $FeCl_3$ uptake by PU foams as function of oxidant concentration for 3 different foam densities: ▲ 0.24 g/cm^3, ● 0.30 g/cm^3, ■ 0.35 g/cm^3.

Source was applied through outer wires, and the voltage drop across inner wires was recorded with a Keithley 197 A Autoranging Microvolt DMM. Thermogravimetric analysis (TGA) was done with a Perkin-Elmer TGA-7 using a nitrogen atmosphere and a heating rate of 10°C/min. Scanning electron microscopy (SEM) was done with an AMR model 1200B microscope equipped with an EDAX detector.

RESULTS AND DISCUSSION

POLYMERIZATION OF PYRROLE

Pyrrole has a relatively low oxidation potential and may be polymerized by oxidants such as $FeCl_3$ which is soluble in methanol (MeOH).

Pyrrole has a relatively high vapor pressure, and the *in situ* vapor phase polymerization of pyrrole can be readily initiated by exposing a polymer containing an oxidant to pyrrole vapor.

The PU foams were first immersed in a $FeCl_3$/MeOH solution, dried in air and then exposed to pyrrole vapor under static vacuum conditions. Figure 1 shows the sorption of $FeCl_3$ by a 0.24 g/cm^3 foam immersed in a 0.5M $FeCl_3$/MeOH solution. Most of the oxidant was absorbed in the first half hour of immersion, and equilibrium was achieved in less than 5 hours.

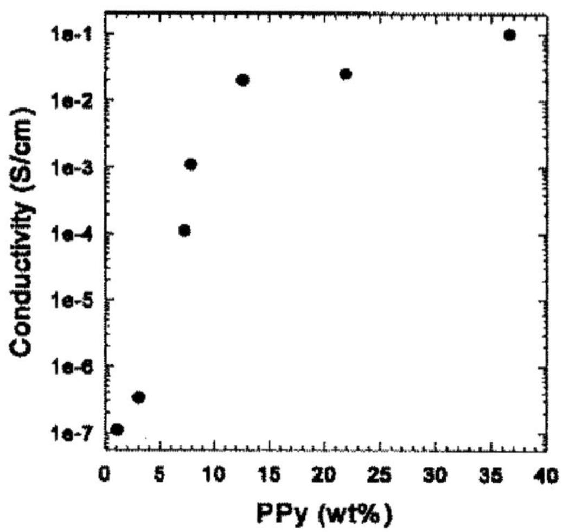

Figure 3. Electrical conductivity of PPy/PU foam composites vs. PPy concentration; the initial foam density was 0.24 g/cm^3.

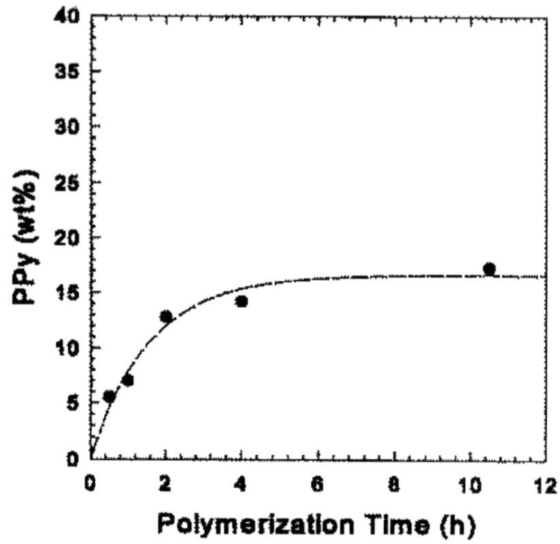

Figure 4. PPy concentration and conductivity of the composite foam as function of polymerization time for an initial foam density of 0.24 g/cm^3 and an oxidant concentration of 70% $FeCl_3$.

The concentration of $FeCl_3$ in the foam was controlled by varying the concentration of the oxidant solution used for swelling step. Figure 2 shows the oxidant uptake after 4 hours immersion a oxidant/methanol solution for three different PU foam densities as a function of the solution concentration. For each foam the $FeCl_3$ uptake increased linearly with increasing oxidant solution concentration, and for a fixed $FeCl_3$ concentration in the swelling solution, the concentration of oxidant incorporated into the foam increased with decreasing foam density.

Once the oxidant was incorporated into the foam, the foam was dried and then exposed to pyrrole vapor to initiate polymerization. The amount of PPy produced was controlled by the $FeCl_3$ concentration in the foam. Figure 3 shows the conductivity of the PPy/PU foams as a function of the PPy concentration. An insulator to conductor transition occurred at a PPy concentration in the foam of about 3-5 wt%, and the conductivity (σ) of the composite foam increased monotonically with PPy concentration. A value of $\sigma = 0.1$ S/cm was achieved for a PPy concentration of 36.6 wt %. Normally, for a dense material filled with conductive particles, a percolation threshold concentration for conductivity is ca. 16 % (vol).[12] For the PPy/polymer composites, vol % and wt% values are about the same, because the densities of most organic polymers are similar (0.9-1.2 g/cm^3). The low percolation threshold concentration shown in Figure 3 occurs because the PPy is incorporated only into the polymer walls, or struts. If the PPy concentration shown on the ordinate in Figure 3 is corrected for the PPy concentration in the polymer phase of the foam, the percolation threshold is close to the theoretical value.

Figure 4 shows the PPy production in a foam and the conductivity as a function of polymerization time for a composite prepared from a foam containing 70% $FeCl_3$. The polymerization was essentially complete after ca. 4 hours. However, most of the conductivity was developed within the first 30 minutes of the polymerization, during which time the PPy content reached about 5 wt %. For 0.5 to 10.5 hours exposure to the pyrrole vapor, the conductivity increased only slightly from 0.01 to ca. 0.03 S/cm. One reason for the relative independence of the conductivity on reaction time is that the 4-point probe measurement is most sensitive to the conductivity near the surface. Since the vapor phase *in situ* polymerization proceeds from the surface of the foam inwards, as a result of the diffusion controlled polymerization process, surface conductivity is established early in the polymerization and does not change substantially as the polymerization front proceeds into the foam specimen.

PYRROLE/N-METHYLPYRROLE COPOLYMERIZATION

Pyrrole and N-methylpyrrole have comparable oxidation potentials and both may be polymerized by either electrochemical or chemical oxidative polymerization. Copolymers have been synthesized by electrochemical methods, and their conductivity varies with the comonomer feed ratio.[6-8]

We used an chemical oxidative, *in situ* copolymerization of pyrrole and N-methylpyrrole vapors in a PU foam to control the conductivity of the resultant composites. The vapor pressures (P_i^*) of pyrrole and N-methylpyrrole at 25°C and 760 mm Hg are 8.25 mm Hg and 21.36 mm Hg, respectively. Whereas for the electrochemical copolymerization, the monomer feed ratio was assumed to be the mole fraction of the monomers in the liquid solution,[10] for the vapor phase polymerization used here, it is the mole fractions in the vapor ($y_i=P_i/P$) that constitute the feed composition. The partial pressure (P_i) of each component in the vapor may be estimated from Raoult's law ($P_i = x_iP_i^*$), where x_i is the mole fraction in the solution. Since previous work showed that the reactivities of the two monomers are essentially the same, we assumed that this also applied for the vapor phase polymerization.

The copolymer composition and the concentration of the conductive copolymer in the composite foam were independently varied by varying the monomer molar feed ratio and the oxidant concentration in the foam, respectively. Figure 5 shows the effects of those two variables on the conductivity of the composite foam.

Conductivity varied over about four orders of magnitude as the composition of the copolymer was changed. The conductivity of the copolymer decreased with increasing PMPy concentration, though the change was non-linear. Above a pyrrole in the copolymer mole fraction of 0.75, the conductivity of the composite was influenced predominantly by the more

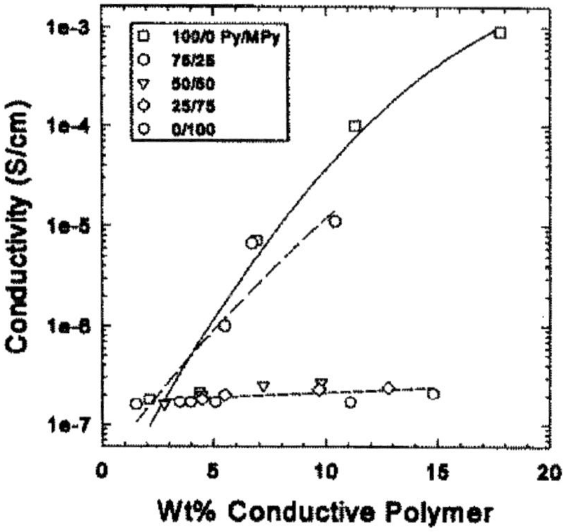

Figure 5. Conductivity of copolymer/PU foam composites vs. conductive copolymer concentration for different copolymer compositions: □ 1.0, ❍ 0.75, ∇ 0.5, ◇ 0.25, ▯ 0.0 mole fraction pyrrole.

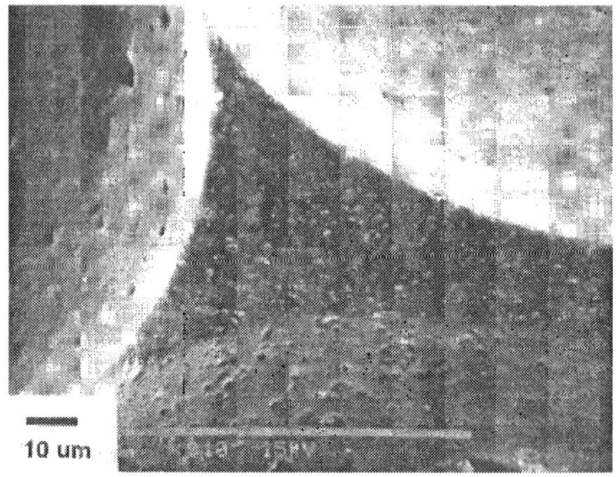

Figure 6. SEM micrograph of cross-section of a cell in a 21.9% PPy/PU foam composite.

conductive pyrrole. N-methylpyrrole decreases the mobility and/or concentration of charge carriers, which are essential for conductivity, and below a pyrrole mole fraction of ca. 0.5, the conductivity of all the composites was ca. 10^{-7} S/cm, regardless of the copolymer or composite compositions. These results demonstrate that the conductivity of the composite foam may be tailored by judicious choice of the copolymer composition and the amount of conductive copolymer produced.

The analysis of copolymer compositions was prevented by the insolubility of the copolymers and the PU foams. FTIR analysis of the composites was not successful at providing the composition, primarily because the spectrum of the PU overlapped the absorption peaks characteristic of pyrrole and N-methylpyrrole unit.

MORPHOLOGY OF THE PPY/PU COMPOSITES

Figure 6 is an SEM micrograph of a cross-section of one of the cells in a composite foam containing 21.9 % PPy. The granular texture of the surface of the cell walls is due to PPy embedded in the polymer phase of the foam; the cell walls of the neat foam were relatively smooth. In general, the composite foams contained a relatively uniform distribution of PPy within the polymeric cell walls and struts. No PPy debris was observed either when the foam was cut for

Table 1. Comparison of mechanical properties of PU and PPy/PU foams

Properties	PU	6.0% PPy/PU
Tensile strength, 10^5 N/m^2	8.1	8.3
Elongation at break, %	143	160
Tear strength, 10^3 N/m	1.8	3.2
Compression set, %	3.5	3.7
Conductivity, S/cm	$<10^{-10}$	10^{-5}

SEM or in the micrograph, which indicates that the polymerization technique effectively isolated the conductive polymer within the dense polymer phase. That was also confirmed by the absence of marking when the conductive foams were rubbed upon a piece of white paper (neat PPy is black and readily marks paper). The PPy had a granular texture with a grain size of about 0.5 μm. EDAX spectra showed three distinct emission peaks: Si K_α (2.3 keV), Cl K_α (2.6 keV) and Fe K_α (6.4 keV). The Si was due to the surfactant used in preparation of the foam. The strong Cl K_α peak was mostly due to Cl^- dopant ions for the PPy and the weak Fe K_α peak was either from dopant ions in the form of $FeCl_4^-$ or residual $FeCl_2$ by-product that was not removed during washing of the product.

THERMAL STABILITY AND MECHANICAL PROPERTIES

TGA thermograms of neat PU foam and the PPy/PU composites were similar between 0-400°C, which indicates that the *in situ* polymerization process did not affect the thermal stability of the PU foam. The mechanical properties of a 6.0 % PPy/PU composite and the parent PU foam are compared in Table 1. Although, the mechanical data are limited, they do demonstrate that incorporation of low concentrations of PPy into the foam was not deleterious to the foam mechanical properties. In addition, the resilience of the composite foam was qualitatively similar to that of the neat PU foam.

CONCLUSIONS

A process was developed for producing conductive elastomeric foams by polymerizing conductive polypyrrole or copolymers of pyrrole and N-methyl pyrrole in the cell walls and struts of a preformed polyurethane foam. The process consisted of first diffusing an oxidant into the dense polymer phase of a solvent-swollen foam and then diffusing a vapor of pyrrole and N-methyl pyrrole into the dried foam. An *in situ* chemical oxidative polymerization of conductive polymer or copolymer occurred where the oxidant was present. Because the conductive polymer was confined to the dense polymer phase, conductivity of the foam composites was achieved at relatively low concentrations of the conductive polymer concentration, ca. 5 wt%. The conductivity of the composite foam was reproducibly controlled between 10^{-7} - 10^{-1}

S/cm by varying either the amount of oxidant used, which controls the amount of conductive polymer produced, and/or the copolymer composition. The conductivity decreases as the concentration of N-methyl pyrrole in the copolymer increases. Preliminary experiments using ferric tosylate as the oxidant and dopant suggested that 1-2 orders of magnitude greater conductivity could be attained at an equivalent PPy concentration compared with using $FeCl_3$.

ACKNOWLEDGMENT

We gratefully acknowledge financial support from Connecticut Innovations, Inc.

REFERENCES

1 D.Moon, A. B. Padias, H. K. Hall, T. Huntoon and P. D. Calvert, *Macromolecules*, **28**, 6205 (1995).
2 H. Masuda, S. Tanaka and K. Kaeriyama, *J. Polym. Sci., Polym. Chem. Ed.*, **28**, 1831 (1990).
3 T Kim and R. L. Elsenbaumer, *Synth. Met.*, **84**(13), 157 (1997).
4 M. Ribo, M. C. Anglada, J. M. Tura and N. Ferrer-Anglada, *Synth. Met.*, **72**, 173 (1995).
5 M. Salmon, A. F. Diaz, A. J. Logan, M. Krounbi and J. Bargon, *Mol. Cryst. Liq. Cryst.*, **83**, 265 (1982).
6 K. K. Kanazawa, A. F. Diaz, M. T. Krounbi and G. B. Street, *Synth. Met.*, **4**, 119 (1981).
7 J. R. Reynolds, P. A. Poropatic and R. L. Toyooka, *Macromolecules*, **20**, 958 (1987).
8 M. S. Kiani and G. R. Mitchell, *Synth. Met.*, **46**, 293 (1992).
9 P. Novak and W. Vielstich, *J. Electroanal. Chem.*, **300**, 99 (1991).
10 M. Nishizawa, T. Sawaguchi, T. Matsue, and 1. Uchida, *Synth. Met.*, **45**, 241 (1991).
11 M. C. DeJesus, Y. Fu and R. A. Weiss, *Polym. Eng. Sci., in press*.
12 H. Scher and R. J. Zallen, *J. Chem. Phys.*, **53**, 3759 (1970).

Neocapacitor. New Tantalum Capacitor with Conducting Polymer

Atsushi Kobayashi, Yoshihiko Saiki
Energy Devices Division, NEC Corporation
Kazuo Watanabe
NEC Toyama Ltd.

INTRODUCTION

Electrolytic capacitors have been widely used in various circuits as one of the key electronic components. There are two electrolytic capacitors: aluminum capacitors and tantalum capacitors. Aluminum capacitors are the most common capacitors today because they are inexpensive. The other electrolytic capacitor is tantalum capacitor. It has many advantages over aluminum capacitors: chemically stable Ta_2O_5 as dielectric film, superior temperature characteristics, lower leakage current and excellent volume efficiency. Therefore, tantalum capacitors have been widely used in highly reliable equipment and portable electronic equipment.

In the last decade, portable electronic equipment has been miniaturized and its energy consumption has been lowered remarkably. This requires much lower noise on power supply line of electronic equipment and lower ESR on capacitors in the line. To meet this demand, NEC has recently developed new tantalum capacitor, NEOCAPACITOR,[1] which uses PPy as its electrolyte (Figure 1). Because of much higher electrical conductivity than that of MnO_2, which is used in conventional tantalum capacitors, PPy provides lower ESR to NEOCAPACITOR (Figure 2). Moreover, thermally degrading property at over 300°C provides healing function to the new capacitor. If a micro defect exists on the dielectric, the current flows at this point. This current generates heat and the heat degrades PPy at the point. Therefore, insulated polymer stops current flow and prevents the capacitor from short-circuiting. In fact, the decrease in leakage current in load life tests has been observed for the new capacitor. This is thought as one of examples of healing function of conducting polymer.

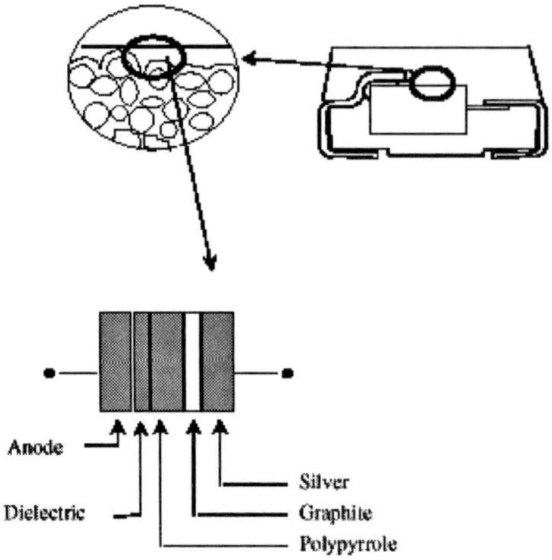

Figure 1. Structure of NEOCAPACITOR.

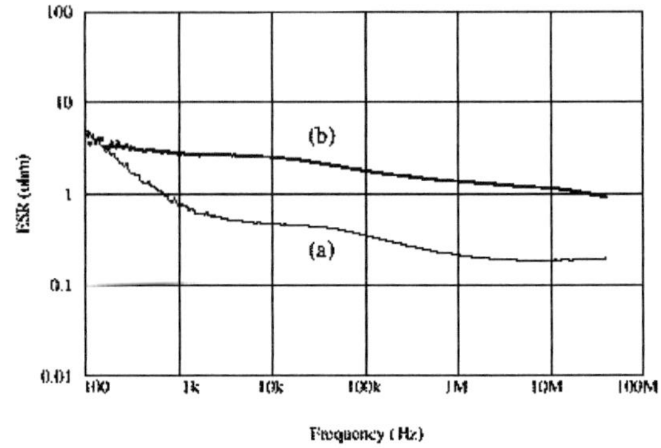

Figure 2. Typical ESR curve for NEOCAPACITOR and conventional tantalum capacitor. (a) NEOCAPACITOR and (b) conventional tantalum capacitor.

However, clearer results and mechanism for the healing function have not shown yet. This paper provides ripple current loading tests and the GC-MS analysis of the polymer.

EXPERIMENT

100uF/10V NEOCAPACITOR (Part No. PSMD1A107M) and R series (Part No. NRD107M010) were examined. They were applied with 10 kHz sign wave ripple voltage. The ripple voltage was generated by Yokogawa Synthesized Function Generator FC110, amplified by Yokogawa Power Amplifier 7058, and monitored with Yokogawa Digital Oscilloscope DL1100. The values of voltage were 1.0, 2.0, 3.0, 3.5 and 4.0V (peak to peak). The temperatures of capacitors were monitored by the thermocouple on the surface of samples. There is no heat-sinking material on the capacitors.

RESULTS

Figure 3 shows the surface temperature of NEOCAPACITOR ripple voltages were applied. The temperature gradually increased and approached the constant values within 100 seconds when ripple voltages were 1.0, 2.0 and 3.0 V. However, the constant value and equilibrium time became larger according to the ripple voltage. In case of 3.5 V, the temperature rapidly increased to 160°C in the first 30 seconds but remarkably decreased and reached the constant

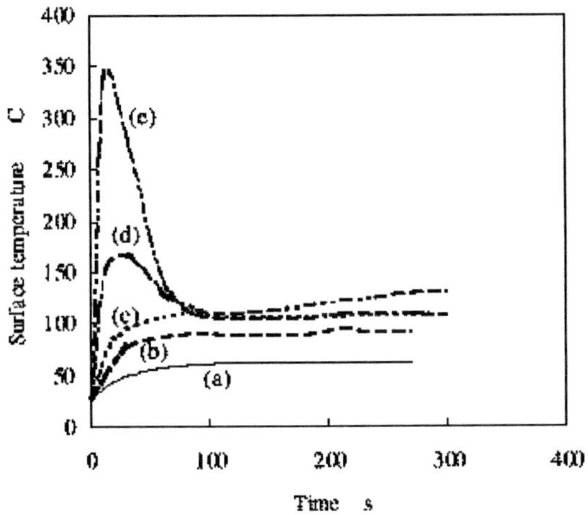

Figure 3. Surface temperature of 10 kHz ripple voltages applied to NEOCAPACITOR. (a) 1.0 V, (b) 2.0 V, (c) 3.0 V, (d) 3.5 V, and (e) 4.0 V.

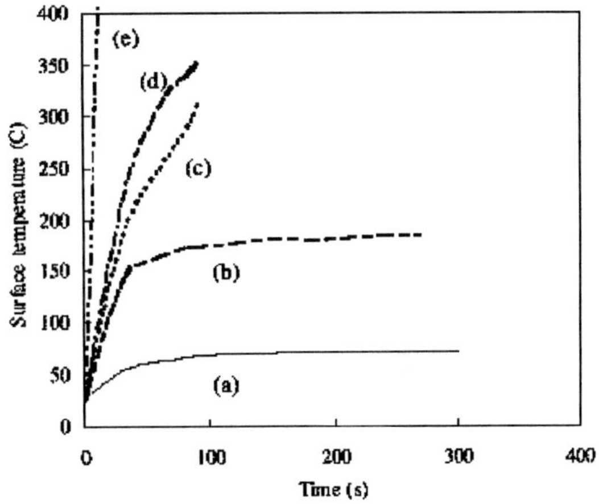

Figure 4. Surface temperature of 10 kHz ripple voltages applied to conventional tantalum capacitor. (a) 1.0 V, (b) 2.0 V, (c) 3.0 V, (d) 3.5 V, and (e) 4.0 V.

value around 100°C. Similar result was observed for 4.0 V. The temperature reached 350°C and it rapidly decreased to 100°C within 50 seconds.

The surface temperatures of conventional capacitors were similar (Figure 4) when the voltages were 1.0 and 2.0 V. The temperature increased gradually and reached the plateau. However, the plateau values for conventional ones were higher than that of NEOCAPACITOR. Moreover, 3.0, 3.5 and 4.0 V ripple voltages significantly increased the temperature and destroyed capacitors catastrophically.

DISCUSSION

As widely know, heat generated by ripple voltages is proportional to square ESR and ripple voltage. The results in the Figure 3 and Figure 4 show the difference in the healing ability. MnO_2 is known that it is decomposed to insulating Mn_2O_3 at over 520°C. However, this temperature is much higher than that of PPy. Its insulating temperature is 300°C or over. Therefore, PPy has superior healing function than MnO_2. However, the process of insulation of PPy has not clearly shown yet. To clarify this process, mass spectrometry (MS) of PPy was measured over the range of m/z from 29 to 650, on a VG-TR10-01 instrument equipped with hand-made temperature controller for the gasses generated from the samples at elevated temperature. PPy has been synthesized as in the literature.[2] Initial temperature was 50°C and the

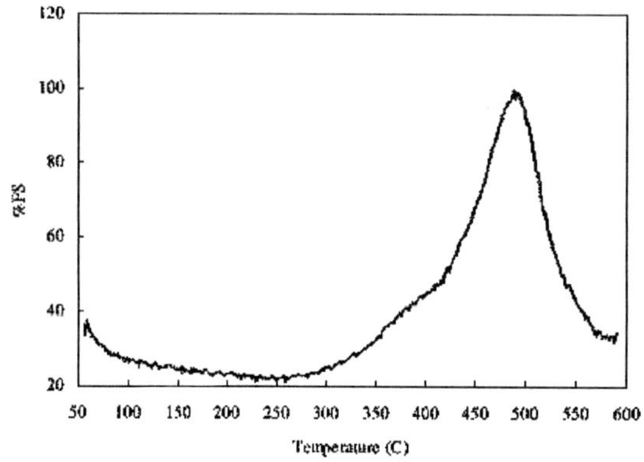

Figure 5. TIC for GC-MS of PPy.

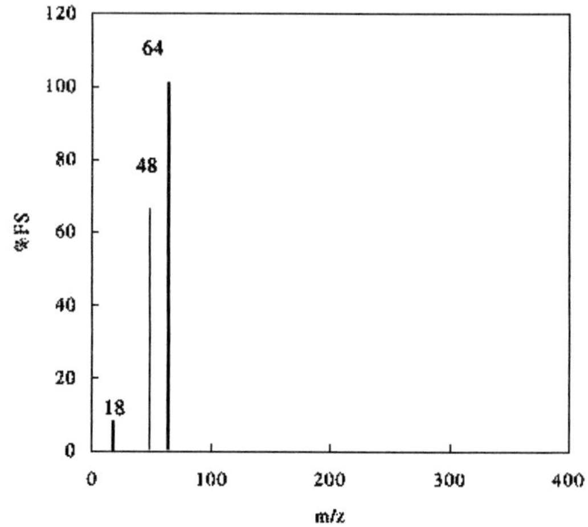

Figure 6. Mass spectrum of PPy at 380°C.

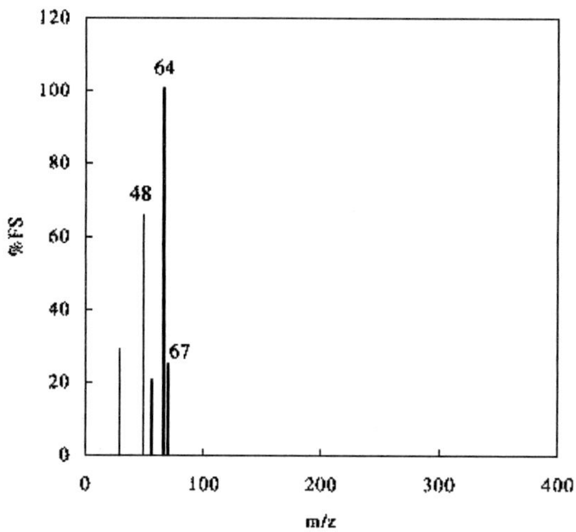

figure 7. Mass spectrum of PPy at 480°C.

temperature was increased by 10°C/min. The generated gasses were also analyzed by gas chromatography-mass spectrometry (GC-MS). Figure 5 shows the results of total ion chromatography (TIC) for GC-MS of PPy. There are a clear peak at 480°C and a shoulder peak at 380°C. Figure 6 and 7 shows the mass spectrum of the peaks at 380°C and 480°C, respectively. Since PPy was doped with sulfonic compound anion, the 48 and 64 m/z in the Figure 6 would be SO and SO_2. In Figure 7, these two are also observed. Moreover, pyrrole, which molecular weight is 67, is detected in Figure 7. These results support that heating PPy decomposes doped anion consisting of sulfonic compound at around 380°C and then polymer backbone at around 480°C.

CONCLUSIONS

NEOCAPACITOR has superior self-healing function than that of conventional tantalum capacitor. This function comes from insulation of PPy through two-step decomposition. At first, doped anion and then the polymer backbone are decomposed.

ACKNOWLEDGMENT

The authors wish to express appreciation to Dr. Masaharu Satoh, Principal Researcher, Functional Devices Laboratories, NEC for GC-MS analysis.

REFERENCES

1 Atsushi Kobayashi, et al., *Denshi Tokyo*, **33**, 153-157, 1994.

2 Masaharu Satoh, et al., *Synthetic Metals*, **72**, 98-105, 1996.

Conductive Polymer-Based Transducers as Vapor-Phase Detectors

Frederick G. Yamagishi, Thomas B. Stanford, Camille I. van Ast, Paul O. Braatz and Leroy J. Miller

Hughes Research Laboratories, 3011 Malibu Canyon Road, Malibu, CA 90265

Harold C. Gilbert

Hughes Aircraft Company Naval and Maritime Systems Business Unit, 1901 W. Malvern Avenue, Fullerton, CA 92634

INTRODUCTION

Conductive polymers have been reported as the active transducer in a number of chemical sensors. They have been used to demonstrate the detection of materials in the vapor state,[1-3] or in aqueous solutions.[4,5] Further, conductive polymer sensors have been reported to detect acid and/or water in nonaqueous and nonpolar media.[6,7] We have developed conductive polymer-based transducers for the detection of volatile organic compounds (VOCs) and other gaseous pollutants for application in environmental monitoring.[8,9]

Our development was through two approaches: 1) to use a composite of two components where one is an attractant material which detects the presence of the substance of interest, and the other is a conductive material to carry the electrical current to the associated electronics, and 2) to vary the counterion associated with the oxidized form of the conductive polymer. In each approach it was felt that one of the components in the transducer (i.e., the attractant polymer or the counterion) would selectively absorb the pollutant of interest while having less attraction for other vapors. With the associated electronics, these sensors are components of a multi-sensor array capable of VOC speciation.

Polyaniline (PANI) and derivatives of polythiophene were chosen as the conductive polymers for these transducers. We believe that the detection of certain chemically inactive pollutants (i.e., those pollutants incapable of oxidizing or reducing the conductive polymer) is a result of a structural perturbation in the conductive polymer caused either by a direct interaction of the conductive polymer with the pollutant, or because of a structural change in the attractant polymer in which the conductive polymer is embedded (e.g., polymer swelling).

It was also found that appropriate combinations of silane surface coupling agents, surfactants, conductive polymer counterions and employing advanced signal processing techniques, sensitivity thresholds of <10 ppm were observed. This combination of transducer components also enhanced the stability and reversibility of these sensors.

EXPERIMENTAL

The sensors were prepared by coating a set of gold interdigitated electrodes with a conductive polymer or a conductive composite film of sufficient thickness to bridge the gap between the sets of digits. The conductivity (or more properly the current flow) was determined by applying a small voltage across the digits (generally 0.200 VDC) and measuring the current.

PANI was prepared both chemically and electrochemically with various conjugate bases (counterions) derived from various acids. The poly(3-alkylthiophenes) were used undoped or doped with anhydrous ferric chloride.

Since some of the conductive polymers used in these tests were not soluble in common organic solvents, cast films were prepared from slurries containing the conductive polymer which has been mixed with a solution of the attractant polymer (e.g., polyisobutylene, PIB) in an appropriate solvent such as N-methylpyrrolidinone. In cases where the conductive polymer was soluble, films were spin-coated from xylene or chloroform solutions.

In some cases, a thin film of a silane coupling agent (e.g., octadecyltrichlorosilane, ODTCS; methacryloxypropyltrimethoxysilane, MTS; or phenylaminopropyltrimethoxysilane, Sylquest Y-9669) was deposited by dip or spin coating from solutions ranging in concentration from 0.1% to 1%. These films were baked for about 30 min at 120°C and were spin coated with solutions containing soluble PANI complexed with dodecylbenzenesulfonic acid (DBSA). Poly(3-hexylthiophene) (PHT) doped with $FeCl_3$ or with undoped regioregular PHT (Rieke Metals, Inc.) were also tested but without silane coupling agent undercoating. In some cases the PANI•DBSA solutions contained a one-to two-part excess of DBSA. Thickness of these composite films was about 500Å.

Sensor testing was done in three different ways: 1) exposure of the sensor in the headspace above the challenge vapor at room temperature; 2) exposure of the sensor to quantities of the challenge vapor in a closed system of known volume; and 3) exposure of the sensor to known quantities of challenge vapor diluted with air or nitrogen at constant humidity, flow and temperature. Thus, measurements were taken at concentrations levels ranging from sub-ppm levels to several percent. Signal processing was done by data detrending and principal component analysis.

RESULTS AND DISCUSSION

SENSITIVITY OF TRANSDUCERS TO CHALLENGE VAPORS

Figure of Merit

For a given test run the sensors were exposed sequentially to increasing concentrations of challenge vapor under flow conditions. In the process of this test data was collected as the current response of the sensor versus the total elapsed time of the run. In certain cases, such as the example shown in Figure 1, the data appeared well-behaved, and could be displayed in a manner that would be expected. However, since each exposure has a different starting baseline, it is difficult to compare the response of the sensor as a function of concentration. We therefore developed a figure of merit (FOM) based on the slope of the response curve, which was plotted as the sensor response versus dosage [(concentration) (time)].

A plot of normalized sensor response versus dosage yielded generally a straight line whose slope is proportional to the response time of the sensor and its sensitivity toward the challenge gas at that concentration. An example of this type of data treatment is shown in Figure 2. In this case, NO_2 is detected rapidly by the sensor and the response is effectively immediate. The slope of this line is the FOM for sensors that showed such a first order dependence.

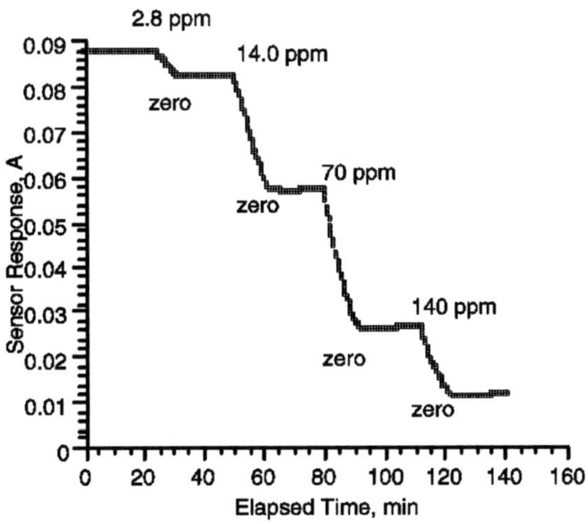

Figure 1. Sensor response vs. time for electrochemically-prepared PANI•TSA, 10.4 μm thick, NO_2.

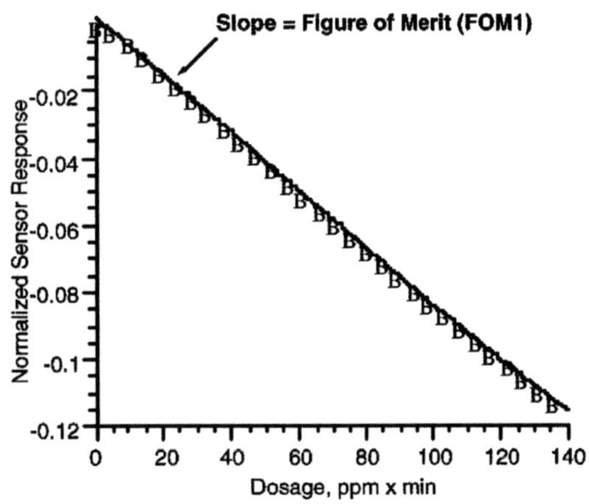

Figure 2. Figure of merit (FOM) derived from the normalized response of a PANI•TSA sensor to 14 ppm of NO_2.

Nitrogen Dioxide

Of the challenge vapors tested, NO_2 showed the greatest and broadest response under a variety of test conditions, and varying physical characteristics of the transducer. Sensors derived from polyaniline complexed with *p*-toluenesulfonic acid (PANI•TSA) were the most sensitive but other sulfonic acid conjugate bases were also very effective. From FOM results, the following observations could be summarized:

- Very high sensitivity to NO_2 concentrations of <140 ppb are detectable;
- Generally, electrochemically-prepared films are more sensitive than cast films, but the cast composite films also respond, and in fact the highest sensitivity was observed with a cast composite film of PANI•TSA in PIB Cast films of PANI with any of the counterions were generally not sensitive;

We suspect that NO_2 is interacting chemically with the transducer, either the polyaniline backbone, or the counterion. If the former is true than the electron withdrawing nature of the NO_2 moiety would significantly modify the band structure of the conductive polymer and would account for the remarkable sensitivity. If the latter is true, the spatial distribution of the counterion would be expected to be significantly modified which would affect the morphology of the conducting chain and modulate the conductivity. Also, the conductivity of the sensor did not return to its original value following the removal of NO_2 vapor around the sensor. This further suggests that the structure of the conductive polymer has been chemically modified to one that has a different physical and electronic structure than the original material.

SILANE COUPLING AGENTS

During the course of this work a need to monitor ketones, esters and aromatic hydrocarbons was defined with concentration levels between 100 ppm to about 1 ppm.

The sensor elements described above were found to be inadequate to meet this need. However, initial testing of sensors composed of PANI complexed with DBSA or polythiophene suggested that these materials showed promise. Further, we found that films composed of PANI•DBSA which contained excess DBSA were much more sensitive to these target challenge vapors than PANI•DBSA alone. However, this new composition lacked long term stability necessary for the desired application.

In an effort to change the surface free energy of the glass substrate surface we coated the interdigitated electrode substrate with three different silanes [e.g., octadecyltrichlorosilane, ODTCS (nonpolar), methacryloxypropyltrimethoxysilane (somewhat polar) and phenylaminopropyltrimethoxysilane, Sylquest Y-9669 (more polar)]. It was anticipated that these agents would enhance the adhesion of the conductive polymer film to the substrate and thereby increase long term stability of the composite. It was found that uniform films of PANI

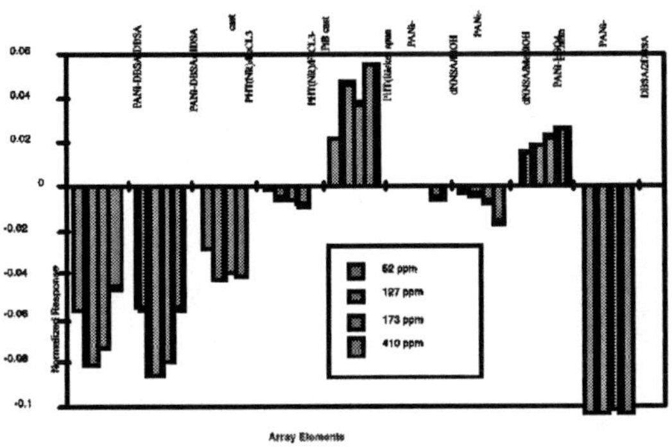

Figure 3. Array response to methyl isoamyl ketone at various concentrations (20% relative humidity).

•DBSA, which is partially charged and therefore slightly polar, could only be prepared on surfaces treated with ODTCS. Substrates treated with the more polar agents MTS and Sylquest Y9669 yielded PANI•DBSA films that were usable for further testing but were not nearly as uniform. PANI•DBSA sensors prepared with very thin films of silane agents showed essentially the same properties as sensors prepared without the surface treatment. However, when thicker coupling agent films were deposited (from higher concentration solutions) followed by depositing a PANI•DBSA layer, high sensitivity to toluene was observed, with rapid return to the original baseline following removal of the challenge vapor. Figure 3 shows the response of the sensors, listed in Table 1, tested against varying concentrations of methyl isoamyl ketone. Sensitivities of <52 ppm are demonstrated routinely. Later results have shown that this sensitivity threshold is actually well below 10 ppm for methyl isoamyl ketone and butyl acetate.

Table 1. Summary of sensor devices

Sensor number	Conductive polymer	Surfactant/ attractant	Dopant	Preparation method
1	PANI-DBSA	DBSA	DBSA	Spin coated
2	PANI-DBSA	HDSA	DBSA	Spin coated
3	PHT (NR)	None	FeCl3	Cast
4	PHT (NR)	PiB	FeCl3	Cast
5	PHT (Rieke)	None	None	Spin coated
6	PANI-dNNSA	Bu Cellos	EtOH	Spin coated
7	PANI-dNNSA	Bu Cellos	2-MeOEtOH	Spin coated
8	PANI-H2SO4	None	H2SO4	E-Chem
9	PANI-DBSA	2DBSA	DBSA	Spin coated

HDSA - hexadecanesulfonic acid, NR - nonregioregular, Rieke - from Rieke Metals Co., regioregular material, dNNSA - dinonylnaphthelenesulfonic acid (PANI-dNNA obtained from Monsanto Co.), PiB - polyisobutylene, Bu Cellos - butyl cellosolve, E-Chem - film deposited electrochemically

The roles of the excess DBSA on the sensitivity and of ODTCS on the reversibility and stability are unknown at this time. However, since the mechanism of detection relies on the physical movement of the conductive polymer morphology, we hypothesize that these excess agents provide a “mobile” phase which facilitates this change in morphology during exposure

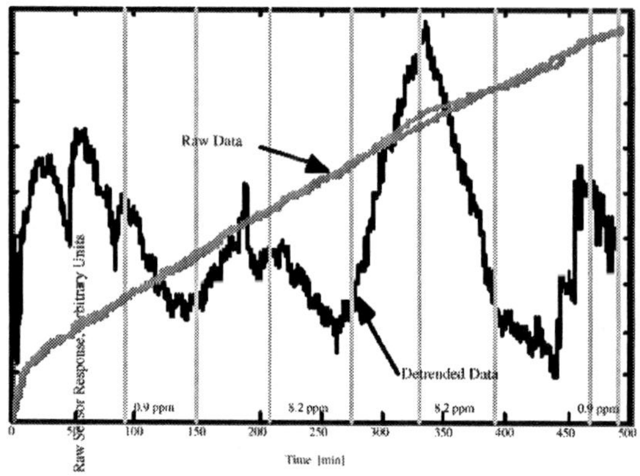

Figure 4. Raw and detrended data for a PANI•DBSA sensor exposed to methyl isoamyl ketone between 8.2 and 0.9 ppm at 20% relative humidity.

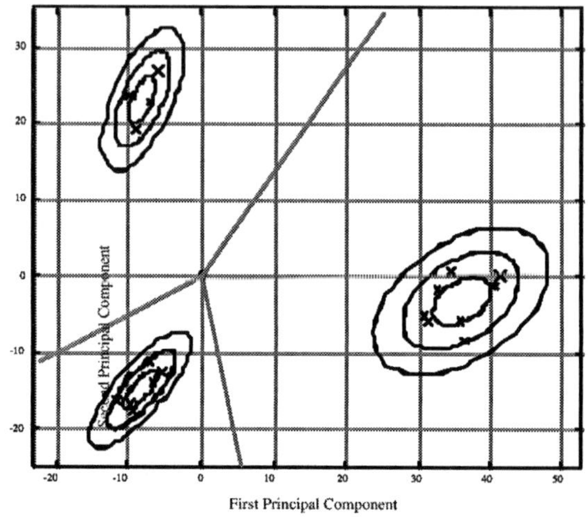

Figure 5. VOC Principal Component Linear Classifier at 20% RH.

to the challenge vapor and allows it to relax back to its original structure once the challenge is removed.

SIGNAL PROCESSING OF SENSOR ARRAYS

In an effort to determine whether our sensors could classify VOC vapors, an array of nine sensors were exposed to each of the vapors of interest at 2 L/min in concentrations ranging from about 500 ppm to about 1 ppm at 20°C and relative humidities of 0.2% to 20%. The active transducers were PANI and PHT with varying counterions, and they were combined with various attractant materials, as shown in Table 1. It was found that each of the nine sensors showed a characteristic response to each vapor. Thus a characteristic signature for that vapor could be generated for that particular array of sensors, as shown in Figure 3.

The sensor response for each individual run to each test condition was analyzed. The raw data was processed to remove obvious bogus data points (outliers) and the resulting data set was normalized with respect to their magnitude. This data set was detrended (removal of background noise) with a simple FIR high-pass filter designed to preserve the frequency components of signal energy while attenuating the low frequency drift in sensor response. Application of this processing technique allowed for the observation of low challenge levels, whereas the raw data appears almost non-variant. An example of the sensitivity of one sensor to about 8 ppm of MIAK is shown in Figure 4, where both the raw data and detrended data are shown superimposed. Exposures down to 0.9 ppm were also done in this run, but the responses were largely obscured by systematic noise. Principal component analysis, as shown in Figure 5, done on data from these array tests demonstrate that the three classes of VOC challenge vapors can be speciated clearly.

CONCLUSIONS

Highly sensitive sensor elements derived from PANI were demonstrated for NO_2 and other VOCs. A figure of merit was defined to allow comparison of the response of different sensors to particular challenge vapors at various concentrations. The use of excess silane coupling agents to modify the surface of our sensor substrates in combination with excess of a sulfonic acid used to convert PANI to its conductive state results in highly sensitive, stable and reversible VOC sensors, along with some polythiophene derivatives, for the detection of ketones, esters and aromatic hydrocarbons. The role of these excess materials is not known at this time, but they may act synergistically to attract VOCs and provide a medium in which the morphology of the conductive polymer chains can be more easily modified. This explanation would account for the increased sensitivity with excess DBSA and for the enhanced reversibility with ODTCS. Individual sensors show different sensitivities to each class of VOC used in this study so that an array of these sensors generated a characteristic signature for each class demonstrating selectivity and classification. These effects can be clearly extracted from the raw data using signal processing techniques.

ACKNOWLEDGMENTS

The authors gratefully acknowledge Mr. L.A. Schatzmann and Dr. J.A. Wurzbach of the Naval and Maritime Systems Division of Hughes Aircraft Company, Prof. Alan G. MacDiarmid of the University of Pennsylvania, Dr. Anthony Guiseppi-Elie of ABTECH Scientific for helpful discussions, Ms. A.M. Lackner and Ms. E. Sherman for assistance in data manipulation, Dr. S.M. Flank, Mr. E.M. Carapezza and Dr. I. Skurnick of DARPA, Drs. D.L. Venezky and S.L. Rose-Pehrsson of the Naval Research Laboratory, Dr. Harold Guard of the Office of Naval Research and Mr. E.M. Monachino of Science and Technology Associates for assistance in contract monitoring. This work was partially supported by DARPA/NRL under Contract No. N66001-94-C-6028 and by DARPA/ONR Technology Reinvestment Project under Agreement No. N00014-95-2-0008.

REFERENCES

1 P.N. Bartlett, P.B.M. Archer and S.K. Ling-Chung, *Sensors and Actuators*, **19**, 125 (1989).
2 P.N. Bartlett and S.K. Ling-Chung, *Ibid.*, **19**, 141 (1989).

3 P.N. Bartlett and S.K. Ling-Chung, *Ibid.*, **20**, 287 (1989).
4 A. Boyle, E.M. Genies and M. Lapkowski, *Synth. Met.*, **28**, C769 (1989).
5 L.D. Couves and S.J. Porter, *Ibid.*, **28**, C761 (1989).
6 F.G. Yamagishi, L.J. Miller and C.I. van Ast, Proc. of the Amer. Chem. Soc. Div. of Polym. Mater.: Sci. and Engineering, Symp. on Transducer-Active Polymers: Components in Sensors and Actuators, Washington DC, 71, 656, August 21-26, 1994.
7 F.G. Yamagishi, L.J. Miller and C.I. van Ast, Proc. Sensors Expo, Symp. on New Innovations in Automotive Sensors, Cleveland, OH, 509, September 20- 22, 1994.
8 F.G. Yamagishi, J. Stanford, Thomas B., C.I. van Ast, L.J. Miller and H.C. Gilbert, Proc. of the Symp. on Chemical and Biological Sensors and Analytical Electrochemical Methods, 1997 Joint International Meeting of the Electrochem. Soc. and The Intern. Soc. of Electrochem., 97-19, 103, August 31 - September 5, 1997.
9 F.G. Yamagishi, T.B. Stanford, C.I. van Ast and L.J. Miller, Abs. of the Electrochem. Soc., San Antonio, TX, 96-2, 1151, October 6-11, 1996.

Conductive Polyphenylene Ether/Polyamide Blends For Electrostatic Painting Applications

J.J. Scobbo, Jr.

INTRODUCTION

Electrostatic painting provides numerous advantages over traditional high pressure, low volume paint processes. These include improved paint transfer efficiency, which translates into lower paint usage. This can prove to be an important economic incentive when one considers that exterior automotive finishes can often cost in excess of $100/gallon.

Electrostatic painting requires that the part which is to be painted is electrically grounded. This is not an issue when one is painting metallic parts. However, many of the automotive exterior trim components are made of engineering resins, which are electrical insulators. In order to take advantage of the efficiencies of electrostatic painting, a plastic part must first be sprayed with a coating of a conductive primer. The particulate metallic constituents of the coating allow it to be grounded, thereby allowing for electrostatic base coat and clear coat deposition. Even with the additional step of conductive priming, there are significant incentives in terms of economics and surface quality/consistency, for electrostatic painting of plastics to be desirable.

An ideal electrostatic painting situation would be to combine the benefits of engineering resins (such as the ability to form parts of complex geometry, economically through injection molding), with the benefits of metallic parts (intrinsic conductivity, precluding the need to deal with a conductive priming step). The solution would be to make an intrinsically conductive thermoplastic resin, that could be injection molded, and would maintain a physical/mechanical performance profile similar to other engineering resins used for the application in question.

To this end, a polyphenylene ether/polyamide engineering resin blend has been recently introduced and is currently in use for electrostatically painted mirror shells. The resin makes use of a graphite nanotube additive to provide sufficient conductivity to allow for electrostatic

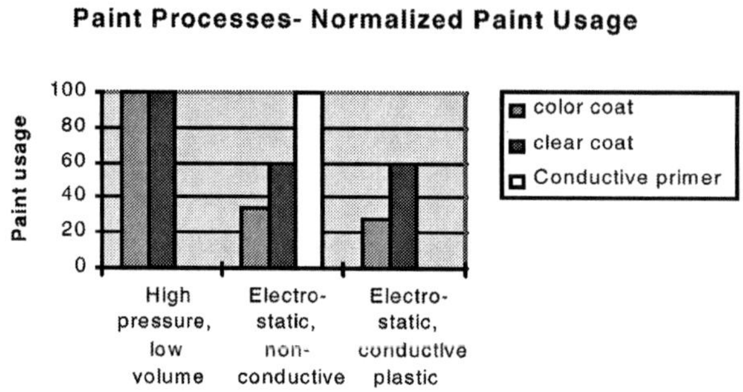

Figure 1. Approximate normalized paint usage for various painting technologies.

painting.[1,2] This paper will discuss the base resin, the benefits of the conductive additive, and impact on the painting process.

ELECTROSTATIC PAINTING

There are significant advantages to electrostatic painting over traditional paint processes, and even further advantages to the electrostatic painting of intrinsically conductive plastics. These advantages are shown graphically in Figure 1. These advantages include:

- Elimination of conductive priming
  - reduced labor
  - reduced materials
  - reduced number of process steps
  - reduced volatile organic compound emissions
- Improved paint transfer efficiency
  - reduced materials (base and clear coats)
- Improvement in first pass yield
  - greater paint wrap
  - uniform coverage

The challenge in resin development is to combine a base resin technology with a conductive technology such that an appropriate balance is struck for economic viability.

PPE/PA BLEND TECHNOLOGY

Polymer blends are formulated to provide a material with an appropriate balance of thermal performance, processability, and toughness, among other properties, that cannot be met with single polymers. Melt compounding, or extrusion, is used to mix the two resins to form the alloy or blend. Because most polymer blend pairs are immiscible and incompatible, they form a two phase system. Furthermore, the phase morphology is unstable and tends toward gross phase separation This makes processing difficult and often results in delamination in injection molded parts. The phase separation also results in poor physical performance.

To circumvent the difficulties encountered in producing useful immiscible blends, generation of compatibilizer molecules by can be achieved during the compounding operation.

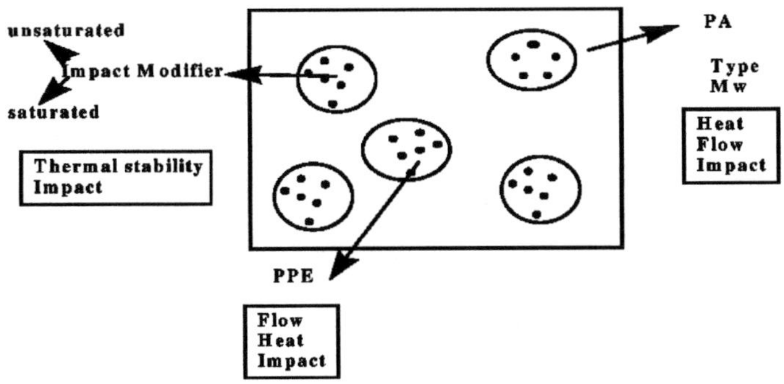

Figure 2. Schematic representation of PPE/PA blend phase morphology.

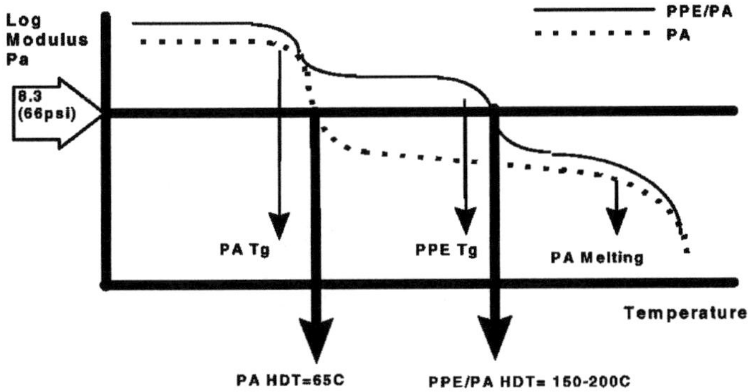

Figure 3. Modulus advantage of PPE/PA over PA at elevated temperature.

These compatibilizers act as macromolecular surfactants, stabilizing the melt phase morphology against coalescence, and provide mechanical adhesion between the dissimilar phases. In this instance, the melt mixing unit operation is combined with that of a chemical reactor. Use of this technology has been applied for many years to the commercial production of polyphenylene ether/polyamide blends (PPE/PA).

The base resin is a compatibilized, impact modified blend of polyphenylene ether (PPE) and polyamide 6,6 (PA). By blending PPE and PA, it is possible to gather the benefits of each, while mitigating the deficiencies. PPE, an amorphous resin with glass transition of about 210°C, brings the following advantages:

- improved dimensional stability (less warp, less shrink, less moisture absorption)
- ease of impact modification
- reduction in density

PA, a semicrystalline resin with glass transition of about 65°C and melting transition of about 260°C, has these benefits:

- chemical resistance
- ease of processability, low melt viscosity
- paintability

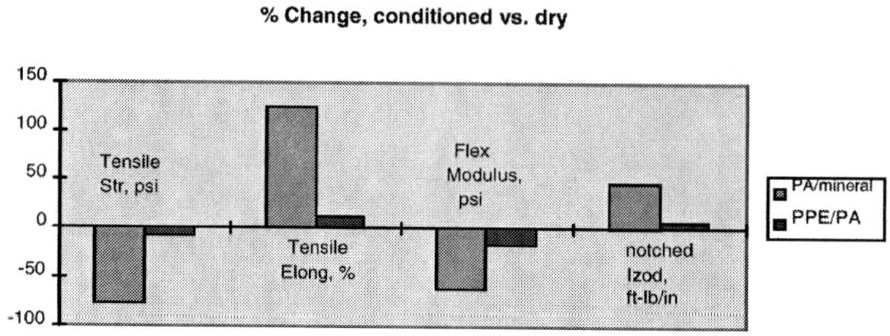

Figure 4. Property retention, conditioned vs. dry as molded (DAM), of mineral filler PA vs. PPE/PA.

Figure 2 shows a diagram of the phase morphology of a typical PPE/PA alloy. PPE is a dispersed phase on the order of a few microns in diameter in the PA matrix. Typically, there is an impact modifier within the PPE phase.

A significant advantage of PPE/PA blends over typical nylon materials is the ability of the material to withstand the elevated temperature of paint bake ovens. This can be seen conceptually as a modulus advantage at elevated temperature, as shown in Figure 3. Or similarly, it is seen in higher heat distortion temperature (HDT, ASTM D648).

In the past mirror shell applications have used mineral filled nylon. The reduced sensitivity to moisture of PPE/PA provide benefits in property retention as a function of environmental conditioning, as seen in Figure 4.

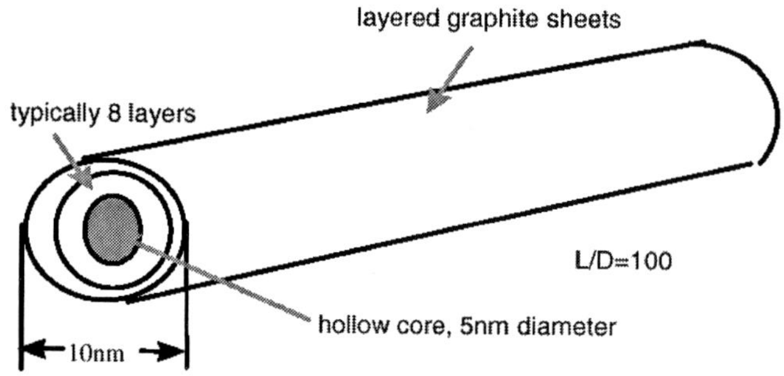

Figure 5. Schematic diagram of graphite nanotube structure.

GRAPHITE NANOTUBE TECHNOLOGY

The challenge has been to find a conductive additive technology that can be used in a PPE/PA resin system without reducing or eliminating the advantages elucidated above. For example, typical graphite fibers can be used. However the loadings required to achieve conductive percolation are so high that embrittlement of the composition occurs. A further complication is the anisotropy that is experienced due to flow fields in injection molding. A second technology worthy of consideration is that of conductive carbon black. In this instance, the loading required for conductivity is significantly less than that for graphite fibers, and isotropy is essentially eliminated. However, mechanical performance can be compromised.

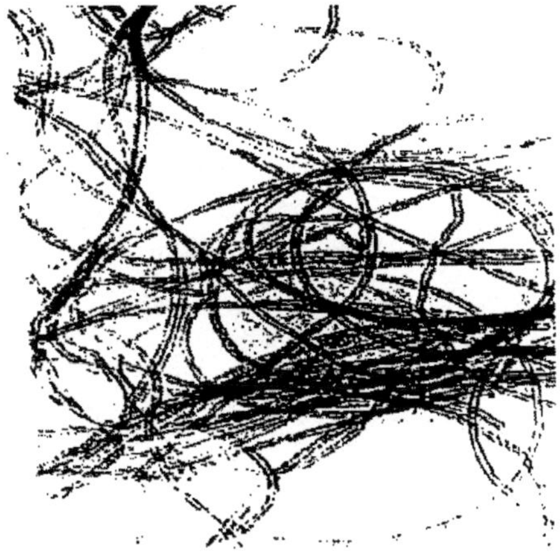

Figure 6. "Bird's nest" aggregate structure of nanotubes as observed by transmission electron microscopy.

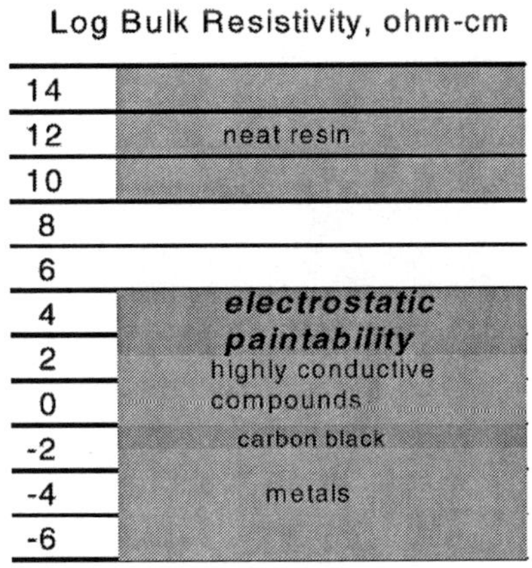

Figure 7. Typical bulk resistivity values required for various applications.

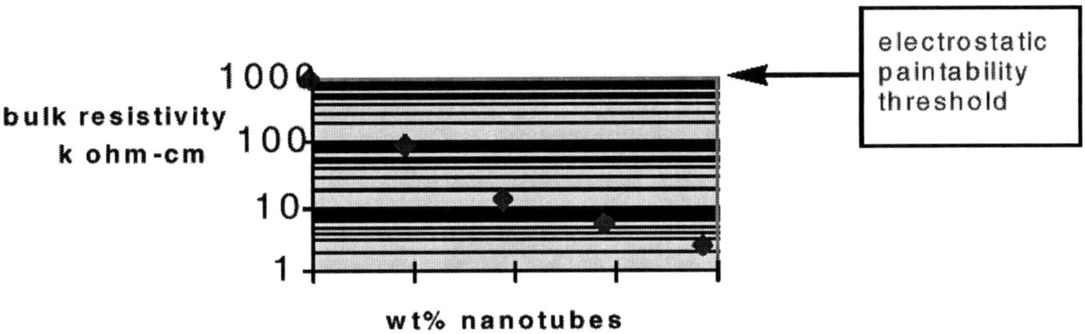

Figure 8. "Percolation behavior" below the electrostatic paintability below the electrostatic paintability threshold.

The next level in conductive additives is graphite nanotubes. These nanotubes are grown catalytically using hydrocarbon gas. A schematic diagram is shown in Figure 5. Figure 6 is a transmission electron micrograph of the "bird's nest" aggregate structure of the nanotubes, which depicts the neat, uncompounded structure. Features of these structures include:

- hollow core, typically 3-5 nanometer diameter
- tubes are concentric layers of rolled-up graphite sheets
- graphite basal plane parallel to cylindrical axis
- 10 to 20 nanometers in diameter
- 2 to 10 microns long

- bulk density less than 0.1g/cc
- “bird’s nest” structure-aggregate of tubes

Because of the large L/D and the tortuous structure of the tubes (they are not straight tubes as depicted in the idealized diagram), they are very efficient at establishing a percolation network at very low loadings. These loadings are less than those required to achieve similar levels of conductivity for carbon black, and therefore have a smaller effect on physical performance.

As previously alluded to, bulk resistivity is a key performance measure to determine electrostatic paintability. The typical threshold for electrostatic paintability is approximately 10^6 ohm-cm. That is, formulations designed with bulk resistivities significantly less than 10^6 ohm-cm ensure good paint transfer through the paint line. A comparison of bulk resistivity values for various applications is shown is Figure 7. An example of the “percolation curve” for a resin system containing nanotubes is shown in Figure 8.

PROPERTIES

Typical physical properties for an example composition for this family of resins are given in Table 1. Such properties were arrived at through a rigorous, statistically-based, design of experiments approach.

Table 1. Physical properties of a typical PPE/PA resin containing nanotubes for electrostatic painting

Property	Value
Specific gravity, g/cm^3	1.08
Heat distortion temperature, 66 psi, 1/4", °C	158
Notched Izod, KJ/m^2	0.6
Flexural modulus (1"/min, 1/4"), MPa	2400
Flexural yield (1"/min, 1/4"), MPa	95
Tensile strength (2"/min), MPa	60
Tensile elongation at break (2"/min), %	22
Bulk resistivity, kΩ-cm	3
Melt flow index (280°C/5 kg)	4.8

SUMMARY

The benefits of conductive plastics for electrostatic painting can be summarized as follows. A significant reduction in base coat and clear coat usage is seen when one changes from traditional painting to electrostatic painting ordinary plastics. However, a conductive priming layer must first be applied. When electrostatic painting of a conductive plastic is employed, excellent paint transfer efficiency is maintained, but the extra process step for conductive priming is eliminated. This means that the paint booths and labor associated with conductive priming can be turned over to base and clear coat operations. Another significant benefit of electrostatic painting of conductive plastics is that the reduction in coatings usage dramatically reduces the emissions of volatile organic compounds (VOCs). If a painting operation is near the legal VOC limits, there is little that can be done to expand capacity. A change in technology to conductive plastics can reduce VOCs to the extent that effectively, capacity may be increased, without significant expense for plant and equipment. Polyphenylene ether/polyamide blends have been developed that provide the benefits described above while also providing the flexibility of injection molding and the physical performance of an engineering resin.

REFERENCES

1 Graphite fiber brings new look to conductive plastics, *Plastics World*, November 1993, 10.
2 Tiny graphite tubes create high efficiency conductive plastics, *Plastics World*, September 1996, 73.
3 News Briefs, Plastics World, May 1997, 1.

Conductive Polymer Films for Improved Poling in Non-Linear Optical Waveguides

James P. Drummond and Stephen J. Clarson
Department of Materials Science and Engineering, University of Cincinnati
Stephen J. Caracci and John S. Zetts
Materials Directorate, Wright Patterson Air Force Base

BACKGROUND

In the past decade, non-linear optical polymers and chromophores have been a topic of intense research. The evidence that these polymers and molecules have the properties that could speed the development of photonic technology[1-4] has researchers continually searching for new materials with better properties.[5,6] Currently, these hopes rest on the use of the second order non-linearity (χ^2) of these materials. In order to realize these non-linear properties, the material used must have a non-centrosymmetric structure. Producing this type of order in a naturally amorphous polymer usually is done through the use of a large electric "poling" field.[7,8] One of the most critical aspects of poling these polymers is achieving the maximum possible electric field without dielectric breakdown. To reduce the voltage needed during poling, the field should optimally be dropped directly across the layer of material to be poled. In optoelectronic devices based on waveguide structures, this is often not the case. In these structures the active layer is generally placed between two highly resistive cladding layers. By replacing these highly resistive layers with those of intermediate resistance (i.e., layers having resistance lower than those of the active EO layer, but still significantly higher than those of the ITO or metallic electrodes), many benefits can be realized. For non-linear optical waveguides, the use of resistive materials such as silicon dioxide (SiO_2), epoxy resin, or polyimide as a cladding layer presents several problems. In attempting to obtain an adequate poling field across the active layer in these resistive triple stacks, external voltages can easily reach into the kilovolt range. Using such high voltages is not always practical, and can lead to undesirable results. Lowering the applied voltage to more reasonable voltages, however, will cause sub-optimal poling of the NLO chromophores and result in lower achievable electro-optic coefficients, and larger switching voltages. Secondly, the mere presence of these insulating layers also requires that devices made from this type of structure will have higher modulation or switching voltages. By replacing these high-resistance claddings with a more highly conductive material, these problems could be avoided.

As stated, the materials used should have a conductivity that is higher than that of the guiding layer. This would therefore drop the majority of the applied poling field across the guiding layer where it is most needed. In other research, attempts have already been made to take advantage of the fact that, in general, polymers undergo an increase in conductivity as they approach their glass transition temperature (T_g).[9] In this method, polymers with T_g lower than that of the guiding layer are chosen as the cladding layers. This approach does have its own problems however. First, the chosen cladding layer may not be stable in the high temperature region near the T_g of the active layer. Since this is the region where the active layer should be poled for maximum poling efficiency, optimal poling may not be achieved in these systems. Second, the benefit of the lower resistance may vanish at lower temperatures, like those encountered where switching/modulation of the structure takes place. Finally, these claddings are chosen to perform with a specific guiding layer, and may not be applicable to other systems. It is therefore logical to look for cladding materials that exhibit enhanced conductivity across the whole temperature range from room temperature to the T_g of the active layer.

In a search for materials to meet the requirements for these conductive cladding layers, the list of possible candidates is small. A very attractive option is the use of inherently conductive polymers. They have many promising properties including high conductivity, simple processing techniques, and the ability to be made into relatively transparent films.[10,11] Practically, however, there are many hurdles to be overcome in their development as usable cladding layers. Here we discuss the successful development of one such conductive polymer system that has shown promise as a conductive cladding layer.

RESULTS AND DISCUSSION

SAMPLE PREPARATION

Initial investigation of conductive cladding layers focused on the conducting polymer poly(ethylene dioxythiophene) (PEDOT). An aqueous solution of PEDOT doped with poly(styrenesulphonic acid) (PSS) was used to allow processing of the polymer to form thin films. To achieve highly transparent conductive films, a blended system of PEDOT/PSS in poly(vinyl alcohol) (PVAl) was chosen. Solutions of 10 wt% PVAl in water were produced by mixing under moderate heat. The PEDOT/PSS and PVAl solutions were then mixed at weight ratios varying from 1:100 to 60:100. Solutions were filtered, and spin cast at 1000 rpm to produce uniform films approximately 2 μm thick. The samples were then dried at 70°C to remove

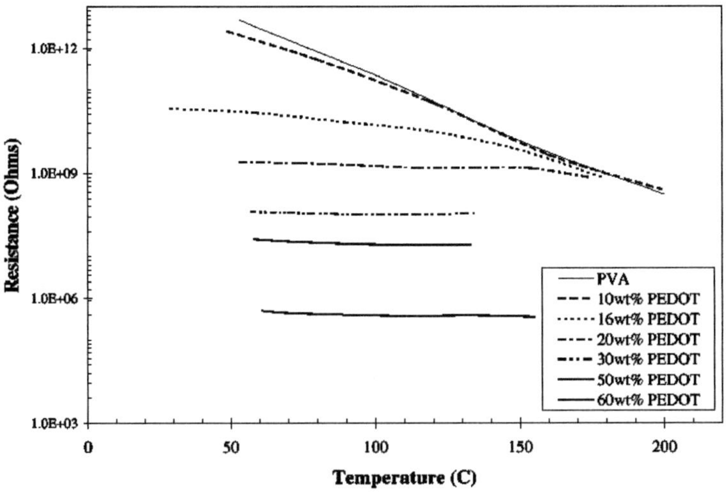

Figure 1. Conductivity measurements for PEDOT/PVAl blends with PEDOT loading percentages from 10 to 60%.

residual solvent. The films formed were transparent with a slight blue tint. Observation of the films indicated that those with up to approximately 30 wt% PEDOT in PVAl had a homogenous distribution of the conducting polymer in the blend. At higher ratios, the PEDOT seemed to form aggregates in the films. For further characterization, these polymer films were cast onto silicon, SiO_2, indium tin oxide (ITO), and glass.

CONDUCTIVITY

Solutions were prepared as above, and were cast onto patterned ITO glass slides. All resistance measurements were taken using a simple two-probe technique, and were conducted in an inert nitrogen atmosphere to prevent any oxidation of the materials tested at high temperatures. Comparative resistance measurements were taken with respect to both voltage and temperature. Resistance values for the blended materials remained constant with voltage for the range of 0 to 200V, but resistance versus temperature graphs revealed some interesting results (Figure 1). When testing samples with various ratios of PEDOT to PVAl, films containing low PEDOT levels (<20 wt%) were found to have decreased low temperature resistance as compared to pure PVAl films. At high temperature these same films took on the resistive characteristics of the pure PVAl films. This result indicates that even low levels of doping in these blended films would enhance the low temperature conductive properties of these polymers and thus enhance their desirability as cladding layers. Overall, it can be seen that the doping of PVAl with PEDOT leads to a gradual increase in the conductive properties of the resulting blends. In this manner, the conductivity of these films could be tuned to fit specified application needs.

ELECTRO-OPTIC POLING

Sample films containing 30 wt% PEDOT in PVAl were prepared for testing as cladding layers for poling of guest-host EO polymer systems. The blended solutions were deposited onto pat-

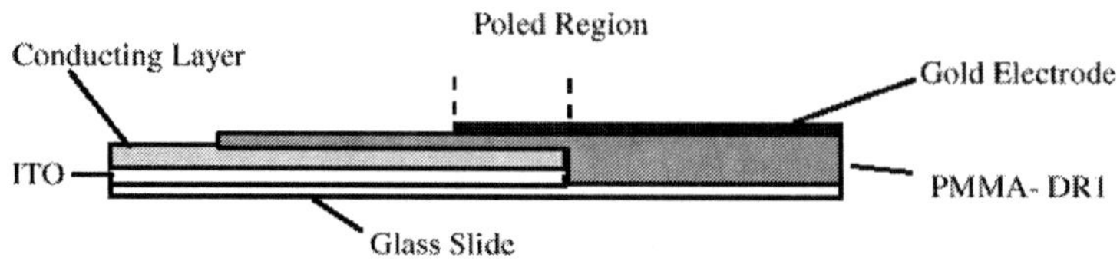

Figure 2. Simple test structure for measuring electro-optic coefficients.

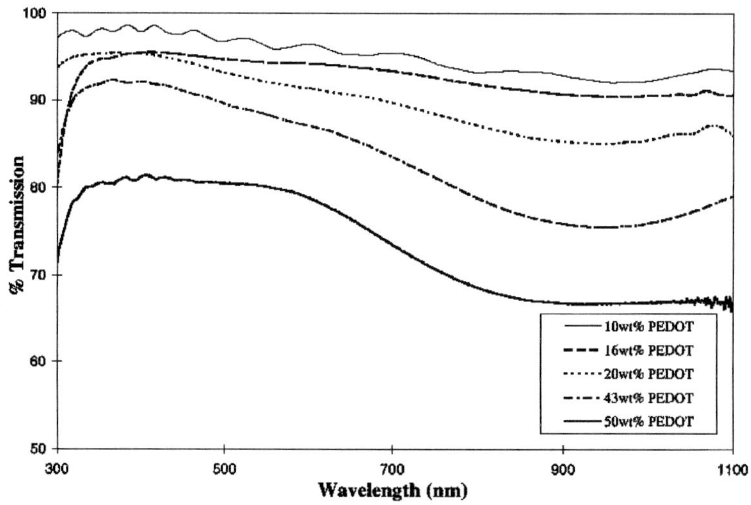

Figure 3. UV/VIS/NIR spectra showing light transmission of PEDOT/PVAl blended films.

terned ITO glass slides, and from these samples, simple EO test structures (Figure 2) were constructed. Testing was performed using an *in situ* poling/EO measurement technique based on the simple ellipsometric method developed by Teng and Mann.[12] A commonly used active layer of poly(methyl methacrylate) (PMMA) doped with 10 wt% disperse red one (DR1) was employed to test the performance of the conductive claddings. A separate group of control samples were also constructed having no conductive layer. Comparable EO testing was performed on both sets of samples. A gradual increase in EO coefficient was seen in both the test and control groups, however it was evident that much larger poling voltages could be attained in the samples with conductive cladding layers (Figure 3). Fields of up to 190 V/m were achieved across the active PMMA/DR1 layer of the test samples. In the control samples, dielectric breakdown was experienced at voltages slightly above 100 V/m. In these experiments, more than a 50% increase in EO coefficient could be achieved through the use of the PEDOT/PVAl layers.

The results seen here seem to indicate that the conductive layers acted as an isolation layer between the active layer and the ITO. They seemed to have the ability to delay the occurrence of catastrophic breakdown by isolating or healing areas where small shorts had occurred and preventing further current leakage. This analysis correlates with previous conducting polymer research, which showed that these layers act as sufficient conducting layers, but also serve to smooth and pacify rough surfaces such as ITO.[13] The addition of the conducting layer therefore served to create a clean, smooth surface for better adhesion and efficient charge transfer.

OPTICAL CHARACTERIZATION

Preliminary optical characterization was done to investigate the value of these blends as cladding layers for optical waveguiding. For these studies, waveguide loss measurements were taken for a group of prospective guiding layers to obtain a guide which had a reproducible loss measurement, and produced high quality waveguides. From the materials investigated, polycarbonate (PC) was chosen to be used as a guiding layer. Samples of PEDOT/PVAl were spin cast onto silicon wafers. Samples were allowed to dry overnight to remove all residual solvent. A top layer of PC was then cast onto the cladding layers. Samples of PC on silicon dioxide were also made for comparison measurements.

All optical loss measurements were performed using an end-fire coupled 632 nm HeNe laser. The optical losses were measured using a video capture method. Optimal coupling of the laser into the guiding layers was achieved in the test and control samples, and well-defined streaks were observable in both cases. These preliminary tests have shown that the two-layered waveguide structures do have increased losses as compared to the PC/SiO_2 sample. The loss values obtained for the two layer structures, however, are still within a useful range. Much of this loss may also be due to surface roughness at the interface, and other adjustable parameters.

Solutions with various levels of doping were also spun on to glass slides to produce samples for optical testing. Transmission measurements were taken using a Hewlett Packard 8453A Spectrophotometer. The PEDOT/PVAl blends were found to have low absorption in the visible region of the spectrum and had a slightly increased absorption in the infrared. Figure 4 shows the decrease in transmis-

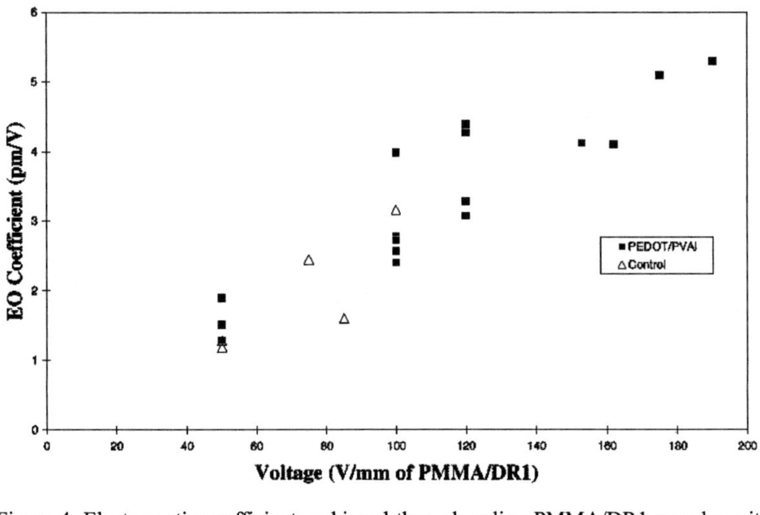

Figure 4. Electro-optic coefficients achieved through poling PMMA/DR1 samples with and without conducting buffer layers.

sion with increasing PEDOT levels in the blended samples. Lightly doped samples had more than 90% transmission across the measured spectrum. The majority of the samples were found to have reasonably good transmission properties with no apparent peaks in transmission loss over the UV/VIS/NIR spectra. This indicates that these blends could be useful as transparent conductive layers in many applications. This absorption data may also help determine how great an affect the absorptive properties of these cladding layers actually have on the optical loss in the waveguide structures.

CONCLUSIONS

The use of blended poly(ethylene dioxythiophene)/poly(vinyl alcohol) films has been shown to improve poling efficiency in electro-optic guest-host systems. In poling these samples, the conductive layer acted as a buffer layer to protect the active layer from catastrophic breakdown. These blends were also used as cladding layers to produce a functional waveguide with a poly(carbonate) guiding layer. In addition, this blended system was shown to have low optical absorption, and a wide range of tunable conductivity.

ACKNOWLEDGMENTS

This research was partially supported by the Air Force Office of Scientific Research. Thanks are also given to Mike Banach and Max Alexander for their support and discussions.

REFERENCES

1 D. Chen, H.R. Fetterman, A.Chen, W.H. Steier, L. Dalton, W. Wang and Y. Shi, *Appl. Phys. Lett.*, **70**, 3335 (1997).
2 N.F. O'Brien, V. Dominic and S.J. Carraci, *J. Appl. Phys.*, **75**, 7493 (1996).
3 S. Kalluri, M. Ziari, A. Chen, V. Chuyanov, W. Steier, D. Chen, B. Jalali, H. Fetterman, and L. Dalton, *IEEE Phot. Tech. Lett.*, **8**, 644 (1996).
4 J. Cites, P.R. Ashley and R.P. Leavitt, *App. Phys. Lett.*, **68**, 1452 (1996).
5 M. Schulze, *Trends Poly. Sci.*, **2**, 123 (1994).
6 D. Gerold, R.T. Chen, W. A. Farone and D. Pelka, *Appl. Phys. Lett.*, **66**, 2631 (1995).
7 M-C. Oh, S-S. Lee and S-Y. Shin, *IEEE J. Quant. Electr.*, **31**, 1698 (1995).
8 P.C. Ray and P.K. Das, *Eur. Polym. J.*, **32**, 51 (1996).
9 D.G. Girton, W.W. Anderson, J.A. Marley, T.E. Van Eck and S. Ermer, Proceedings of the Organic Thin Films Conference (OSA, Portland, OR, 1995).
10 Y. Cao, G. Treacy, P. Smith and A.J. Heeger, *Appl. Phys. Lett.*, **60**, 2713 (1992).
11 Y.Z. Wang, J. Joo, C-H. Hsu and A.J. Epstein, *Synth. Met.*, **68**, 207 (1995).
12 C.C. Teng and H.T. Man, *Appl. Phys. Lett.*, **56**, 1734 (1990).
13 S.A. Carter, M. Angelopoulos, S. Karg, P.J. Brock and J.C. Scott, *Appl. Phys. Lett.*, **70**, 2067 (1997).

The Corrosion Protection of Metals by Conductive Polymers. II. Pitting Corrosion

Wei-Kang Lu
Materials Science and Engineering, The University of Texas at Arlington
Ronald L. Elsenbaumer
Department of Chemistry, The University of Texas at Arlington

INTRODUCTION

It is well know that a sheet of mild steel exposed to a moisture environment within several days will rust badly with pits covered by corroded products. The most common pitting is the selective attack of surface scratch or induced breakdown of the protection film. The pitting mechanisms of aluminum and copper alloys may differ but the basic features are similar. Comparative pitting results of these three kinds of metal alloys will be made and presented. Electrochemical techniques can be used to investigate the passive film breakdown to study pitting propensities. Aluminum always undergoes a pitting problem in sodium chloride. Since oxides always exist on the surface, aluminum alloys and surface treatments may alleviate the degree of localized corrosion attack.

Film-forming polymerization of conductive polymers on electroactive metals[1] and conductive polymers formulated with other polymers which have good adhesion properties to metal surfaces have been used for recent corrosion research in last decade.[2,3] Among all the conductive polymers studied so far, Conquest® of DSM and Ormecon's Corrpassiv® are the first two to achieve commercial availability. Pitting corrosion happens on aluminum, steels and copper commonly and also affects the utilities lifetime tremendously due to aggressive growing pits with damaging species concentrated inside the pits. Furthermore, the mechanical properties of metallic materials can be changed in a short period. So far, no other experimental results in the area of applying the intrinsically conductive polymers to avoid or at least lessen the pitting corrosion on metals was specifically reported. The possibility of whether or not the conductive polymers prevent the pit propagation and growth in the areas of electrolyte-exposed areas or metals under closed end pinholes of protective film is the key motive of this paper.

EXPERIMENTAL

The electrochemical cell setup can be seen elsewhere.[4,5] The test equipment is a Gamry CMS120 software-controlled, automated digital ECN system and a Gamry PC3 potentiostat/zero resistance ammeter was used for both CP and ECN measurements. The panels were received as 2 by 2-in. 2024, 6061 and 7075 aluminum alloys, C1010 grade carbon steel and A316 stainless steel. After appropriate surface cleaning and polishing, coupons were coated with the PANI-PET blend with a certain binder supplied from Americhem Inc. after reformulation by authors. Those compounds were used for coating application by hot dip or spray methods. Those sample materials of polyaniline blends were named AC1, AC3 and AC7 respectively. Corrpassive (zk) is a polyaniline PANI-PMMA mixture supplied by Ormecon company. Conquest (py) is solution of polypyrrole dispersion in polyurethane made by DSM Chemicals. The deaeration tests were conducted in a two neck flask with a corrosion resistant purge tube inside the used electrolyte.

RESULTS AND DISCUSSION

According to corrosion rate determination data in Table 1, AC1 coating material has a far lower corrosion rate through whole immersion time compared to other commercial and control sample sets. The initial and final stages of corrosion rate of purposed uncovered area for AC1 increment is negligible. Zk kept a stable corrosion tendency that is at least 100 times faster than AC1. Figures 1 and 2 indicate that there is a high agreement of pitting tendencies between ECN and CP results for AC1, AC3 and AC7. AC1 shows an extremely low pitting current density (about 10^5 less) compared to the control. AC3 reveals metastable pitting prevention at the initial stage and a subsequent pit propagation and growth pattern shows a sudden spike on the ECN spectrum which was also confirmed by microscopic examination of the inside dent region. AC7 did not produce any pitting prevention at all. The hysteresis loop and

Table 1. Calculated corrosion rates for drilled epoxy top coat conductive polymers coated C101 mild steels in 3.5% NaCl (unit: mpy)

Time/sample	Control	AC1	Zk	Py
day 1	0.018	0.003	1.402	0.006
day 7	0.037	0.008	0.817	0.013
day 28	0.125	0.008	0.722	0.009
day 56	0.208	0.009	0.906	0.082

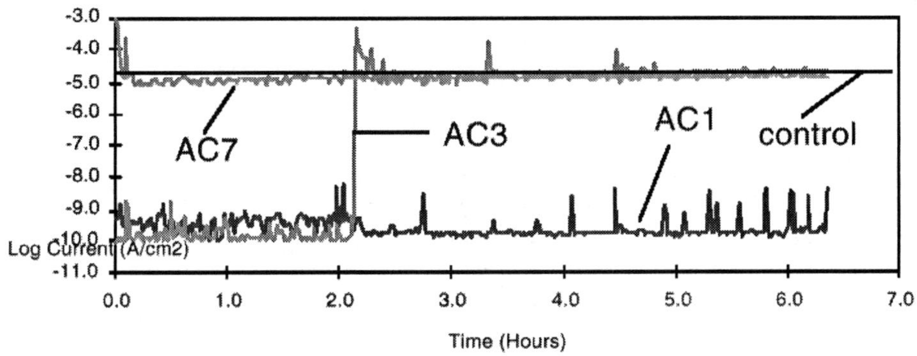

Figure 1. ECN results of drilled epoxy/PANI blends/C101 steel in 3.5% NaCl.

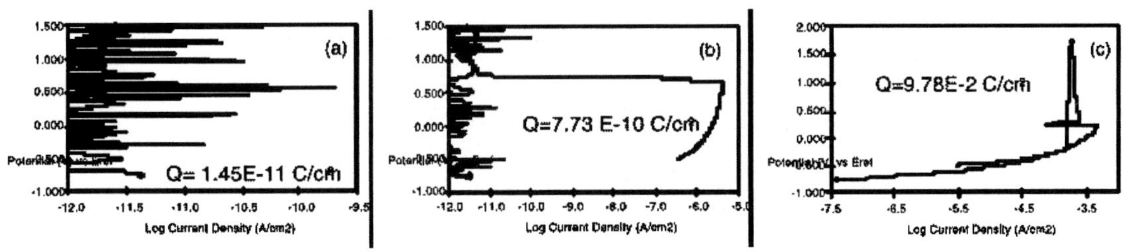

Figure 2. Cyclic polarization curves of dented epoxy/PANI blends/C1010 steel in 3.5% NaCl (a) AC1, (b) AC3, (c) AC7.

the total accumulated charge of pitting-repassivation cycles of these three coatings exhibit the same tendencies. Both AC1 and AC3 appeared with low corresponding pitting current within narrow ranges but AC3 had been identified as passivity breakdown at reversed scan with a 10 fold amount of cycle charge during the same scale measurement. AC7 expressed a high pitting current and a tremendous amount of dissolution charge with a tendency of at least 10^5 times greater. Both ECN and CP appear similarity in the pitting description of the AC series compounds studied. It is believed that the AC1 polyaniline blend has excellent "extended" pitting protection ability to exposed dent surface as compare to the formulations of AC6 and AC7 in the saline environment.

The pitting effects of zk and py as primer layer for drilled exposed area to brine are shown in Figure 3. Without doubt, no primer layer sample demonstrated a bigger hysteresis loop region at the highest measured pitting current. It also has largest potential gap between pitting potential and repassivation potential (protection potential). Generally, zk has a best pit

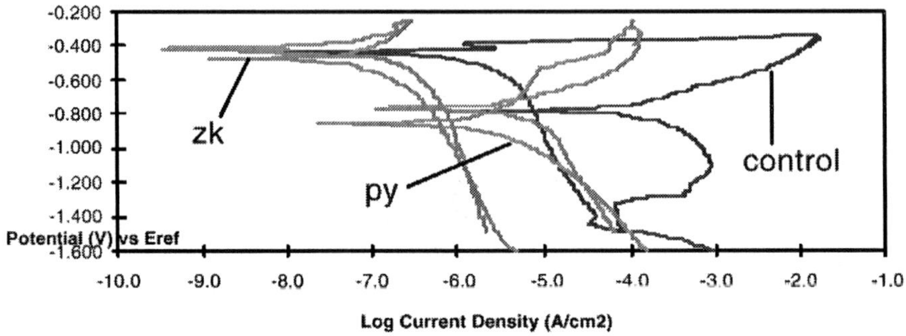

Figure 3. Cyclic polarization comparison among conquest and corrpasive as primer layer with epoxy top coat/C1010 steel and no prime layer sample.

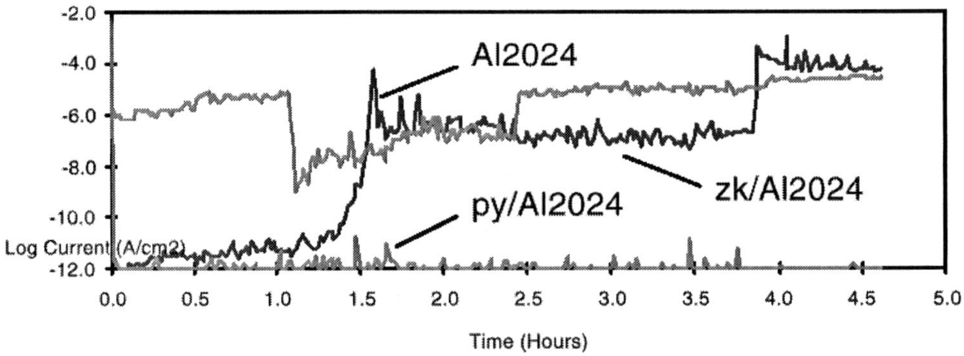

Figure 4. ECN results of conductive polymer coated aluminum alloys in 3.5% NaCl.

prevention performance with a defined small loop at relative positive potential with a narrow potential difference. This means pitting potential and repassivation potential are close each other which has the possibility to halt pit propagation even with initial pit formation on steel surface. Figure 4 shows py had an excellent pitting passivation property even with existing pinholes through cast films covered on aluminum alloy. The bare 2024 aluminum alloy showed the onset of pitting after 1.5 hours of immersion then a plateau region. Meanwhile, zk coated 2024 aluminum alloy shows quite a variation on pitting expression: surfaces will passivate for a short time then go up to keep the original scale of measured current.

The constant phase element (CPE) can be seen from Figure 5. CPE behavior is observed from most of kinds of protective films in impedance response.[6] One can consider the film and

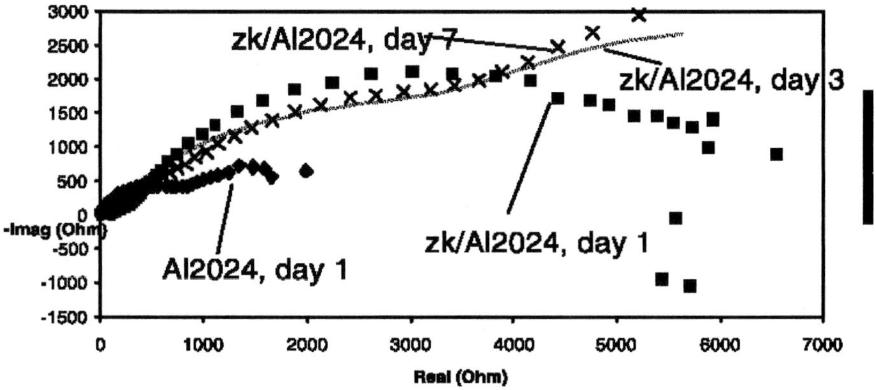

Figure 5. EIS results of conductive polymer coated Al2024 alloys in 3.5% NaCl.

dielectric interface of electrolyte/film to act like a series of complex non-ideal capacitors. Somehow, the impedance spectra correspond to porosity and localized pit activities that can be monitored. From figure 5, a random distribution of low frequency data points for zk coated 2024 Al alloy during the first day of submersion mean that active pit growth unstabilize the signal lock, and the shield effect of Al_2O_3 passivation film on aluminum alloy is not as good as the zk coated one, but surface coverage and cathode/anode separation also need to be considered. However, the zk-aluminum oxides complex can increase the corrosion resistance of covered passivated films by time and charge transfer resistance maintaining an almost a fixed value and polarization resistance increased from time to time. The porosity of zk film exhibits no change and still has open-end pinholes. In order to observe the actual pitting, the peeling of protective films was done showing sparsely scattered hollow pit sites on both the aluminum alloy and carbon steel surfaces. The major difference is the refill-by-products degree of mild steel seems higher than Al. For the purposed drilled hole, because of the rougher surface at the sidewall, more pits can be noticed around the inner bottom surface with some tangling fractual sidewall and not at the edge of the film-metal interface. It is suggested that the control momentum of pitting is also the potential gradient and metallurgical preparation.

The inert gas purge tests can be helpful to understand oxygen participation in the corrosion mechanism and pitting formation mechanism. Microscopic examination was performed during momentary interruption tests for comparative samples. Visual confirmation of early stages occurred with no difference. The change of color and aggregation of passivation film on original alloy surfaces were varied compared to ambient sets in the same acidic environment. Due to the hydrogen embrittlement in oxygen-free environment, the argon gas inlet samples seem to develop some sort of erosion top layer and loosely laid on surface. After cleaning the surface, several severe deep columnar pits were spotted. However, the passivation film of oxygen-rich (ambient) immersion samples look the same before the intermediate period then formed a more compact oxidation film to shield the carbon steel. With the comparison of surface pits at the dented area, oxygen-deficient samples show more corrosion than oxygen exposes ones. We conclude that corrosion protection of mild steel by conductive polymers in 0.1 M HCl requires oxygen input in the protection mechanism when the system turns into anodic protection mode.

CONCLUSIONS

The corrosion protection of carbon steel by conductive polymers in acidic environment proved to be efficient but cannot reduce much of the pitting trend. Furthermore, most of conductive polymers cannot achieve successful long-term corrosion prevention in artificial seawater with pitting inhibition. However, existing pitting suppression by almost all conductive polymer materials is obvious in sodium chloride solution. The newly innovated Americhem formulated polyaniline (AC1) had demonstrated very strong anticorrosion properties in both general corrosion and local corrosion in 3.5 % NaCl. On the contrary, most of commercially available materials lack protection efficiency under the same experimental condition. Considerable pitting current in drilled hole can be reduced by the application of AC1 and are technologically important because the easy processing and coating procedures lend to satisfactory mechanical strength. However, reformulation of PANI mixed with binder show the pitting behaviors of ferrous and nonferrous alloys influenced by intrinsically conductive polymers can be studied by ECN, CP and EIS techniques providing more understanding of the corrosion and pitting mechanisms.

Preliminary inert gas aeration results indicate oxygen is not a factor in the early immersion period but will become important when the protection mode moves to anodic protection which was confirmed by previous Tafel slope interpretations.

ACKNOWLEDGMENTS

The authors wish to thank Dr. V. G. Kulkarni and Mr. Tim Chen of Americhem Inc. for great help in providing technical assistance and the partial financial support for this project.

REFERENCES

1 W. Su and J. Iroh, *Electrochim. Acta*, **42**, 2685 (1997).
2 V. G. Kulkarni, *private communication.*
3 S. P. Sitaram, J. O. Stoffer and T. J. O'Keefe, *J. Coat.*
4 W. K. Lu, S. Basak and R. L. Elsenbaumer, **Handbook of Conducting Polymers**, 2nd edition, (Stotheim, ed.), *Marcel Dekker,* New York, Ch31, p881-920 (1997).
5 D. J. Mills, G. P. Bierwagen, B. Skerry and D. Tallman, MP, 33, May 1995.
6 L. M. Calle and L. G. MacDowell III, paper No. 97268, Corrosion '97, New Orleans, LA. Tech., 69, 65 (1997).

Studies of Electronically Conducting Polymers for Corrosion Inhibition of Aluminum and Steel

Dennis E. Tallman,[1] Youngun Pae,[2] Guoliang Chen,[1] Gordon P. Bierwagen,[2] Brent Reems[1] and Victoria Johnston Gelling[1]

Departments of Chemistry[1] and Polymers and Coatings,[2] North Dakota State University, Fargo, ND 58105-5516

BACKGROUND

Electronically conducting polymers (ECP's), such as polyaniline, polypyrrole and polythiophene, continue to be the subjects of intensive research. The electrical, electrochemical and/or optical properties of these polymers make them potentially useful for a number of commercial applications, including sensors, rechargeable batteries, electrochromic displays, selective membranes, charge dissipative coatings, corrosion resistant coatings, etc. Since many of these applications require casting the polymers as films or coatings, solution processibility is a key issue. One potential application of ECP's now receiving considerable attention is in corrosion resistant coatings. Since the 1981 paper by Mengoli *et al.*,[1] a number of reports have appeared describing studies of the corrosion inhibiting properties of various conducting polymers. A recent review appeared in 1997.[2]

In this report we describe ongoing work in our laboratory involving several strategies for addressing the processibility issue. Preliminary results of immersion testing using electrochemical impedance spectroscopy and electrochemical noise methods on conducting polymer coated steel and aluminum alloys are presented. The polymers currently under study include an organic soluble polyaniline, a water soluble polyaniline rendered insoluble through polymer-polymer complex formation, and an organic soluble alkyl substituted polypyrrole.

EXPERIMENTAL DETAILS

MATERIALS

The metal panels used in this work were cold-rolled steel (Bonderite 1000) and aluminum alloy (7075 T6 and 2024 T3). The organic soluble polyaniline(PANDA, M_W 70,400) was obtained from Monsanto (St. Louis, MO), contained dinonyl naphthalene sulfonic acid (DNSA) as counterion and was dissolved in xylene. The water soluble polyaniline (SPANI, M_W 10,000) was a sulfonated polymer obtained from Nitto Chemical Industry Co. (Tokyo, Japan). Poly(4-vinylpyridine) (PVP, M_W 50,000) was obtained from Polysciences, Inc. (Warrington, PA). The poly(3-octylpyrrole) (POP) was synthesized electrochemically by the Intelligent Polymer Research Institute (University of Wollongong, Australia) and contained a mixture of perchlorate and p-toluenesulfonate counterions.

PREPARATION OF SAMPLES

The Bonderite 1000 steel substrates were prepared for coating by washing with acetone or hexane. The aluminum alloys were prepared by polishing on 600 grit emery paper followed by an acetone or hexane wash. Conducting polymer films ranging from 10 to 50 microns were cast on the metal substrates using either a draw bar coater or a solvent casting technique. Thinner coatings were prepared by dip coating or spin coating. Exposure of aluminum samples was by immersion in dilute Harrison solution (0.35% $(NH_4)_2SO_4$, 0.05% NaCl). Exposure of steel samples was by immersion in 3%NaCl.

INSTRUMENTATION

Electrochemical current and potential noise measurements (ENM) were performed on a DENIS Corrosion Characterization System (CML, Ltd., Manchester, UK). Electrochemical impedance spectroscopy (EIS) measurements were performed at the open circuit potential on an EIS-900 from Gamry Instruments, Inc. (Warminster, PA). Atomic force microscopy (AFM) images were obtained in the contact mode using a Nanoscope IIIa from Digital Instruments, Inc. (Santa Barbara, California).

RESULTS AND DISCUSSION

ELECTROCHEMICAL IMPEDANCE SPECTROSCOPY

EIS was used to investigate the electrochemical response of the Monsanto (PANDA) polyaniline-coated steel (40-50 micron coatings) as a function of immersion time in 3% NaCl. The Nyquist plot in Figure 1 shows two regions of distinct electrochemical response. At low frequencies (lower limit = 0.01 Hz), the response is limited either by the diffusion of counterions within the polyaniline film or by diffusion of oxygen through solution to the

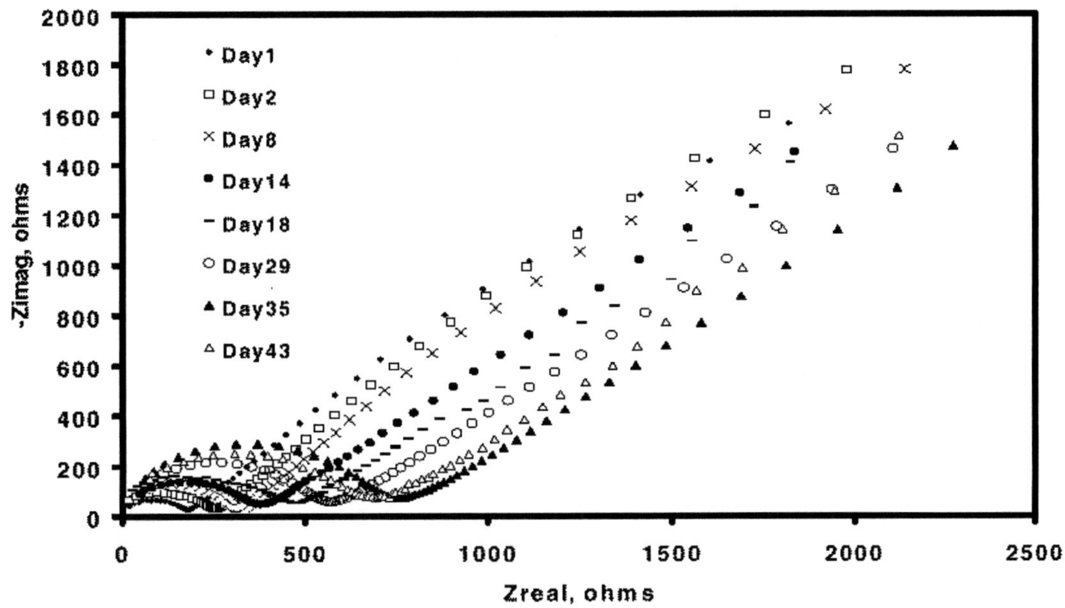

Figure 1. Electrochemical impedance spectra for polyaniline, PANDA, coated steel as a function of immersion time in 3% NaCl. Frequency range: 0.01 Hz to 5 kHz at the open circuit potential.

polymer/solution interface, giving rise to the long diffusion tail represented by the straight line with 45° slope. At higher frequencies, (upper limit = 5 kHz), the kinetics of a charge transfer process at the electrode surface combined with an interfacial capacitance leads to the semicircle, from which the value for the charge transfer resistance (R_{ct}), solution resistance (R_s), and double layer capacitance (C_{dl}) can be extracted. We attribute the charge transfer resistance to electron transfer between the metal and the polyaniline coating, and we conjecture that it is this process which is responsible for formation and stabilization of a passive oxide layer on the metal surface. Separate EIS experiments carried out over the range 10^{-4} Hz to 100 kHz revealed no additional arcs at high frequencies and little deviation from linearity at low frequencies. This low frequency behavior is consistent with semi-infinite oxygen (or perhaps cation) diffusion through solution. Diffusion through the polymer film would be expected to show finite diffusion effects (even for the rather thick films of this study) which would lead to capacitive behavior at low frequency, manifested by a bending of the diffusion tail upward toward a vertical limit. The DNSA counterion in the polyaniline film is not expected to be mobile, and it is likely that cations must move into (or out of) the film during reduction (or oxidation). Experiments now in progress employing control of solution oxygen concentration should permit a more definitive assignment of this diffusion process. The preliminary results from these experiments indicate that the charge transfer resistance responds to the concentration of oxygen in the bulk solution.

Non-linear least-squares fits of the impedance data to equivalent circuit models permitted estimation of the above three parameters. R_{ct} is observed to increase with immersion time, reflected in the increasing radius of the semicircle (Figure 1). The maximum value (ca. 600 Ω) is reached after 35 days of immersion. After ca. 50 days the coating fails (blistering and corrosion products were visible) and R_{ct} falls to ca. 200 Ω. We attribute the increasing value of R_{ct} to the formation and increasing coverage (and/or thickness) of a passive oxide layer which, in turn, reduces the rate of charge transfer between metal and polymer. This interpretation is consistent with passivation mechanisms described by Elsenbaumer *et al.*,[3] by Wessling *et al.*,[4] and by Kinlen and coworkers.[5]

An identical EIS immersion experiment in which the steel substrate was replaced by platinum also exhibited a single arc and a diffusion tail. However, R_{ct} remained essentially constant over the same period of immersion. Thus, the observed changes in R_{ct} with steel do not appear to be attributable to changes occurring within the bulk of the polymer but rather are associated with the active metal/polymer interface.

The electrochemical impedance spectrum of PANDA polyaniline coated aluminum 7075 T6 also showed a similar trend of increasing R_{ct} as a function of immersion time (data not shown). The rate of increase in R_{ct} was larger than that of the steel sample, consistent with Al being more active than Fe. The resistance decreased drastically after 26 days of immersion at which time visible delamination of the coating was observed. For comparison, the impedance spectra of alodine treated aluminum 7075 T6 were obtained. The Nyquist plots exhibited a large capacitive arc on the first day of immersion, indicating a large R_{ct} value (> 200 kΩ), followed by a dramatic decrease in R_{ct} on the second day of immersion (to ca. 20 kΩ), and then a slow rise in R_{ct} to ca. 50 kΩ by day 27. The early behavior may reflect a hydration of the oxide layer upon first immersion. The higher R_{ct} values of the alodine system probably reflects the rather thick oxide layer which forms during this surface treatment.

During the EIS experiments, the open circuit potential of each substrate was monitored. The open circuit potentials of all polyaniline-coated substrates (steel, aluminum 7075 and platinum) were ca. 0.2 V (vs. SCE) and were stable up to the point of coating failure. However, the alodine treated aluminum 7075 displayed an open circuit potential of ca. –0.6 to –0.7 V, considerably more active than the polyaniline coated substrates. The large difference in open circuit potential between the polyaniline samples and the alodine treated aluminum sample may reflect differences in corrosion protection mechanisms of these two systems.

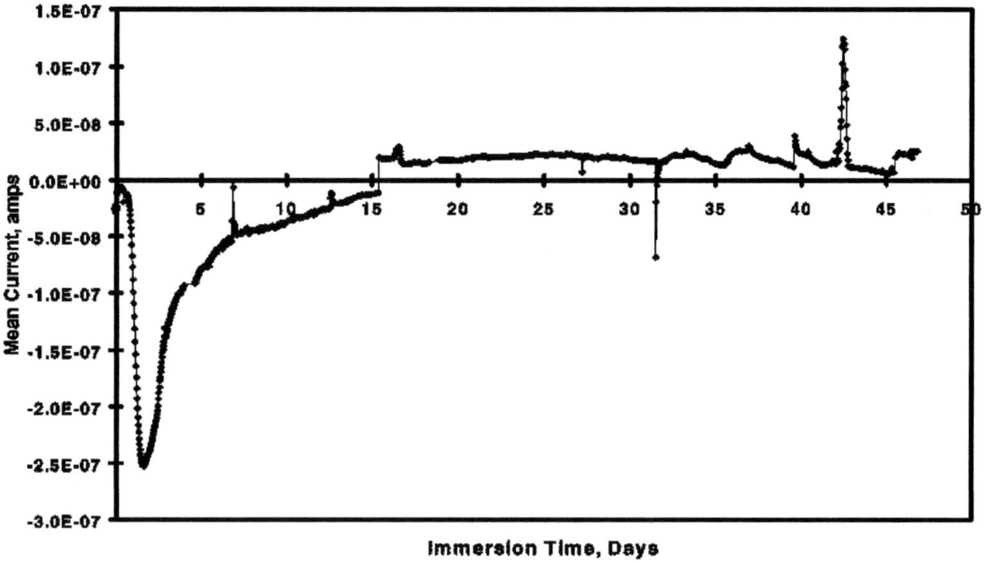

Figure 2. Mean current from electrochemical noise measurement for polyaniline, PANDA, coated steel as a function of immersion time in 3% NaCl.

ELECTROCHEMICAL NOISE

The mean current, mean potential and the noise in these quantities were monitored as a function of immersion time for the polyaniline-coated steel. The mean current is plotted in Figure 2. At the beginning of immersion, a significant current transient is observed, approaching zero after ca. 5 days. We conjecture that this current transient reflects the difference in the rate of formation of the passive layer on the two "nominally identical" panels employed for measurement. From this mean current data, we conclude that the passivation process is most active during the initial 5 days of immersion. A plot of mean potential vs. immersion time (not shown) reveals some oscillatory behavior over the first two days followed by a gradual increase from ca. 0.15 V to ca. 0.27 V by day 30. The development of a more noble potential suggests that the system is becoming increasingly passivating as the immersion time increases. Finally, a plot of noise resistance, (R_n, defined as σ_V / σ_I, where σ_V, σ_I are the voltage and current noise, respectively) vs. immersion time for the polyaniline-coated steel sample exhibits some initial oscillatory behavior, but varies little from between 10^4 and $10^5\ \Omega$. The polyaniline sample appears to exhibit a markedly different behavior from barrier type coatings. Good barrier coating systems have high R_n values, typically above $10^6\ \Omega$. Therefore, polyaniline does not appear to function as a particularly good barrier coating, not surprising since polyaniline is a polyelectrolyte with charges that would allow facile penetration of water and ions throughout the coating.

ATOMIC FORCE MICROSCOPY

The surface morphology of the bare steel was imaged by AFM and is characterized by a surface rms roughness of 61 nm. After coating with the polyaniline (all coatings were applied using a draw bar), the surface roughness is reduced to 1.4 nm, indicating the very smooth nature of the films. After a dry contact time (i.e., no immersion) of 18 days, the polyaniline film was removed and the steel surface was again imaged, exhibiting a rms of 61 nm, identical to that of steel never in contact with polyaniline. Thus, it appears that surface modification of the steel (namely, passive oxide formation) does not occur (or occurs very slowly) in the dry state. After immersion of a freshly prepared polyaniline-coated steel sample in 3% NaCl for 1 week, the film surface increased in rms roughness to 27nm, probably reflecting solvent uptake and concomitant swelling of the film. After 10 days immersion, the polyaniline film was detached and the steel surface was again imaged, yielding a rms roughness of 93 nm. These results suggest that modification of the polyaniline-coated steel surface has occurred, but that solvent is required within the film and perhaps at the polymer/metal interface for oxide formation to take place.

WATER SOLUBLE POLYANILINE

A polymer complex between a sulfonated water soluble polyaniline (SPANI, Nitto Chemical Industry Co., Ltd., Tokyo, Japan) and poly(4-vinylpyridine) (PVP) is formed by mixing an aqueous solution of the protonated (hydrogen chloride) form of PVP, a cationic polyelectrolyte, with an aqueous solution of SPANI.[6] A gel-like precipitate forms which has limited solubility in many common solvents. Thus, the approach represents a possible route to the aqueous solution processing of polyaniline. The nitrogen-to-sulfur ratio of the complex indicates approximately a 1:1stoichiometry between PVP and SPANI monomer units. The complex exhibits modest conductivity (3.3×10^{-5} S/cm) and is electroactive when immobilized on carbon or platinum electrodes. The swellability of the gel form of the complex is characterized by a solvent content of 16 grams per gram of dry material. Thermal analysis of the dry complex indicates stability to 225°C. Films of the complex have been prepared on aluminum alloy (2024T3) by casting a sonicated suspension of the complex(colloidal in nature) using a draw down bar. The Al panels were then baked at 70°C for four hours. Films prepared in this way exhibit good adhesion to the Al surface. Films have also been cast on steel by sequentially dip coating in separate solutions of PVP and SPANI (work performed at IPRI, Wollongong). Immersion studies of these films using EIS are in progress and results will be reported.

POLY(3-OCTYLPYRROLE)

Work is underway to explore the corrosion properties of poly(3-octylpyrrole) (POP) on aluminum alloys. This polymer (prepared at IPRI, Wollongong) is soluble in several organic solvents. Films prepared by solvent casting from solutions of POP dissolved in CH_2Cl_2/CCl_4 exhibit good adhesion to aluminum using the tape pull-off test. Thin films prepared by spin coating techniques will also be evaluated. These substrate/film interfaces are currently being probed using EIS and the effects of prolonged immersion in dilute Harrison solution will be assessed. Immersion experiments in which epoxy and urethane top coats are applied over either a PVP/SPANI or a POP film are planned, and results of this work will be reported in due course.

ACKNOWLEDGMENTS

We gratefully acknowledge the Intelligent Polymer Research Institute of the University of Wollongong, Dr. Gordon G. Wallace, Director, for collaboration on portions of this work.

REFERENCES

1 G. Mengoli et al., *J. Applied Polymer Science*, **26** (1981) 4247.
2 S.P. Sitaram, J.O. Stoffer and T.J. O'Keefe, *J. Coatings Technology*, **69** (1997) 65-69.
3 W-K. Lu, R. L. Elsenbaumer and B. Wessling, *Synthetic Metals*, **71** (1995) 2163-2166.
4 B. Wessling, S. Schroder, S. Gleeson, H. Merkle, S. Schroder and F. Baron, *Materials and Corrosion*, **47** (1996) 439-445.
5 P.J. Kinlen, D.C. Silverman and C.R. Jeffreys, *Synthetic Metals*, **85** (1997) 1327-1332.
6 D.E. Tallman and G.G. Wallace, *Synthetic Metals*, **90** (1997) 13-18.

Novel Electrically Conductive Injection Moldable Thermoplastic Composites For ESD Applications

Moshe Narkis
Department of Chemical Engineering, Technion - Institute of Technology, Haifa, Israel
Gershon Lidor, Anita Vaxman and Limor Zuri,
Carmel Olefins Ltd, Haifa, Israel

INTRODUCTION

Electronic components are susceptible to damage from electrostatic discharge (ESD). The annual losses in products containing sensitive electronic components and subassemblies due to ESD during manufacturing, assembly, storage and shipping has been estimated in billions of dollars.[1]

A variety of materials has been developed to package sensitive electronic devices and prevent damage during storage and shipping. The Electronic Industry Association (EIA) classifies packaging materials according to their surface resistivity as being either conductive, dissipative, or insulative. According to EIA standards, conductive materials have a surface resistivity of less than $1.0\text{x}10^5$ ohms/sq, dissipative materials have a surface resistivity from $1.0\text{x}10^5$ to $1.0\text{x}10^{12}$ ohms/sq and insulative materials have a surface resistivity greater than $1.0\text{x}10^{12}$ ohms/sq. For many articles in ESD protected environments the optimal surface resistivity is in the range of 10^6 - 10^9 ohms/sq.[2,3] Even higher values may be accepted if the article is capable of dissipating charge fast enough. Too high surface resistivity results in an uncontrolled discharge.[2,3]

There is a number of mechanisms by which a polymeric material can be made conductive, static dissipating or antistatic. The conventional methods are through painting or coating, the addition (internal or external) of hygroscopic materials, or conductive fillers. Electrostatic dissipating thermoplastic compounds have successfully eliminated electrostatic discharge failures in many applications in the electronics industry. A variety of conductive fillers is presently available to material engineers, including carbon blacks (CB), carbon fibers (CF), metallic powders, flakes or fibers, and glass spheres or glass fibers coated with metals. For a given polymeric compound, electrical conductivity is determined by the amount, type and shape of the conductive fillers.[4,5,6] The critical amount of filler necessary to initiate a continuous conductive network is referred to as the percolation threshold, which varies from polymer to polymer for a given CB type. A small increase of the filler concentration has a much smaller effect and subsequently a plateau is reached. Although percolative systems can easily give highly conductive compounds, it is difficult to reliably attain desired intermediate conductivities levels required for ESD applications, due to the steepness of the resistivity vs. CB concentration curve. The conductivity of such compounds depends not only on the CB concentration and morphology, but also on the specific polymer matrix used and the generated morphology.[7,8,9,10]

Carbon black loaded static controlling products usually contain 15 to 20 wt% CB by weight. Local variation of the concentration of the conductive additives can result in conductive and insulative regions in the same product. Contamination is also an important issue since, in highly filled CB compounds, the carbon powder tends to slough and thus contaminate the environment. There is a challenge in developing cleaner injection moldable compounds with consistent surface resistivity in the static dissipative range.

The CarmelStat new technology is based on combining a number of polymeric materials with glass fibers and CB to produce a uniform, multi-phase thermoplastic composite with consistent electrical and mechanical properties.[11] When compared with the currently available static dissipative plastics, based on CB, CF, or surface coating of molded parts, this new technology offers many advantages, i.e., a combination of permanent and consistent resistivity levels of 10^6 -10^9 ohms/sq, achieved with about 1 wt% CB, a controlled stiffness/impact balance, and a high heat deflection temperature.

This chapter describes new thermoplastic composites for injection molding, containing CB or CB/CF in relation to their electrical and mechanical behavior, as compared to the existing CB filled thermoplastic compounds.

EXPERIMENTAL

The conductive composites were prepared in a co-rotating twin screw extruder (Berstorff, 25 mm, L:D = 28:1) and subsequently injection molded (Battenfeld, 80 ton).

Commercial grades of polypropylene (PP), polyethylene (PE) and polystyrene (PS) (Carmel Olefins, Israel) and Noryl (GE, USA) were used in this study. The following carbon blacks were used: Ketjenblack EC 600 JD (Akzo, The Netherlands), Printex XE 2 (Degussa, Germany), Black Pearl 2000 (Cabot) and Conductex 975 (Columbian Chemical). The short glass fibers used had a length of 3 mm and a diameter of 10 microns (Vetrotex, Owen Corning). The carbon fibers used had a length of 6 mm (Tenax, Germany).

The surface and volume resistivity of injection molded samples (discs or bars) was tested using procedures described in EOS/ESD S11.11 and EIA 541, based on ASTM D 257, using a Keithley 6517 or 2400 Electrometer which was connected to a concentric (guarded ring) fixture (Keithley, Model 8009), or a 4- point-probe (in the latter method silver paint was used to eliminate the contact resistance). Each reported value is an average of six test specimens.

The ASTM test methods D638, D790 and D256 were used to determine mechanical properties, ASTM D648 for the thermal properties and ASTM D792 and D570 for density measurements. Each reported value is an average of five test specimens.

The Taber abrasion test was used to characterize the abraded average volume loss. This test (ASTM D1044) involves subjecting a molded plaque to contact with a rotating abrasive wheel (CS-10) under a 1 kg load for 1000 cycles.

The material's contamination potential was evaluated by extraction in deionized water at 80°C. The water was then analyzed for leachable anions by an inductively coupled mass spectroscopy (IC/MS) (Dionex 4500i). Total metals was detected by atomic absorption spectrometer (Perkin Elmer AAS 3100).

RESULTS AND DISCUSSION

ELECTRICAL PROPERTIES

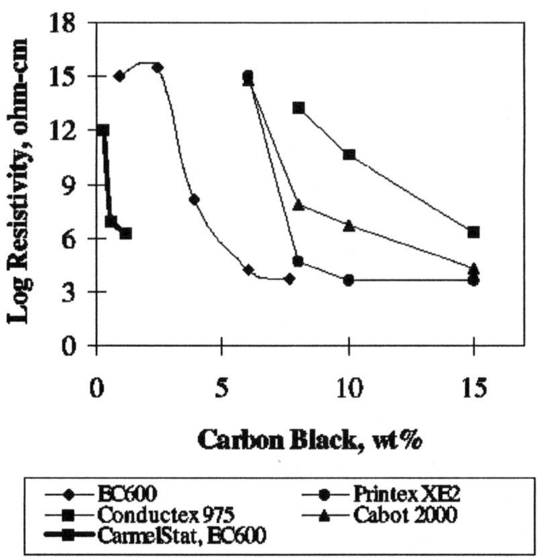

Figure 1. Resistivity vs. carbon black content of an injection molded CarmelStat composites and polypropylene compounds containing different carbon black types.

The influence of CB content on volume resistivity (EOS/ESD S11.11) of an injection molded PP composite, compared with the resistivity of reference unreinforced PP compounds containing different CB grades, is presented in Figure 1. The characteristic insulating to conducting transition is observed for all the systems studied. The percolation concentration for the CarmelStat composite occurs at about 1 wt% CB, significantly lower compared with the reference CB filled PP compounds. From Figure 1 it is evident that the resistivity values of the new CarmelStat material are well within the ESD range (10^6 - 10^9 ohms/sq) at a CB concentration of 1-1.5 wt%. The resistivity of all the other unreinforced compounds, at 1-1.5 wt% CB exceeds 10^{12} ohms/sq. Figure 1 shows that the

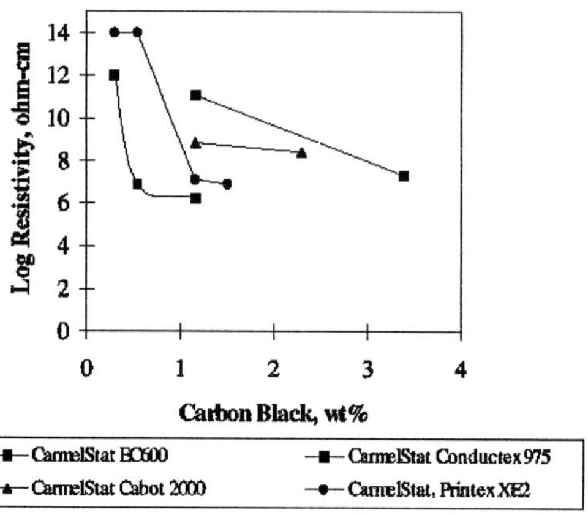

Figure 2. Resistivity vs. carbon black content of injection molded CarmelStat composites containing various carbon black types.

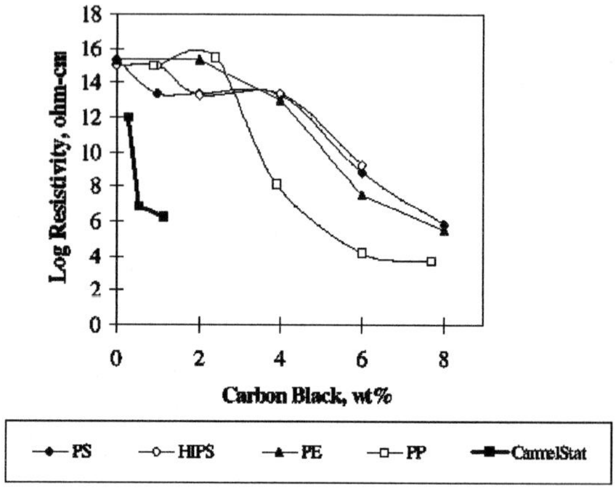

Figure 3. Resistivity as function of EC 600 loading for different matrices.

general shape of the resistivity-concentration curves for the CB grades studied is similar and the critical CB concentrations fall below 10 wt %. Thus, it has been confirmed that in CB filled polymers, at low CB concentrations the CB particles are isolated and the electrical resistance is high, while beyond a critical loading (greater than 10 wt%), particles form structures which provide an electrical network through the insulative polymer matrix. Small changes in filler concentration correspond to multiple order of magnitude changes in the electrical resistivity. The lowest concentration was found for the filled PP compounds containing high structure EC 600 carbon black (about 4 wt%). The superior behavior of EC 600 is also indicated in Figure 2, where the influence of several CB grades on the volume resistivity of CarmelStat composites is presented. To achieve the ESD range (10^6 - 10^9 ohms/sq) higher concentrations of Printex XE2, Cabot 2000 and Conductex 975, are required compared with EC 600. EC 600 was thus chosen for the further experiments because it provides resistivity levels in the 10^6 -10^9 ohms/sq range at the lowest CB contents.

Figure 3 shows the percolation behavior of different polymer matrices containing EC 600. It is evident that the PP compound needs the lowest amount of CB compared with PE, PS or HIPS to generate resistivities in the 10^6 -10^9 ohms/sq range. Table 1 summarizes the properties of different conductive thermoplastic composites based on the CarmelStat technology.

Table 1. Properties of injection molded CarmelStat composites containing about 1 wt% carbon black, based on various matrices

Property	Base resin				
	PP	PS	HIPS	PE	Noryl
Specific gravity, kg/m^3	1.082	1.22	1.21	1.07	1.24
Water absorption (24 h, room temp), %	0.04	0.13	0.14	0.06	0.17
Tensile modulus, MPa	2634	3650	2630	746	3382
Tensile strength, MPa	56.2	66	61	19.3	64
Flexural modulus, MPa	4684	7100	6410	1663	6501
Izod impact, notched, J/m	77	51	47	80	38
Volume resistivity, ohm-cm	1.6×10^6	7.0×10^7	3.1×10^9	2.6×10^9	2.0×10^{11}
Surface resistivity, ohm/sq	1.1×10^7	7.3×10^7	2.8×10^7	2.8×10^8	4.2×10^{10}

Table 2. Properties of CarmelStat PP composites containing 1 - 1.5 wt% carbon black as function of glass fiber content

Property	% GF by weight			
	10	15	21	25
Specific gravity, kg/m^3	0.981	1.025	1.082	1.212
Tensile modulus, MPa	1771	2030	2634	3133
Tensile strength, MPa	39.4	44	56.2	58
Flexural modulus, MPa	3054	3801	4684	5394
Izod impact, notched, J/m	64	69	77	71
Volume resistivity, ohm-cm	1.9×10^7	1.7×10^7	1.6×10^6	2.0×10^6
Surface resistivity, ohm/sq	2.7×10^7	1.3×10^7	1.1×10^7	5.7×10^6

MECHANICAL PROPERTIES

Table 2 depicts mechanical properties of PP composites with 1 - 1.5 wt% CB as a function of glass fiber content. Tensile strength and modulus, and the flexural modulus increase with fiber content. Table 2 also shows that CarmelStat exhibits a consistent surface resistivity in the static dissipating range. The consistency occurs over a wide range of glass fiber concentrations thus also assuring control of the other material properties.

Table 3 compares commercial conductive PP compounds with CarmelStat materials. The electrical properties data indicate that the other materials are in the conductive or static dissipative category. The mechanical properties data shown indicate that the CarmelStat composite is significantly stiffer and stronger than the other PP compounds, which results in

Table 3. Properties of CarmelStat composites compared with commercial compounds

Property	Carmel Olefins Carmel Stat CS1015	Carmel Olefins Carmel Stat CS1115	Premix Pre-Elec PP1370	Lati Latistat 57/7-02	Cabot Cabelec CA3839	RTP ESD-A -100	LNP Stat-Kon M HI
Specific gravity, kg/m^3	1.01	1.01	0.98	.94-.96	1.05	.97-1.01	1.00
Tensile strength, MPa	55	20	18	30	20	23.4	37
Flexural modulus, MPa	3300	2300	1200	2100	865	1241	1100
Izod impact, notched, J/m	57	125	24	25	18	425	500-700
Heat distortion temp. (045 MN/m^2), °C	160	155	90	115	56	127	87
Volume resistivity, ohm-cm	10^6-10^7	10^7-10^8	$<10^3$	10^2	10^4	10^3-10^9	10^2-10^4
Surface resistivity, ohm/sq	10^6-10^7	10^7-10^8	$<10^4$	10^2	10^7	10^3-10^{12}	10^2-10^4

lower particle shedding and better dimensional stability. Carbon powder is a particulate filler which reduces the mechanical strength of a thermoplastic base resin compound and thus lower CB concentrations are an advantage. Moreover, at high CB concentrations, as in the commercial compounds, the release of carbon particles against a counter-face, commonly called "sloughing", may make carbon powder compounds unsuitable for some applications.

INFLUENCE OF CARBON FIBERS (CF)

Addition of CF to the polymer/CB/glass fiber composites results in higher conductivities and enhancement of some mechanical properties.[12] Figure 4 shows a significant decrease in resistivity by incorporation of up to 17 wt% CF, to CarmelStat composites containing about 1.2 wt% CB and 20 wt% glass fibers.

INFLUENCE OF POLYPROPYLENE GRADE

To determine if changes in melt viscosity of the base polymer affect the composite conductivity, three melt flow index (MFI) grades of PP were compounded with 25 wt% glass fibers and 1 wt% CB. Figure 5 reveals that a higher MFI is useful for achieving lower resistivity. Figure 5 also shows that the flexural modulus is practically independent of MFI. CarmelStat composites were prepared with different PP types (homopolymer and copolymer). Table 4 shows that a conductive composite based on PP copolymer, demands a higher CB loading to reach a resistivity level similar to the homopolymer based composites. Thus, the presence of rubber, as shown in Table 4, increases the impact resistance, however the conductivity of the composite decreases. Carbon black in the rubber containing composites tends to concentrate more in

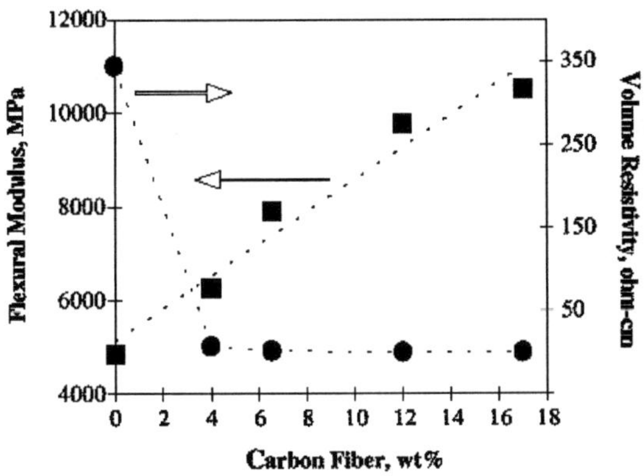

Figure 4. Resistivity and flexural modulus vs. carbon fiber content, for CarmelStat composites containing 1.2 wt% carbon black and 20 wt% glass fibers (4-point probe method).

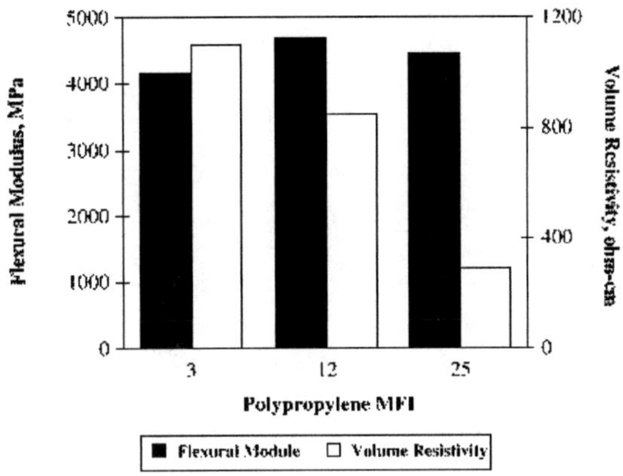

Figure 5. Resistivity and flexural modulus of CarmelStat composites containing 1.2 wt% carbon black and 25 wt% glass fiber for PP with different melt flow index (MFI).

Table 4. The influence of polypropylene type (homopolymer vs. copolymer) on surface resistivity and impact resistance of CarmelStat composites

	Homopolymer	Copolymer				
Carbon black, wt%	1.2	1.2	3.7	4.6	4.8	5.1
Glass fibers, wt%	15.4	15.4	10	15.4	10	10
Surface resistivity, ohm/sq	10^6	$>10^{12}$	10^8	10^6	10^5	10^4
Izod, notched, J/m	57	117	140	115	140	135

the elastomeric phase rather than forming conductive paths. The distribution of CB in PP/rubber compounds (40 wt% polymer, 20 wt% rubber, 40 wt% conductive filler) was studied by Haddadi.[13] This study has shown that part of the conductive filler transferred into the rubber phase and thus about 19 wt% CB was located in the rubbery phase of the compound.

CLEAN ROOM APPLICATION

Free particles adversely affect semiconductor performance. Particles added to the wafers during fabrication are the result of contact between the wafers and the wafer carriers. Reduction

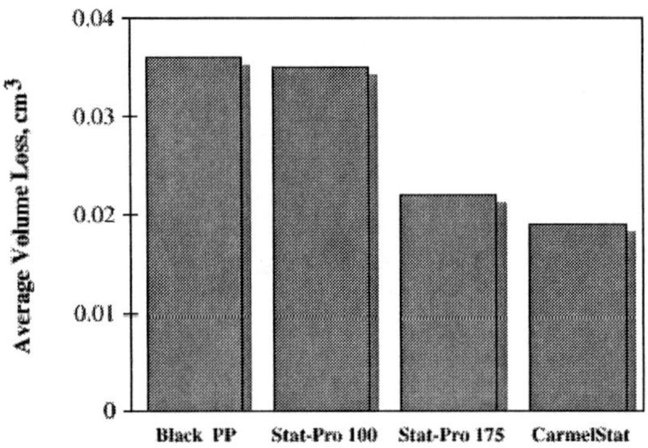

Figure 6. The volume loss, Taber abrasion, of a CarmelStat composite material compared with commercial compounds.

Table 5. Contamination levels of CarmelStat composites, PP containing 15 wt% glass fibers and 1 wt% carbon black (surface resistivity of 10^6 ohm/sq)

Parameter	Different composites		
	I	II	III
Leachable anions			
Cl, ppb	1680	280	120
SO_4, ppb	260	120	840
Metal composition			
Al, ppm	3900	470	53
Ca, ppm	9700	3200	280
Mg, ppm	150	45	8
Na. ppm	840	540	69

of the number of such released particles is obviously achieved by using wafer carriers that generate fewer particles. In our study, the Taber abrasion test was used to characterize the particle shedding performance of the CarmelStat materials. Figure 6 shows the volume loss of a CarmelStat material compared with some other conductive compounds for clean room applications.[14] The data from this Figure reveal that CarmelStat shows better abrasion resistance than the CB filled PP (black PP and Stat-Pro 100) compounds and has similar low particle generation as carbon fiber filled PP (Stat-Pro 175).

Clean room materials utilized in storage boxes and wafer carriers, should have the potential of least ion contamination. The concentrations of leachable anions and metal were investigated in different CarmelStat PP. From Table 5 it is obvious that compound III is the most suitable for clean room applications, since it has the lowest contamination level compared with the other composites.

CONCLUSIONS

Static dissipative injection moldable polymer composite materials have been characterized. Composites with consistent resistivity in the range of 10^6 - 10^9 ohms/sq can be prepared based on combining a number of polymeric materials with glass fibers and about 1 wt% CB.

Mechanical properties of CB filled thermoplastics can be tailored by reinforcing the polymer with glass fibers and by modifying the polymer with rubber. Such materials can be designed for applications in a variety of industries where control of static and contamination is required.

REFERENCES

1 S. Paul Singh and H. El-Khateeb, *Packaging Technology and Science,* 7, (1994).
2 K. Vakiparta, EOS/ESD Symposium, 229 (1995), Phoenix.
3 R. W. Cambell and W. Tan, EOS/ESD Symposium, 218 (1995), Phoenix.
4 M. Narkis, A. Ram and F. Flashner, *J. Appl. Polym. Sci.*, **22**, 1163 (1978).
5 M. Narkis, A. Ram and Z. Stein, *J. Appl. Polym. Sci.*, **25**, 1515 (1980).
6 M. Narkis and A. Vaxman, *J. Appl. Polym. Sci.*, **29**, 1639 (1984).
7 Carmona, The Second International Conference on Carbon Black, 213 (1993), Mulhouse.
8 Lee, *Journal of Vinyl Technology*, **15**, 173 (1993).
9 S. Petrovic, B. Martinovic, V. Divjakovic and J. Budinski-Simendic, *J. Appl. Polym. Sci.*, **49**, 1659 (1993).
10 R. Tchoudakov, O. Breuer, M. Narkis and A. Siegmann, *Polym. Networks and Blends*, **6**, 1 (1996).
11 M. Narkis, R. Tchoudakov, A. Siegmann and A. Vaxman, "Electrically Conductive Compositions and Methods for Producing Same", *USA patent application (1996).*
12 P. Johnes, I. Emami, J. Goodman and K. Mikkelsen, Microcontamination, January (1993).
13 V. Haddadi-Asl, *Iranian Polymer Journal*, **5**, 75 (1996).
14 J. Mikkelsen, Microcontamination, March, (1996).

Electrical Properties of Carbon Black-Filled Polypropylene/Ultra-High Molecular Weight Polyethylene Composites

Jiyun Feng and Chi-Ming Chan
Department of Chemical Engineering, The Hong Kong University of Science and Technology, Clear Water Bay, Kowloon, Hong Kong

INTRODUCTION

An important issue considered in the preparation of carbon black (CB)-filled conductive polymer composites is the CB content. To maintain balanced properties of the composites, the CB content must be kept as low as possible otherwise the processing of the composite is difficult, the mechanical properties are very poor, and the total cost is higher. Thus, the method of reducing CB content is a major research subject in this field.

It has been found that an effective way to reduce CB content in CB-filled conductive polymer composites is the use of binary immiscible polymer blends as polymer matrix.[1-8] In CB-filled immiscible polymer blends, CB particles can be controlled to be selectively localized in one polymer phase or at the interface of immiscible polymer blends due to different affinities of CB for different polymers during the processing of the composites. For instance, Gubbels[5] studied CB-filled PE/PS blends and selectively localized CB particles at the interface of PE/PS blends by kinetic control. The percolation threshold for the CB-filled PE/PS(45/55) composites is only 0.002 volume fraction of CB, which is much lower than that of CB-filled conventional polymer composites. This result also indicates that the use of immiscible polymer blends is a promising approach to prepare CB-filled conductive polymer composites with a very low percolation threshold.

In this investigation, CB-filled PP/UHMWPE composites were prepared by the conventional melt mixing method, the electrical properties of the composites including the PTC effect and the CB distributions in the PP/UHMWPE matrix were studied.

EXPERIMENTAL

The polymers used in this investigation were PP (Profax PD382 from Himont) and UHMWPE (Hifax1900 from Montell). The CB was V-XC72 from Cabot. The CB-filled PP/UHMWPE composites were prepared using a Haake mixer at 200°C and 30 rpm for 15 minutes. The mixed materials were further compressed into 2 mm thick sheets using a hot press at 200°C and 16 MPa.

The resistivity of the composites was determined by a multimeter. Prior to the measurements, the sample surfaces were coated with a layer of silver paint to eliminate the contact resistance. The resistivity of the composites versus temperature was determined by a computerized system, which comprises a multimeter, a computer and a programmable oven. The heating rate used was 2°C/min. The CB distributions in the composites were studied by optical microscopy using the transmission mode. Thin sections of 1 μm in thickness were cut by a cryomicrotome at -100°C.

RESULTS AND DISCUSSION

ELECTRICAL RESISTIVITY

The electrical resistivity of CB-filled PP/UHMWPE composites with a fixed CB content (10 wt%) as a function of UHMWPE content in the PP/UHMWPE matrix is depicted in Figure 1. Obviously, the UHMWPE content significantly influences the resistivity of the composites. As the UHMWPE content increases, the resistivity of the composites decreases remarkably, reaching a minimum value when the UHMWPE content attains 50 wt%, then increases when the UHMWPE content further increases. Basically, the addition of UHMWPE particles to the PP matrix influences the electrical resistivity of the composites in two aspects. Due to the fixed CB content and extremely high viscosity of UHMWPE particles, CB particles can only be localized at the interface between PP matrix and UHMWPE particles or in the PP matrix. Hence, the addition of UHMWPE in the composites results in an increase in the CB content in the PP matrix which is the continuous phase, leading to a decrease in resistivity. On the other hand, the addition of UHMWPE particles gives rise to a significant increase in viscosity and shear strength of the polymer melt. The strong shear strength can break down the secondary structure of CB, resulting in an increase in resistivity. Accordingly, the higher the amount of UHMWPE in the composites, the higher is the resistivity of the composites. This is consistent with the results when the UHMWPE content is higher than 50 wt% as shown in Figure 2. Thus, by the combination of the two effects, when the UHMWPE content is lower than or equals 50 wt% in the polymer matrix, the increase in CB content in the PP matrix dominates the electrical resistivity of the composites while when the UHMWPE content is higher than

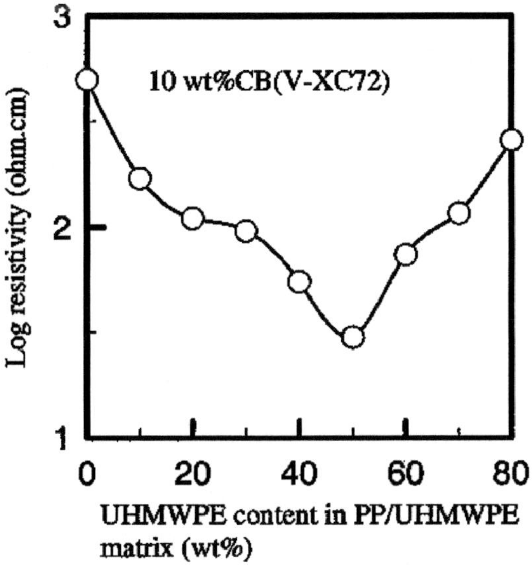

Figure 1. Log resistivity of CB-filled PP/UHMWPE composites vs. UHMWPE content in PP/UHMWPE matrix.

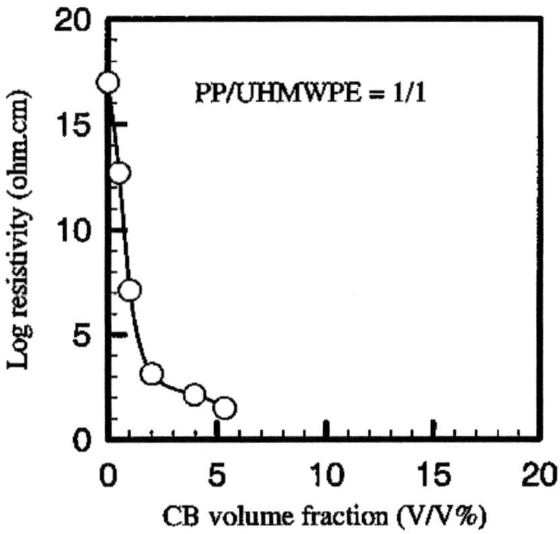

Figure 2. Log resistivity of CB-filled PP/UHMWPE composites vs. CB volume fraction.

50wt%, the break down of the secondary structure of CB dominates the electrical resistivity of the composites.

The effect of the CB content on the electrical resistivity of CB-filled PP/UHMWPE(1/1) composites is illustrated in Figure 2. Apparently, when the CB volume fraction reaches about 1 v%, a sharp decrease in resistivity occurs. Using the percolation theory, the percolation threshold of the composites is determined to be 0.8% volume fraction of CB with the best fit. This value is much lower than that of conventional CB-filled single polymer composites. Thus, PP/UHMWPE (1/1) matrix is a good polymer matrix for the preparation of CB-filled polymer composites with a very low percolation threshold by the conventional melt-mixing method.

PTC EFFECT

In addition to electrical resistivity, the PTC effect is another important electrical property of CB-filled semicrystalline composites. Figure 3 shows the log resistivity of CB-filled PP/UHMWPE (1/1) composites with various CB contents as a function of temperature. Clearly, two sharp jumps in resistivity are observed at about 140 and 170°C. This is a novel physical phenomenon, which is different from the PTC behavior of CB-filled single semicrystalline polymer composites. This effect is referred to as the double-PTC effect. The

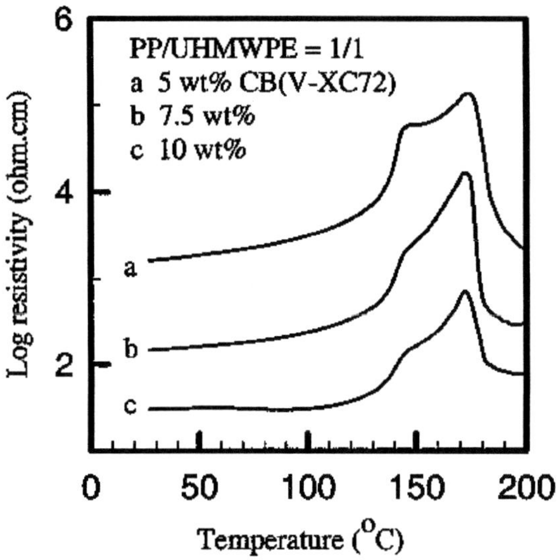

Figure 3. Log resistivity of CB-filled PP/UHMWPE (1/1) composites vs. temperature.

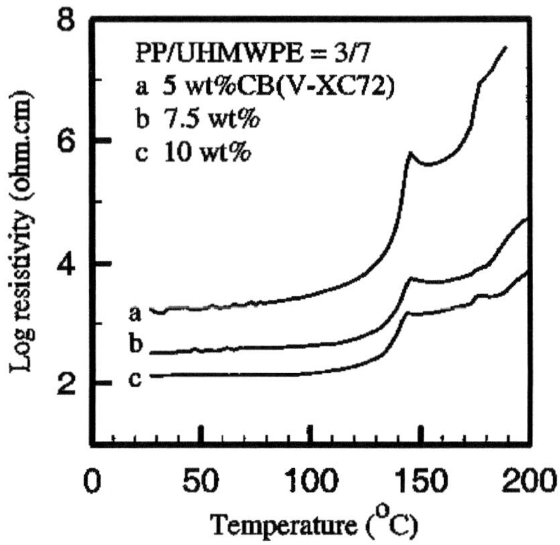

Figure 4. Log resistivity of CB-filled PP/UHMWPE (3/7) composites vs. temperature.

first jump in resistivity is defined as the first PTC effect and the second jump in resistivity as the second PTC effect. According to the mechanism of the PTC effect of CB-filled single semicrystalline polymer composites, the first PTC effect and the second PTC effect are attributed to the large thermal expansion as a result of the melting of UHMWPE and PP crystallites, respectively. These results also illustrate that the formation of conductive networks is related to the UHMWPE particles in the composites. One possibility is that CB particles may accumulate around the UHMWPE particles especially when the CB content is low due to the thermodynamic driving force.[5] In this case, as the temperature increases to the melting point of UHMWPE, the UHMWPE particles melt and expand significantly, breaking down most conductive pathways around them and resulting in a sharp jump in resistivity, as shown in Figure 3. As CB content increases, CB particles are forced to be dispersed in the PP matrix when all interfaces between PP and UHMWPE particles are occupied by CB particles. In this case, most conductive pathways are formed in the PP matrix. This is the reason for the fact that as CB content increases, the first PTC effect of the composites becomes weaker. It should be noted that after the second PTC effect, a sharp negative temperature coefficient (NTC) effect occurs. This NTC effect is due to the formation of new conductive pathways which is similar to that of CB-filled single semicrystalline polymer composites without crosslinking.

To study the effect of the PP/UHMWPE weight ratio on the PTC effect of CB -filled PP/UHMWPE composites, the composites with a low PP/UHMWPE weight ratio (3/7) and

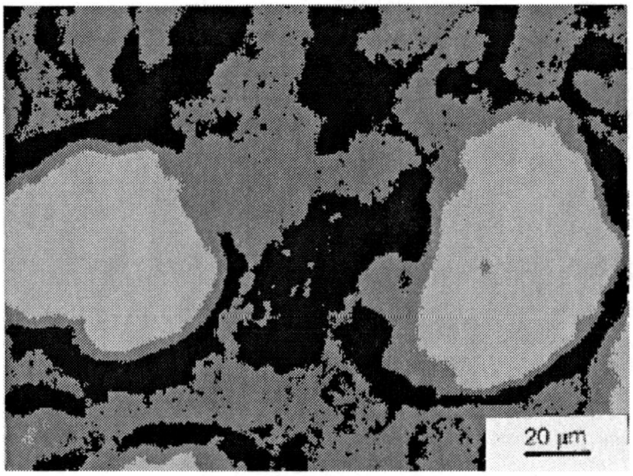

Figure 5. An optical micrograph of 1.0 wt% CB-filled PP/UHMWPE (1/1) composite.

various CB contents were studied. Figure 4 depicts the log resistivity of the composites versus temperature. Clearly, the double-PTC effect is also observed even though the second PTC effect becomes weaker as the CB content increases. The reason for this fact is that as the PP/UHMWPE weight ratio significantly decreases, the PP content also decreases. Hence, the thermal expansion due to the melting of PP crystallites decreases, resulting in a weaker second PTC effect. A new and important observation for the composites lies in the fact that they do not exhibit any NTC effects, which is different from that of the composites with a high PP/UHMWPE weight ratio. The reason for the elimination of NTC effect may be due to the very high concentration of CB in the PP matrix as the PP/UHMWPE weight ratio is lower, which results in a high viscosity of the composites, hindering the formation of new conductive pathways and eliminating the NTC effect.

CB DISTRIBUTIONS IN THE PP/UHMWPE MATRIX

To determine the CB distributions in the PP/UHMWPE matrix, optical microscopy was used. Figure 5 displays the optical micrograph of 1.0 wt% CB-filled PP/UHMWPE (1/1) composites. Clearly, a two-phase structure is observed. The light areas are identified as the UHMWPE particles due to the fact that CB particles cannot enter UHMWPE particles as a result of extremely high viscosity of the UHMWPE. These UHMWPE particles form the dispersed phase. While the grey areas are identified as the PP matrix and they form the continuous phase. It is interesting that around the UHMWPE particles, dark rings are observed, indicating that CB particles accumulate at the interface between PP matrix and UHMWPE particles. This observation is in agreement with the above discussions. Thus, this result confirms that CB particles are not only localized in the PP matrix but also accumulate at the interface between PP matrix and UHMWPE particles.

According to Sumita's results,[3,4] when CB particles are mixed with PP/HDPE blends, they are selectively localized in the HDPE phase. Thermodynamically, CB particles have stronger affinity for HDPE than for PP. In the case of CB-filled PP/UHMWPE composites, although thermodynamically CB particles preferentially enter UHMWPE particle, in practice they can only reach and accumulate at the interface of UHMWPE particles and cannot go inside the UHMWPE particles due to their extremely high viscosity of UHMWPE. Finally, CB-rich regions are formed at the interface between PP matrix and UHMWPE particles. The results suggest that by the combination of kinetic and thermodynamic factors, the selective localization of CB particles in immiscible polymer blends can be tailored and obtained.

CONCLUSIONS

Our results indicate that the PP/UHMWPE weight ratio and CB content are the two main factors that can significantly influence the electrical resistivity of the CB-filled PP/UHMWPE composites. At a fixed CB content (10 wt%), a minimum value in resistivity is observed at the PP/UHMWPE weight ratio of 1/1. At a fixed PP/UHMWPE weight ratio (1/1), a percolation threshold of 0.008 volume fraction of CB is observed which is much lower than that of conventional CB-filled polymer composites.

The measurements of resistivity of the composites versus temperature depict that the stronger double-PTC effects are observed at low CB content and low PP/UHMWPE weight ratio. The melting of UHMWPE crystallites at about 140°C and the melting of PP crystallites at about 170°C are responsible for the first PTC effect and the second PTC effect, respectively.

The results of the optical microscopy studies illustrate that CB particles are selectively localized at the interface between PP matrix and UHMWPE particles or in the PP matrix due to the extremely high viscosity of UHMWPE.

ACKNOWLEDGMENT

This work was supported by UPGC Research Infrastructure Grant under Grant No. RI93/94E.

REFERENCES

1 B. Wessling, *Kunststoffe*, **76**, 930 (1986).
2 G. Genskens, E. De Kezel, S. Blacher and F. Brouers, *Eur. Polym. J.*, **27**, 1261 (199 1).
3 M. Sumita, K. Sakata, S. Asai, K. Miyasaka, and H. Nakagawa, *Polym. Bull.*, **25**, 265 (1991).
4 M. Sumiata, K. Sakata, H. Nakagawa, S. Asai, K. Miyasaka, and M. Tnaernura, *Colloid Polym. Sci.*, **270**, 134 (1992).
5 F. Gubbels, R. Jermore, Ph. Tessie, E. Vanlathern, R. Detour, A. Calderon, V. Parente, and J. L. Breads, *Macromolecules*, **27**, 1972(1994).
6 M. Narkis, R. Tchoudakov, and O. Breuer, ANTEC'95 P1343-1346.
7 R. Tchoudakov, O. Breuer, M. Narkis, and A. Siegaman, *Polym. Eng. Sci.*, **36**, 1336(1996).
8 J. Feng and C-.M Chan, *Polym. Eng. Sci.* In press.

The Use of Conducting Polymer Composites in Thermoplastics for Tuning Surface Resistivity

Sam J. Dahman
RTP Company, Winona, MN
Jamshid Avlyanov
Eeonyx Corporation, Pinole, CA

INTRODUCTION

Electrostatic discharge (ESD) is an issue well known to those in the electronics industry.[1] ESD can lead to expensive catastrophic and latent failures in sensitive electronic devices. Losses due to ESD events have been estimated into the billions of dollars.[2]

With rapid miniaturization, electronic components have become increasingly susceptible to ESD. Furthermore, plastics have become the materials of choice in the manufacturing and packaging of electronics. Unfortunately plastics are insulators, which increases the likelihood of ESD events. Sufficient charge is then able to accumulate with the possibility of a sudden discharge causing damage to sensitive electronic components. Thus, there is a need for plastics that control ESD.

Static dissipative thermoplastics have successfully met the requirements for ESD control in a number of applications for the electronics industry. These applications include but are not limited to tote boxes, matrix trays, chip carriers, cassettes, card enclosures, fibers, etc. Static dissipative thermoplastics offer many advantages for ESD protection over other materials such as metals. For instance, finished devices are typically lighter in weight, more aesthetic, easier to fabricate, and less expensive.

The ideal static dissipative thermoplastic dissipates charges as they are generated in a controlled manner. At any point on the surface there is no accumulation of charge, which could cause hazardous discharge. This means that the thermoplastic should be at least somewhat conductive. However, very high surface conductivities lead to uncontrolled discharges. For many ESD applications the optimal surface resistivity resides from 10^6 to 10^9 ohms/square.[3]

A static dissipative thermoplastic is a material in which a conductive additive is incorporated into a thermoplastic resin matrix. Conventional conductive additives for ESD applications have included antistatic agents, carbon black, carbon or graphite fibers, and metal powders or fibers. Unfortunately, it is difficult to provide consistent surface resistivity in the ideal range utilizing traditional additives. These compounds are either not conductive enough or exhibit a sudden increase in conductivity that extends beyond the ideal range once a critical concentration is reached. An example of this type of behavior is shown in Figure 1.

A new class of conductive additives has been developed through *in-situ* deposition of intrinsically conducting polymers (ICP) onto carbon black substrates. The resulting ICP composites are thermally stable. Conductive thermoplastics were produced using the ICP composites in a number of matrix resins without degradation or conductivity loss.[4]

A significant advantage of the ICP composites is their inherent flexibility. The type of ICP, degree of deposition, dopant, pH, and carbon black substrate may all be varied. This flexibility allows selected properties to be tuned to desired values. Through this flexibility, it is possible to produce static dissipative thermoplastics with surface resistivities tuned into the ideal range for ESD applications.

EXPERIMENTAL

Various ICP composites were prepared as described elsewhere.[5] Commercially available thermoplastics were used as matrix resins. Conductive thermoplastics were manufactured by first physically mixing the ICP composites with matrix resin pellets and then compounding using a twin-screw compounding extruder. Compound melt flow behavior was analyzed with a capillary rheometer.

Test specimens were produced via injection molding. The mechanical and electrical properties of the specimens were tested according to ASTM test methods D-638 for tensile strength and flexural modulus, D-256 for Izod impact, and D-257 for volume and surface resistivity. Matrix resins and compounds were dried according to established resin guidelines prior to compounding, rheometry, and molding.

Thermal stability was investigated by aging at elevated temperatures. Test specimens were aged using standard air circulating ovens at 80, 100, and 130°C. Specimens were equilibrated at 25°C and 50% relative humidity prior to testing.

RESULTS AND DISCUSSION

A mechanical performance comparison of various conductive nylon-6 compounds relative to unfilled resin is given in Table 1. Trends in mechanical properties often associated with particulate filled materials are lower tensile strengths, greater stiffness, and decreased impact resistance. These trends are also observed with these conductive compounds. However, the

Table 1. Performance comparison of conductive nylon-6 compounds (mechanical properties relative to neat resin)

	Carbon black	ICP composite		
		A	B	C
Tensile strength, %	45.5	62.2	74.9	69.3
Flexural modulus, %	92.0	126	108	116
Izod impact, notched, %	43.2	64.8	46.4	74.4
Volume resistivity, ohm-cm	4.12×10^1	8.92×10^0	1.38×10^5	2.59×10^9

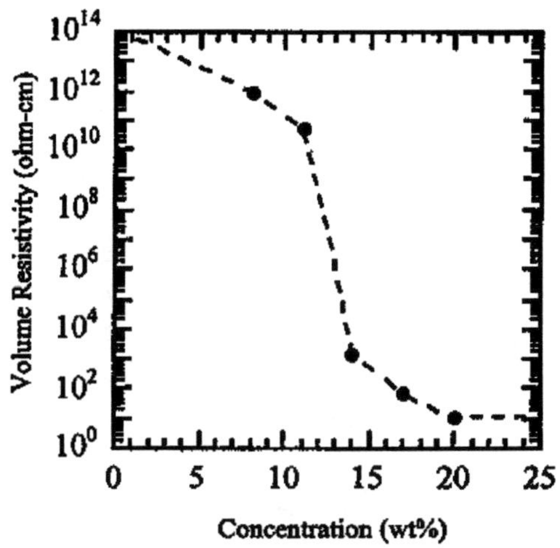

Figure 1. Volume resistivity as a function of carbon black concentration in nylon.

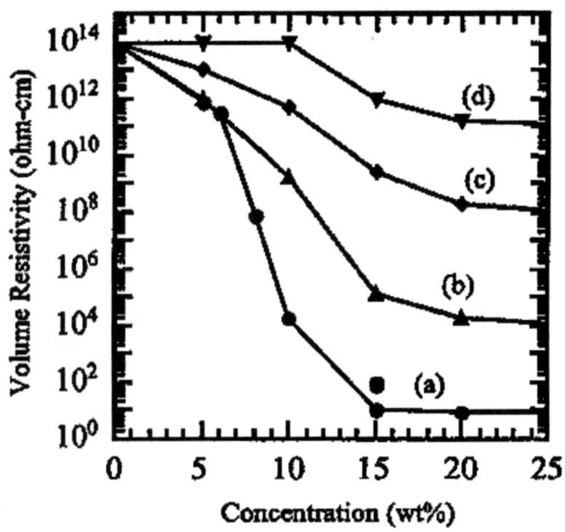

Figure 2. Volume resistivity of nylon-6 compounds as a function of ICP composite concentration. ICP composite (a) A, (b) B, (c) C, (d) D.

compounds containing ICP composites outperform the unmodified carbon black material. The increase in mechanical performance is most likely due to a more complete wetout and greater adhesion between the conductive particles and the matrix polymer. This is also indicated by the observation that compounds containing ICP composites exhibit reduced particle generation or "sloughing" compared to standard carbon blacks when exposed to abrasion.

It is widely known that the resistivity does not decrease with carbon black concentration linearly. Generally, what is observed is a sudden decrease in the resistivity once a critical concentration is reached. On either side of the discontinuity, only gradual decreases in the resistivity are seen. It is difficult to provide controlled and consistent resistivities in the criti-

Table 2. Surface resistivities of nylon-6 compounds

	Surface resistivity, ohm/sq
ICP composite A	$<10^4$
ICP composite B	2.1×10^6
ICP composite C	1.0×10^8
ICP composite D	1.7×10^{10}

cal region since minor changes in bulk additive levels and local fluctuations can cause large shifts in the resistivity.

Volume resistivities for the conductive compounds are also listed in Table 1. Note that the volume resistivities vary for different ICP composites. Volume resistivities of various ICP composites in nylon-6 as a function of concentration are given in Figure 2. ICP composite A is the most conductive and displays similar behavior to traditional carbon blacks with a steep percolation curve. ICP composites B, C, and D become less conductive and demonstrate a more controllable behavior since the percolation curves are less steep over a wider range of additive concentration.

The surface resistivities of certain nylon-6 compounds with ICP composites are given in Table 2. Usual definitions to characterize materials based on surface resistivity are: antistatic (10^9 to 10^{12} ohm/sq), static dissipative (10^5 to 10^{12} ohm/sq), and conductive (less than 10^5 ohm/sq). With the flexibility of ICP composites, it is possible to tune the surface resistivity of the compound into any desired classification. It is also possible to achieve the optimal surface resistivity range (10^6 to 10^9 ohm/sq) required for many ESD applications.

It is important to note that the surface resistivity of the static dissipative compounds with ICP composites is consistent from specimen to specimen and from location to location on a specimen as observed during spot checking. This is partly due to the reduced steepness of the percolation curves. This minimizes variation caused from bulk or local concentration differences. The consistent behavior is also a result of the additives themselves and not the additive level being adjusted to provide the desired surface resistivity. Thus, a well-defined network throughout the matrix can be achieved. This reduces fluctuations caused by local conductive and insulative regions.

It is generally accepted that conductive compounds filled with carbon black exhibit higher viscosities than with other types of fillers. This increase in viscosity is due to the higher surface area of carbon blacks. As a result, it is often difficult to compound carbon blacks with high surface area into thermoplastics. A key side effect of the deposition of ICPs onto carbon black surfaces is the reduction of surface area and pore volume.[6]

Compounds incorporating ICP composites were found to process much easier than compounds containing standard carbon blacks. Evidence of easier processing can be found in Figure 3. An ICP composite has a lower viscosity than the corresponding untreated carbon black. Figure 3 shows that the viscosity is lower even when the ICP composite is used at higher loading levels than carbon black.

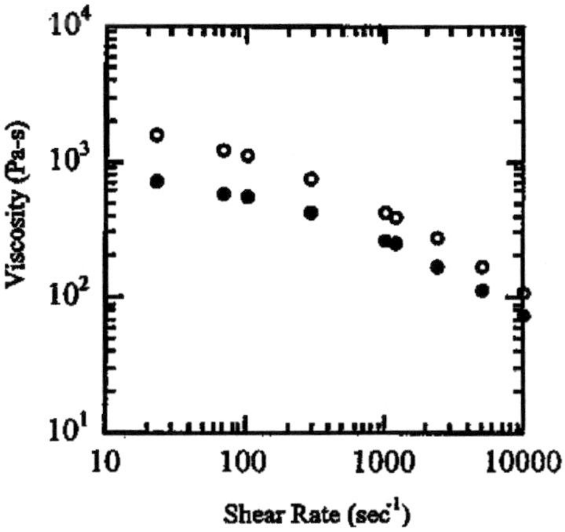

Figure 3. Melt flow behavior of conductive nylon-6 compounds: O 5 wt% high structure carbon black, ● 7.5 wt% of ICP composite A.

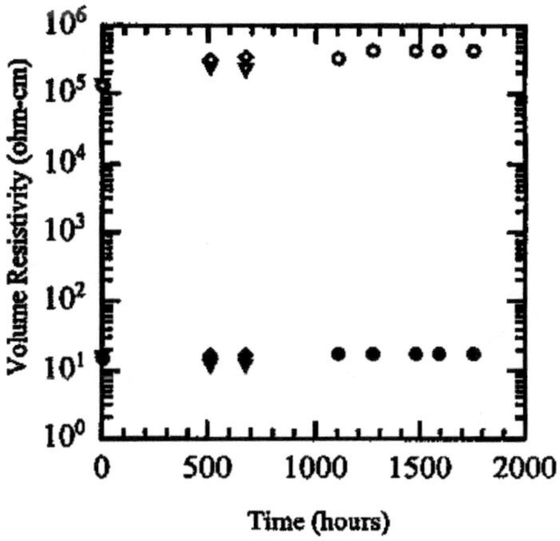

Figure 4. Volume resistivity of nylon-6 compounds as a function of time for (filled) ICP composite A and (unfilled) ICP composite B at elevated temperatures: (O) 80°C, (∇) 100°C, and (◇) 130°C.

One of the main limitations of ICPs is poor stability especially at elevated temperatures. This is generally seen as a reduction in conductivity with increased time of storage or use, particularly in air. It has been shown that the conductivity of polyaniline and polypyrrole can deteriorate by more than 50% within 24 hours at 120°C.[7,8]

ICP-carbon black composites have shown excellent thermal stability in both TGA analysis and when heated in air at 300°C.[4] Specimens molded from compounds containing ICP composites also display excellent thermal stability as shown in Figure 4. The resistivity has not changed in over 1700 hours for compounds with ICP composites at 80°C. Similar behavior has been observed for both 100 and 130°C. At these temperatures, the thermal aging has been proceeding for over 600 hours.

CONCLUSIONS

A new class of conductive additives for thermoplastics has been developed through *in-situ* deposition of ICPs onto carbon black substrates. A major advantage of these ICP composites is their inherent flexibility. This flexibility was utilized to obtain consistent surface resistivity and to tune the surface resistivity into the optimal range of many ESD applications. These materials also display improved mechanical properties over unmodified carbon black filled compounds. An additional benefit of using ICP composites is the ease of processing as compared to conventional conductive additives. Furthermore, these compounds exhibited excellent thermal stability.

REFERENCES

1 J. L. Sproston, Electrostatic Damage in the Electronics Industry, Proceedings of the Static Electrification Group of the Institute of Physics, IOP Publishing Ltd., England, 1987.
2 S. P. Singh and H. El-Khateeb, *Packaging Tech. Sci.*, **7**, (1994).
3 K. Vakiparta et. al., EOS/ESD Symposium, 229, (1995).
4 S. Dahman and J. Avlyanov, ANTEC Proceedings, 1313, (1998).
5 U.S. Patent 5,498,372, (1996).
6 J. Avlyanov and S. Dahman, ACS Symposium Series, Dallas, (1998).
7 L. Olmedo, P. Hourquebie, and P. Buvat, ANTEC Proceedings, 952, (1997).
8 H. Kuhn and A. Child, **Handbook of Conducting Polymers**, 2nd Ed., T. Skotheim, R. Elsenbaumber, and I Reynolds, eds., 993, *Marcel Dekker*, New York, 1998.

Monosandwich Injection Molding: Skin-Core-Structure and Properties of Sandwich-Molded Anti-electrostatic Components

K. Kuhmann, G.W. Ehrenstein
Lehrstuhl für Kunststofftechnik, Am Weichselgarten 9, D-91058 Erlangen, Germany

INTRODUCTION

With the sandwich injection molding process the combination of different thermoplastic materials in sandwich structures is possible. One material forms the outer skin layer while a second material forms the inner core layer of the sandwich part.

In the beginning the sandwich injection molding technology was used to combine a foamed core component and compact skin layers to thick walled parts.[1] In recent years further applications have been developed, using the skin material for sufficient surface properties (e.g. appearance, hardness, thermal or chemical resistance, soft touch) while the core material may contribute to sufficient mechanical properties or lead to cost reduction (use of recycled materials).[2,3]

A further application potential of the sandwich molding process is represented by the production of anti-electrostatic sandwich parts. The use of conductive, anti-electrostatic surfaces is essential in the production of electronic parts e.g. for anti-electrostatic chip carriers.[4] In this case the sandwich-molding technology allows to combine a conductive filled skin material with a cheaper unfilled core material based on the same material for good skin-core-adhesion.

The anti-electrostatic sandwich parts are characterized by the distribution of the unfilled core component and the filler distribution in the skin layer. In order to achieve an anti-electrostatic surface the conductance has to reach approx. 10^{-10} S and must not exceed approx. 10^{-7} S.[4]

MONOSANDWICH INJECTION MOLDING OF ANTI-ELECTROSTATIC SANDWICH PARTS

STANDARD SANDWICH AND MONOSANDWICH PROCESS

The standard sandwich molding process is performed by two-component injection molding machines, using separate injection units for each of the two components. This process is characterized by a sequential injection of the two components using the same gating system. After filling the mold partially with the skin matcrial with one injection unit the core component will be injected by the second injection unit. In order to avoid flow marks on the part surface a simultaneous phase is inserted between the injection of skin and core components. The final packing phase may be performed either with the core or the skin component.

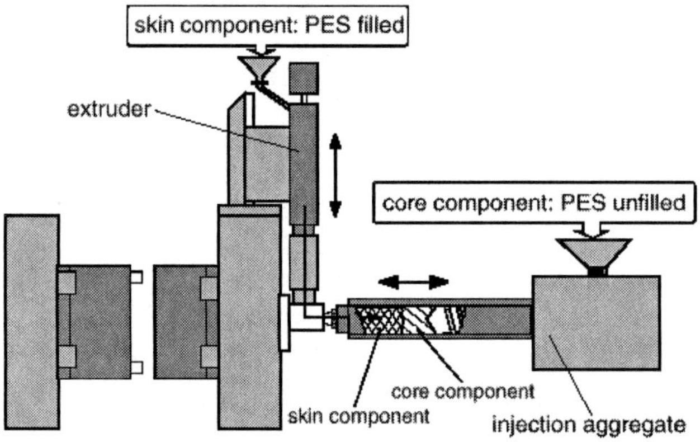

Figure 1. Sketch of a monosandwich injection unit.

In 1992, a new sandwich injection molding concept, called monosandwich, was developed by Jaroschek and Thoma (Ferromatic Milacron).[5] In contrast to the standard sandwich molding process, this technology uses only one injection unit for both skin and core components, Figure 1. The cycle starts with the plastification of the core component in the injection unit. Then the extruder moves to the bottom position, the injection unit moves forward to the extruder nozzle to link the nozzles of the extruder and the injection unit. The extruder starts plastification of the skin component and extrudes the melted skin component into the screw antechamber of the injection unit. Thus the skin and core components are located one after the other in the screw antechamber. After the extruder moved back to the top position, the injection unit moves forward to the mold followed by a conventional filling phase. Due to the fountain flow effect the first injected material forms the skin layer followed by the second component forming the core. Compared to the standard sandwich process the injection phase of the monosandwich process is less complicated as it is identical to the conventional injection molding process.

PARAMETERS INFLUENCING THE SKIN-CORE DISTRIBUTION

The distribution of the skin and core components is influenced by the material properties, the injection or process parameters and the part and gate geometry. These parameters have been examined for the standard sandwich molding process in different works [e.g.[3,6]]. Especially the viscosity ratio (core component viscosity/skin component viscosity) influences the core distribution. Viscosity ratios below 1 may tend to two-component flow instabilities resulting in uneven distribution of the core component. In addition the process parameters in sandwich molding e.g. melt and mold temperatures and the injection velocity influence the material flow behavior and the cooling conditions in the mold and thus influence the core thickness distribution as well as the shape of the skin-core-interface. Regarding the part geometry especially the filling of ribs and gaps has to be considered.[3]

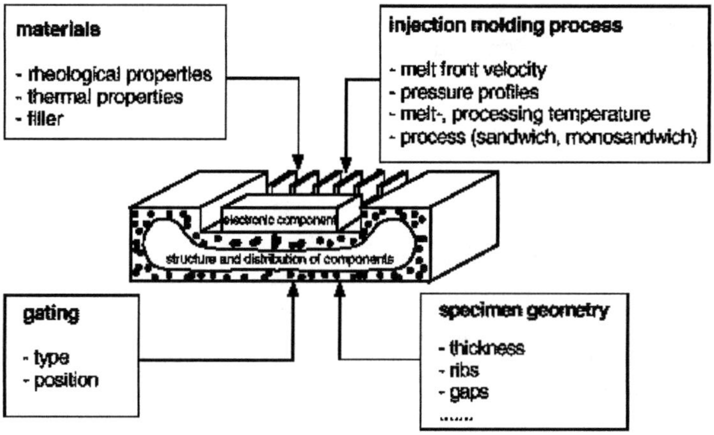

Figure 2. Parameter influencing sandwich molding.

In this work the influence of these parameters is examined for the monosandwich process using a conductive filled skin material and unfilled core material to find limiting construction, process and material parameters for anti-electrostatic sandwich parts. Figure 2 indicates a schematic sandwich structure to carry electronic components sensitive to electrostatic charges.

EXPERIMENTAL

The examination of the monosandwich molding process was carried out with a high temperature resistant thermoplast combination of two different viscous PES grades. The skin component Ultrsaon® E1010 is a low viscous PES (melt volume rate MVR 360/10 = 140 cm^3/10 min), the core component Ultrason E3010 is a higher viscous PES grade. This material combination was selected in order to avoid viscosity ratios to far below 1 as the viscosity of the skin component increases rapidly when adding the conductive filler.

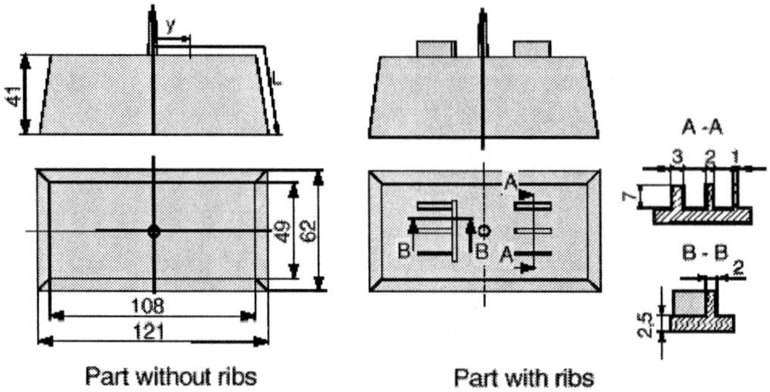

Figure 3. Part geometry.

Table 1. Variation of parameters (parts without ribs)

	Low settings	Reference settings	High settings
Filler content, wt%	30 15	45	-
Mold temperature, °C	118	140	158
Injection speed, mm/s	25	50	75 100
Melt temperature, °C	-	360	-

Table 2. Variation of parameters (parts with ribs)

	Low settings	Reference settings	High settings
Filler content, wt%	-	45	-
Core part, %	42	47	52
Mold temperature, °C	-	140	-
Injection speed, mm/s	25	50	100
Melt temperature, °C	340	360	-

The conductive filler consists of small mica plates (diameters 5 to 15 μm, thickness about 0.5 μm) coated with metaloxides based on tin, silicon and titanium. This filler type is especially suitable to reach defined, low conductance for anti-electrostatic applications.[4]

Two testing geometries (parts with and without ribs) were used given in Figure 3. Non-ribbed parts were used to examine different filler amounts. The surface conductance of the parts has been measured using plate electrodes according to DIN VDE 0303 (Voltage 100 V).

The variation of the parameters for monosandwich injection molding was conducted according to settings given in Tables 1 and 2 (monosandwich injection molding machine by Ferromatic Milacron, screw diameter of the injection unit 35 mm). The melt temperatures of the skin and core components were kept equal. Proceeding from the reference setting the parameters were varied separately to higher and lower settings. Different filler contents were mixed into the PES skin component using a twin screw extruder (screw diameter 34 mm) Leistriz LSM 3034. In addition, the amount of the core component was varied using the geometry with ribs.

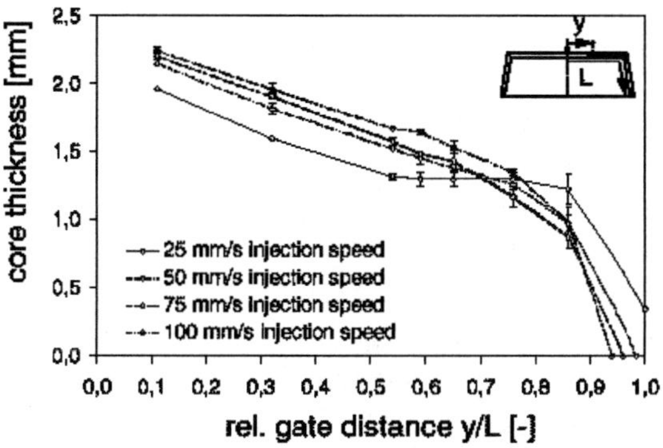

Figure 4. Distribution of core thickness as a function of injection speed.

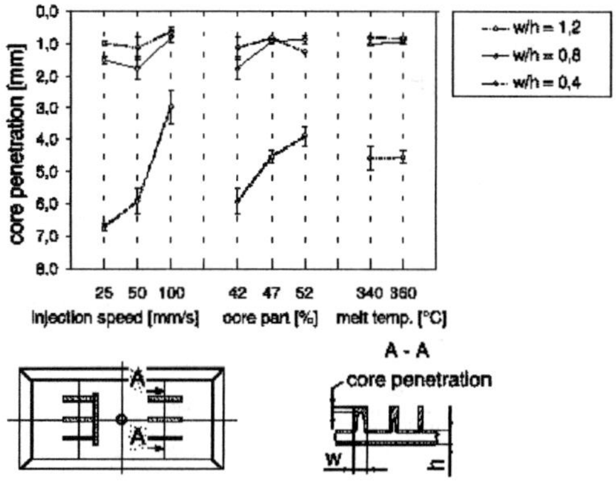

Figure 5. Distribution of core component in single ribs as a variation of molding parameters (cross-section A-A).

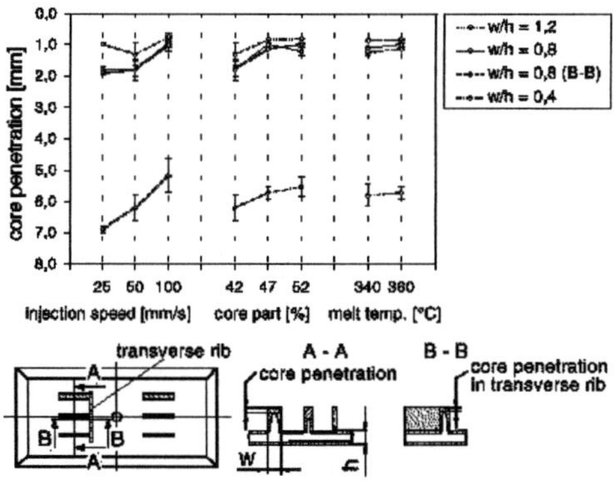

Figure 6. Distribution of core component in connected ribs at a variation of molding parameters.

The core thickness distribution in the part, the core penetration in the ribs and the filler distribution in the skin layer were evaluated by microscopical inspection of fracture and ground surfaces.

RESULTS AND DISCUSSION

CORE THICKNESS DISTRIBUTION AND REACH

The distribution of the core thickness observed with the non-ribbed test geometry is given in Figure 4 for different injection velocities. With decreasing injection velocity the core thickness distribution becomes more even and the core material reach increases. The core thickness decreases near the gate and increases in the area far from the gate. At the lowest injection speed of 25 mm/s the core reaches the maximum flow length and penetrates the skin component. Similar to observations in the standard sandwich molding process other process parameters e.g. melt or mold temperatures show only slight influences on the core thickness distribution compared to the effect of the injection velocity.

Decreasing injection velocities lead to increasing growth of the frozen skin layer during the filling process resulting in increasing consumption of skin component along the flow path. This effect is superposed by the change of the viscosity ratio because the filled skin

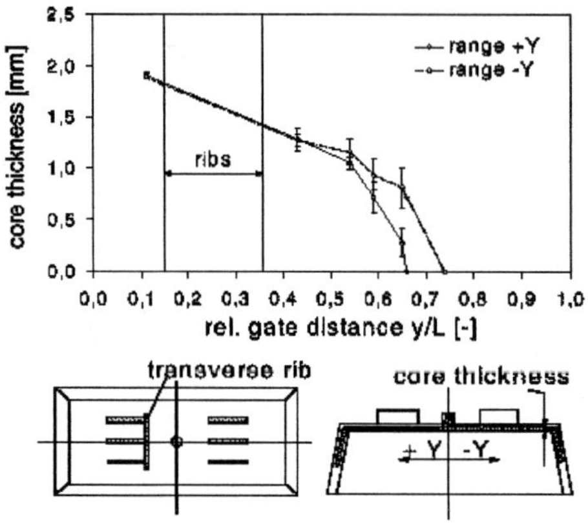

Figure 7. Influence of ribs on the distribution of core component.

component and the unfilled core component show different shear thinning behavior. Viscosity measurements in a capillary rheometer revealed decreasing viscosity ratios below 1 with increasing shear velocity for the examined material combination (45 wt% filler in skin component). However these effects cannot be experimentally separated. In spite of the low viscosity ratios no instable flow effects could be observed in the mono-sandwich experiments.

The core penetration in ribs parallel and perpendicular to the flow direction is illustrated in Figures 5 and 6. It becomes obvious, that the core penetration reaches a maximum at the 2 mm thick ribs (ratio w/h=0.8). Using 3 mm thick ribs (w/h=1.2) does not improve the core penetration. Thin ribs (w/h=0.4) contain nearly no core component at low injection velocities. However the core penetration in the single thin rib is significantly improved by increasing injection velocities. The prefilling of the thin rib with skin component seems to be reduced by the increasing injection velocity.

When the ribs are connected by a transverse rib the core penetration into the thin rib becomes worse, Figure 6. Hardly the filling of the thicker ribs is influenced. It is remarkable, that the core penetration into the transverse rib (perpendicular to the flow direction) is nearly the same compared with the rib (w/h=0.8) parallel to the flow direction, Figure 6. Finally the transverse rib has negative influence on the core penetration into the area behind the ribs, Figure 7. The core thickness and the reach decrease along the flow path compared with the area behind the single ribs.

SURFACE CONDUCTANCE AND FILLER DISTRIBUTION

In order to determine the necessary filler amount to achieve a sufficient surface conductance, different amounts of filler have been added to the skin component. As shown in Figure 8 a distinct increasing conductance of the sandwich molded parts is achieved at filler contents above 30 wt%. Additional experiments with different filler contents from 40 wt% up to 50 wt% indicate the development of a percolation filler network responsible for the conductance. A filler content of 45 wt% leads to the desired conductance above 10^{-10} S hitting the target range of 10^{-10} S to 10^{-7} S.

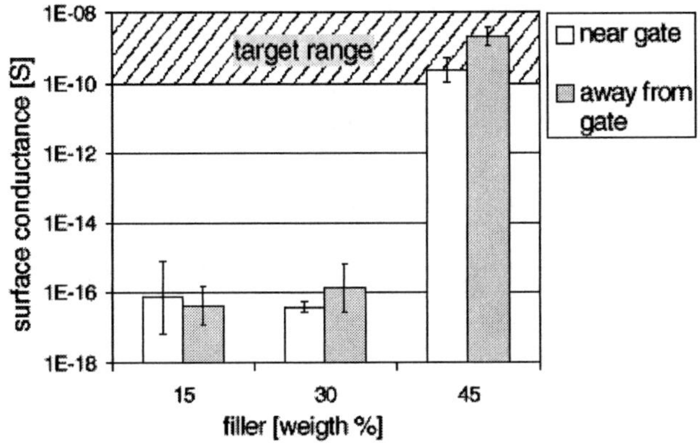

Figure 8. Surface conductance of sandwich molded parts as a function of filler content.

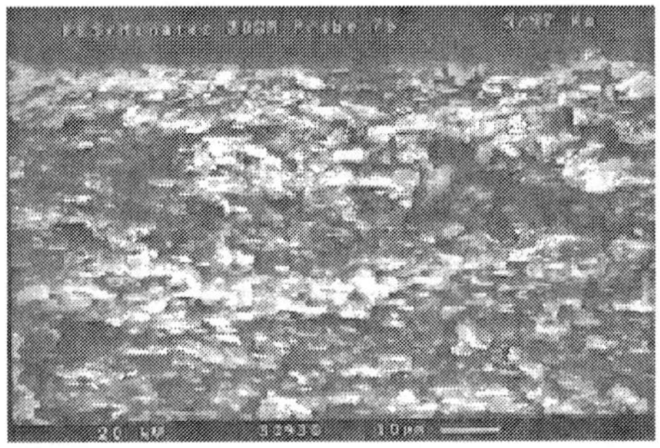

Figure 9. Filler distribution on the part surface (fracture surface, magnification 1000).

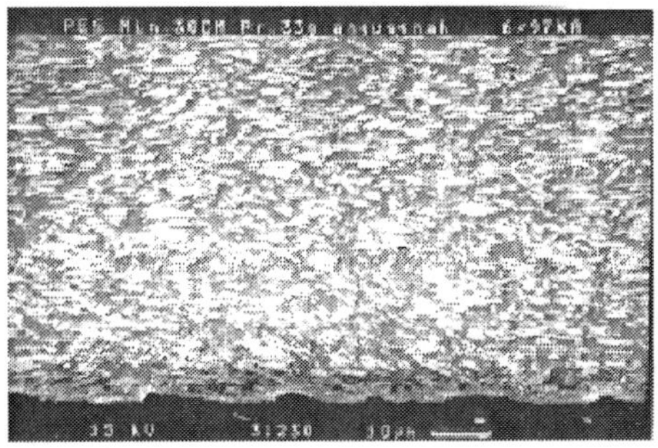

Figure 10. Filler distribution on the part surface (ground surface, magnification 1000).

The filler distribution of the percolation network in the outer skin layer at 45 wt% filler content is shown in Figures 9 and 10. One realizes a slight wavelined filler orientation with predominating orientation into flow direction. Variations of the process parameters showed nearly no influence on this characteristic orientation distribution near the part surface. The fracture surface in Figure 9 reveals a poor filler matrix adhesion leading to brittle failure of the skin component under mechanical load.

CONCLUSIONS

The monosandwich injection molding experiments showed the PES-combination with 45 wt% filler content being suitable to produce anti-electrostatic sandwich components due to development of a low conductive filler percolation network in the skin layers. In accordance with the standard sandwich process the injection velocity reveals the most important process parameter influencing the core distribution, the surface-near filler distribution is hardly influenced. The resulting small viscosity ratios do not cause flow instabilities effecting the core distribution. In addition the rib geometry has to be considered for sandwich part design, thin ribs (w/h << 0.8) show poor core penetration.

ACKNOWLEDGMENTS

The authors would like to thank the German Federation for Research (DFG) for supporting this research project. The support with materials by BASF AG and Merck and with machinery equipment by Ferromatic Milacron GmbH is gratefully acknowledged.

REFERENCES

1 E. Escales, Das ICI-Sandwich-Spritzgießverfahren. *Kunststoffe*, **60** (1970) 11, S. 847-852.
2 A. Sternfield, Where they are using sandwich molding and why. *Modern Plastics International*, (1983) 6, pp. 25-27.
3 T. Zipp, Fließverhalten beim 2-Komponenten-Spritzgießen. Dissertation RWTH Aachen, 1992.
4 H. Münstedt, U. Brendel, Permanent antielektrosta-tische Kunststoffe. *Kunststoffe*, **86** (1996) 1, S. 73-78.
5 C. Jaroschek, Spritzgießen von Formteilen aus meh-reren Komponenten. Dissertation RWTH Aachen, 1994.
6 F.A. Eigel, G.R. Langecker, The Sandwich Injection Moulding Technology - Effects and Limiting Quantities for the Spatial Distribution of the Components. SPE Proceedings, 55th Annual Technical Conference, Toronto 1996, pp. 456-460.

Thermoformed Containers for Electrostatic Sensitive Devices

Walter E. Gately
Ex-Tech Plastics, Incorporated

BACKGROUND

People and objects in motion can generate surprisingly large and damaging static charges. If you can feel the zap when you touch a doorknob, the charge must be in the three to four thousand volt range. AT&T cautions technicians and visitors to their assembly operations that under the right conditions the human body can charge up to as much as 35,000 volts by merely walking on a synthetic carpet.[1] Since some sensitive electronics can be damaged or destroyed at well under 1000 volts, uncontrolled static can be extremely expensive.

In the middle 1960's Dan Anderson of Richmond Industries became famous as an extremely vocal proponent of pink plastic bags for electronics packaging. In this product a fatty amine is added to low density polyethylene with a red dye for identification. As the amine migrates to the surface, it allows the formation of a layer of atmospheric moisture on the bag making it slightly electrically conductive. This provides a path for static charges to flow safely to ground, affording some protection to the bag contents.

The first clamshell container for an electrostatic sensitive application was developed in 1969 at Hewlett/Packard by Richard Moss. It consisted of two molded plastic hemispheres joined by a leather hinge with a metallic coating and it was used to contain a semiconductor device.

During the 1970's, Life-Line Products in California produced the first modern appearing thermoformed clamshell containers for the electronics industry. They were treated with a commercial chemical antistat after forming. Plastic bags, however, were to remain the most widely used static control packaging medium.

Electrostatic treatment of effective static controlled clamshell containers by hand application was slow and cumbersome. Clearly, a thermoformable sheet with built-in electrostatic properties would have to be developed in order to address the market.

In fact, several manufacturers brought pre-treated PVC, PET and PETG based products to market in the 1980's with considerable success. That success was only partially due to product quality and marketing zeal. To a much larger degree, the success of the thermoforming concept in the electronics market was a matter of being in the right place at the right time.

LATENT DEFECTS

At the 1982 EOS/ESD Symposium, McAteer, Twist and Walker presented a federally funded paper.[2] In the paper, the authors presented conclusive evidence that electronic devices could be damaged by static electricity even at levels below the design threshold. Further, the damage was subtle enough to escape detection in quality screening operations. Devices would be manufactured, shipped, received and installed. And then they would fail. As the failures were separated from the cause by time and distance, it was impossible to assess the blame or properly allocate the cost. Whoever owned the devices at the time of failure suffered the loss.

Since you could not identify latent-damaged parts, the only possible course of action was to initiate disciplines and strategies to prevent the damage from happening. Static control programs defining work surfaces, flooring, clothing and packaging became a necessary part of manufacturing operations - especially if you planned to sell to the government.

Static electricity is for the most part invisible and the effects of static are not immediately apparent. While some people believed in static control and some did not, the installation of a fully qualified, documented and audited static control program was no longer a matter of choice. For many it became a contractual obligation rather than a voluntary safeguard.

POLYCARBONATE COMPATIBILITY

In 1985 Patrick Guinn and Bill Roderick at Hewlett/Packard in Loveland, Colorado discovered that the amines in conventional pink poly bags caused stress cracking in polycarbonate parts. After verifying this with General Electric, the resin supplier,[3] they took steps to eliminate the use of pink bags in their operation. After twenty years of market domination, pink poly bags were being replaced by new materials and new packaging techniques.

Manufacturers of the suspect bags quickly reformulated the active ingredient in pink poly but ran head first into another disaster.

THE GIDEP ALERT

The Government/Industry Data Exchange Program was set up so that the two cooperating bodies could have a means to alert all concerned when a potentially hazardous or costly situation was developing. In 1987, Eldon Cady at Lockheed in Sunnyvale, California published a GIDEP alert[4] specifying that the use of pink bags from a specified manufacturer caused corrosion damage to resistors and reduced the solderability of electrical component leads. The critical ingredient was a hydrocarbon, N-octanoic acid, and the report stated that "an effort to clean up the formulation…was not successful". Copies of the alert were circulated widely in the electronics industry with an obvious effect on the sales of pink poly when competing with the newly introduced thermoformed clamshells.

BLACK CONDUCTIVE POLY

People committed to bagging operations had one alternative when the decision to replace pink bags was unavoidable. Black, carbon loaded LDPE bags were opaque and somewhat dirty to handle, but they were inexpensive. Accordingly, the black bag business enjoyed a definite uptick until 1988 when the Department of Defense published MIL-HDBK-7735.[5] The document stated "In all cases the first contacting layer (with bare item) of the selected protective material will be static dissipative and not conductive."

It had been determined that a static sensitive device could generate a charge by rubbing against the container. If the container was a conductor, there could be a sudden discharge of sufficient voltage to damage the device. Static dissipative materials are very slightly conductive so that charges flow to ground slowly and safely.

While the events described did not strike a death blow to either pink or black poly, they caused major consternation within the industry especially when some of the larger companies banned the use of specific products unconditionally.[6,7] This concern helped to open the doors of many decision makers who would not have seriously considered the conversion to a thermoformed container system a very few years before.

RETAIL ELECTRONICS

By far, the biggest boost to the development of thermoformed electronics packaging was the emergence of the retail electronic parts market. The personal computer business created a demand for "add-on" circuit boards for increased memory and networking applications. Mall-based retail outlets provided a large, new marketing opportunity to circuit board manufacturers but a few headaches as well.

Prospective buyers would walk into stores wearing well insulated sneakers, shuffle across the rug and ask to handle the circuit boards. From a static control viewpoint, this was a potentially disastrous situation. Thermoformed clamshells made from static dissipative, clear plastics were the only apparent answer. The buyer now could pick up the circuit board and inspect it closely without ever removing it from the protective package.

PROBLEMS

ELECTROSTATIC PERFORMANCE

Many of the plastic bags sold utilize a metallic layer, ostensibly to prevent the static sensitive contents from electrostatic fields. There has been a problem in convincing some people that thermoformed containers lacking the metallic layer would offer significant electrostatic protection. It became necessary to develop a body of information that would reliably address these concerns.

In 1992 Arch Warnecke and Cecil Deish presented a milestone paper[8] during the Technical Program at NEPCON West, a major electronics exposition. In the study, subsequently reprinted in EOS/ESD Technology Magazine, it was established that a static dissipative clamshell incorporating an air gap of 0.25 inches between the contained device and the inner wall of the container accomplishes by isolation all that a metallic layer does through conductivity. The ESD Association incorporates this information in the Basic ESD Tutorial[9] offered at their annual symposium while stretching the gap to 0.375 inches. A more recent study by Stephen Halperin and Associates[10] published in Evaluation Engineering in 1994 further supports the air gap isolation concept.

While organizations such as AT&T publicly support the use of air gaps over conductive materials,[11] the strongest argument might be that the technology is no longer new and there have been no reported failures despite widespread adoption of the packaging concept.

COST

Inevitably, someone will ask if a thermoformed clamshell costs more than a plastic bag. Of course it does. If one were to ask if a packaging system using clamshells is more costly than a packaging system using plastic bags, the very opposite is true. Most companies today converting from bagging operations to clamshells do it to reduce costs.

Western Digital converted from bags to clamshells and studied the impact on costs of inventories, labor, freight and material costs. They published[12] that their original packaging cost of $4.80 per unit for bag-in-a-box dropped to $2.05 with the adoption of a packaging system using clamshells. Generally speaking, an electronics firm converting to clamshell containers never reverts to bags.

THERMOFORMING ADVANTAGES

PHYSICAL PROTECTION

In a properly designed clamshell, the sensitive part snaps into a positive location and the lid also has a positive closure. This provides superior physical protection and makes the package automation compatible by allowing the use of non-intrusive bar code scanning. It is important that there is no need to expose the device to the outside world until it is brought into a static safe area for installation.

Plastic bags are notoriously susceptible to punctures from the sharp leads protruding from circuit boards while clamshells are essentially puncture proof. An exposed lead can provide a conductive path to serious electrostatic damage. Further, a circuit board securely held in a clamshell cannot flex and damage the solder pads of surface mounted devices.

ECOLOGICAL IMPACT

The use of a clamshell packaging system typically enables the end user to reduce the bulk, weight and variety of packaging materials used in his operation. Clamshells are made from a single resin and are usually embossed with the appropriate symbol for directing the used package to the proper waste stream. In addition, used packages and trim waste have cash value. Material suppliers are more than willing to buy back these materials to add to the recycled content of their products. Some major electronics firms have active programs to encourage the use of ecologically appropriate products and will favor their use in bidding situations.

THE FUTURE

Electronic devices continue to become smaller, more powerful, more expensive and incredibly more sensitive to damage. Responding to this continually changing technology, modern packaging materials for use with electronics must meet formal specifications relating to electrical resistance, charge generation, oven aging, contact corrosivity, polycarbonate compatibility, outgassing and scannability. There are certainly more hurdles to come.

The major sheet material suppliers have kept in step with the requirements of the electronics industry and will continue to do so. Consequently, it is imperative that the development of business in this area should be conducted as a three way partnership involving the exchange of technical assistance between the material supplier, thermoformer and end user.

CONCLUSIONS

While the need for static control packaging was apparent to some people as early as the 1960's, the establishment of latent damage as a real and costly threat made static control programs compulsory.

A series of well-publicized problems with established packaging materials in the mid 1980's opened the door to new packaging techniques, materials and procedures.

The initial skepticism regarding the electrostatic performance and cost of thermoformed packaging has been addressed in published and available documentation.

A conventional comparison of available packaging for electronics that conscientiously considers the mechanical, chemical, electrostatic, economical and ecological needs of the application will inevitably favor the use of thermoformed containers for electrostatic sensitive devices.

REFERENCES

1 AT&T Technologies Employee Information Bulletin (unnumbered).

2 O.J. McAteer, R.E. Twist and R.C. Walker, "Latent ESD Failures", Electrical Overstress/ Electrostatic Discharge Symposium Proceedings 1982.

3 Renee A. Miller, "Amine Degradation of Lexan® Resin Components", Lexan® Technifacts T-81, General Electric Company.

4 Eldon Cady, "Materials, Plastic, Antistatic", Government Industry Data Exchange Program ALERT, 1987 Jan 23.

5 MIL-HDBK-773, 1 April 1988, "Electrostatic Discharge Protective Packaging", Department of Defense, Washington, D.C. 20301.

6 DWG NO. 025-0174, "Bag, Antistatic, Plastic" Collins Transmission Systems Division, 1200 N.Alma Rd., Richardson, TX 75081.

7 F.L. Springer, D.R. Baugh, W.C. Miller, "Clarification of Fab Packaging Requirements/Revision", letter to Rockwell International Suppliers, Rockwell International, Dallas, Texas, May 15, 1987.

8 A.E. Warnecke, Cecil Deisch, "Shielding Evaluation of Semi-Rigid Plastic Trays", Proceedings of the Technical Program, National Electronic Packaging and Production Conference, Anaheim, CA February 25-27, 1992.

9 Burt Unger, "Basic ESD Seminar" sponsored by Electrostatic Discharge Association, Inc., Rome, NY.

10 Stephen Halperin, ESD Shielding Without Metallization or Conductive Fibers, Stephen Halperin and Associates, Ltd., Bensenville, IL 60106 as published in Evaluation Engineering - Sept 1994.

11 G. Theodore Dangelmayer, AT&T Network Systems, **ESD Program Management**, *Van Nostrand Reinhold*, New York 1990.

12 Troy Merrell, Senior Engineer, Western Digital Corporation, "Redesigned Package Saves Time and Money" Case History, EOS/ESD Technology Magazine, reprint distributed by ITW Electronic Component Packaging Systems.

Electronic Packaging for the Next Century

Steve Fowler
Fowler Associates, Inc.

HISTORY

One may assume that the Chinese had problems with fireworks explosions due to static discharge centuries ago. The munitions industry has always had to be careful of static discharges. The APOLLO command module fire in 1967 was thought to be static related. Several premature or unintentional rocket ignitions have been blamed on static discharges. Today we may hear of gasoline pumps exploding due to static. Workers in detonator fuse assembly areas know too well the problems with static. These problems are known and assumed to be real. In the electronics industry the problems of static damage are more recent and less visible. In the early days of electronics, items such as vacuum tube devices and other extremely robust products from an ESD point of view had little or no need for static control in the materials used for packaging. Then in 1947 with the invention of the transistor the course of packaging for electronics was beginning to change. The first transistors were very insensitive to static charges because of their size and sheer bulk. As the size of these components came down and speed became a factor in their operation, the idea of static sensitivity became known but not yet fully necessary. The progression of transistors to metal oxide versions such as MOSFET's on to integrated circuits with smaller and smaller sizes opened the industry's eyes to static failures which began to affect production yields. With the invention of the true microprocessor and the explosion of computer technology, static control became a very real issue which could not only affect the yields and profitability of circuit manufacturers but also keep latent failures from causing failures later in the product life with even safety consequences. During the late 1970's and 1980's many major companies developed well organized ESD control programs with major benefits to yields and profitability. Today the emphasis has shifted to one less concerned with preventative measures. Many companies have abandoned the gains realized in the early days of static control and believe the technology has developed to a point where concern is reduced to the lowest price per package. In the field of packaging, there are two types of "chips": IC chips and potato chips. A $500 IC chip may be packaged in a 3 cents bag while I cent worth of potato chips will be packaged in a 10 cents bag. The food industry does not normally underestimate the value of its packaging while the electronics industry allows its products to constantly be underpackaged. If food spoils, the consumer may not buy the product or may be affected by pathogens grown in poorly packaged products. The results are disastrous when botulism or salmonella affects someone or some group. In electronics the failure of high dollar products due to ESD may not necessarily cause sickness (pace makers, respirators, refrigeration controls, etc) however, they may ruin the day of an airplane crew and passengers when navigation equipment goes out (Flight 007 KAL). Automatic pilot systems could send planes into uncontrolled rolls. Premature firing of air-to-air missiles in helicopter #6 could make the flight leader miss dinner. Electronics is considered the highest technology industry. It is the most archaic in its packaging technology. This paper will help inform those involved in this industry in the advantages of value packaging of its products.

ESD PACKAGES OVERVIEW

The ESD packaging industry has grown up to support the desires of the electronics industry. As stated above the industry has been made up of flat films, bags and boxes. Not too much innovation; hardly any automation. The bags began as black carbon loaded polyethylene until the advent of "pink poly" and other static dissipative films. Shielding bags were first foil laminates mainly used for moisture barrier. Now they are typically metallized polyester laminates. The driving forces of this industry has become costs not function. Most electronic industry end users believe that the present technologies are OK and that cost is all important. This brings the need for more automation closer than ever to favor. Cross sections in Figures 1 through 10 describe the typical materials used today in ESD packaging.

CHEMISTRY OF DISSIPATIVE PLASTICS

There are two dominate chemical technologies that achieve static dissipation as Type 11 electronics packaging films, as defined by EIA 541 or MIL-B-81705-C (the "Mil-Spec"), and as the static dissipative heat sealing layer of many EMI shielding Type I and static shielding Type III films. An amine is a long chain of carbon atoms surrounded by hydrogen atoms. This hydrocarbon chain is derived from a fatty acid and can vary from a few carbon atoms long to twenty or more carbon atoms. The last carbon is connected to a nitrogen atom which has two hydrogens linked on the sides. Because simple fatty amines need a higher affinity for water to perform well as antistatic additives, ethylene oxide is reacted with them to make ethoxylated fatty amines. This gives the molecules two polar end groups: OH or alcohol groups. Ethoxylated fatty amines are the additives used in dissipative polymer amine technology.

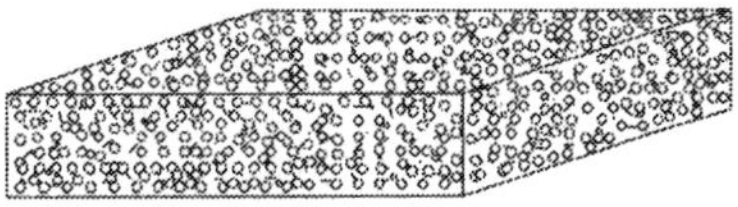

Figure 1. Black poly.

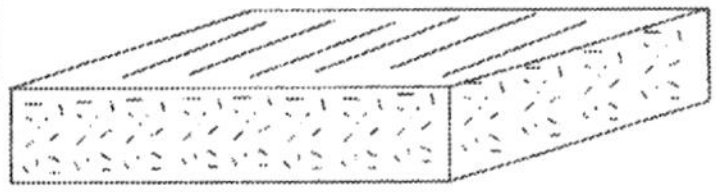

Figure 2. Pink poly.

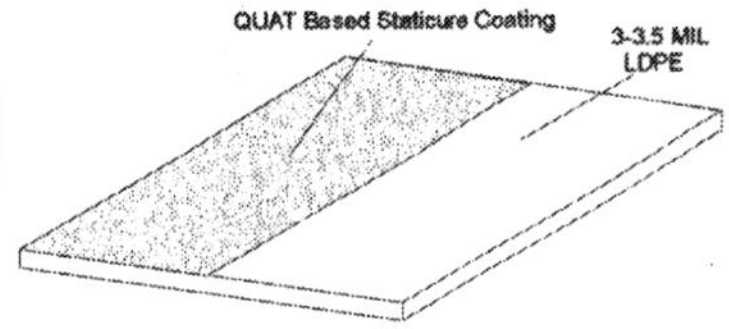

Figure 3. Staticure.

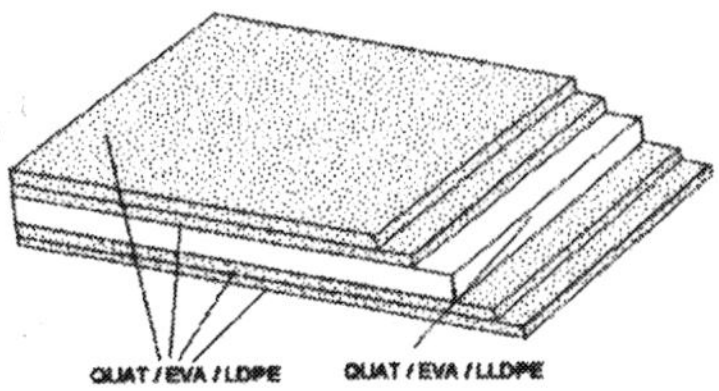

Figure 4. Dissipative coextrusion.

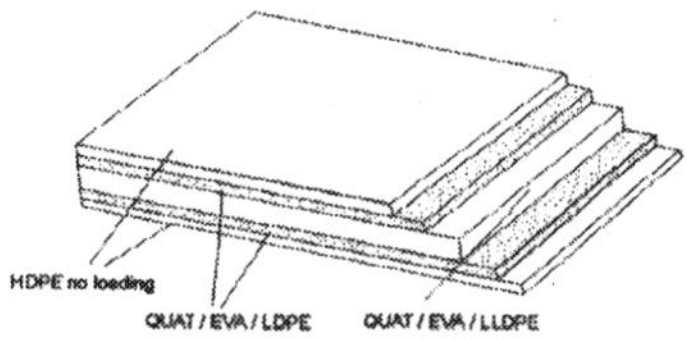

Figure 5. Clean skin coextrusion.

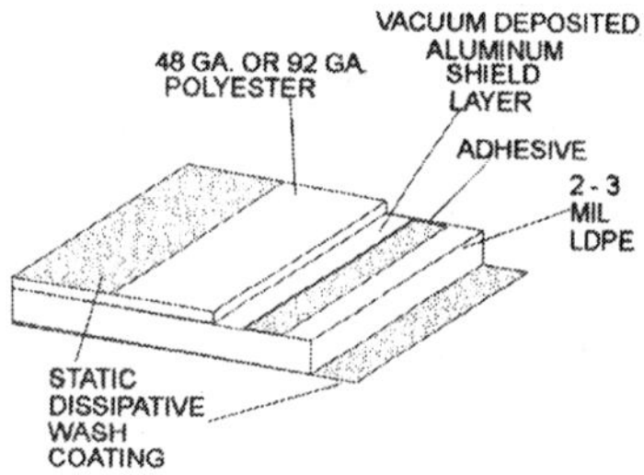

Figure 6. Metal-out shielding.

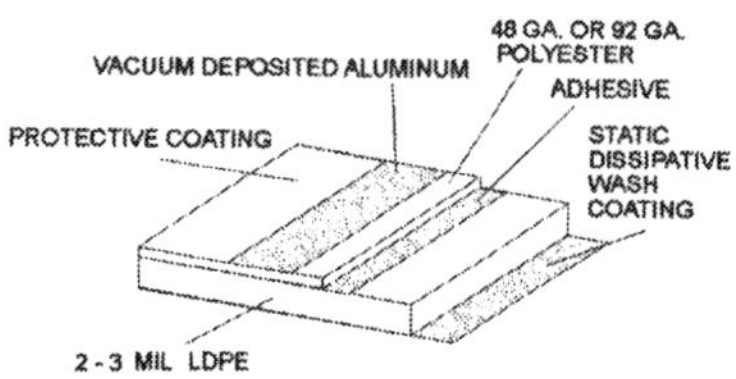

Figure 7. Metal-in shielding.

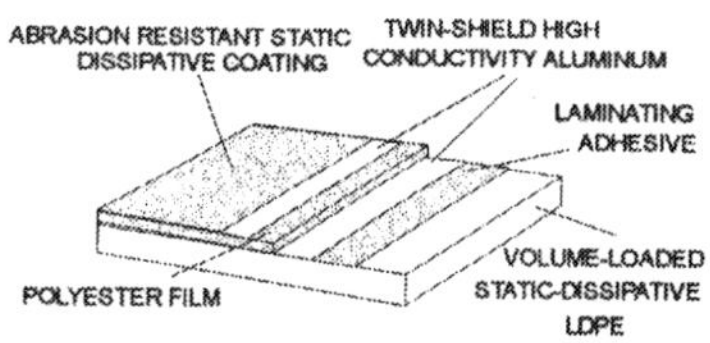

Figure 8. Twin shielding.

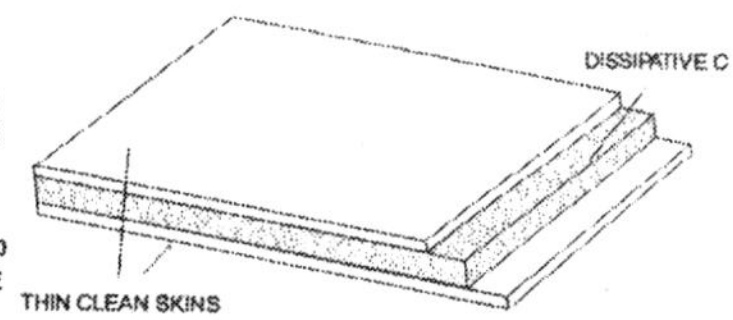

Figure 9. Coextrusion shrink film.

Amide additives used in dissipative polymers are similarly based on a fatty acid attached to nitrogen which is also reacted with ethylene oxide. However, the carbon atom next to the nitrogen has a double bond attachment to oxygen instead of two single bonds to two hydrogens.

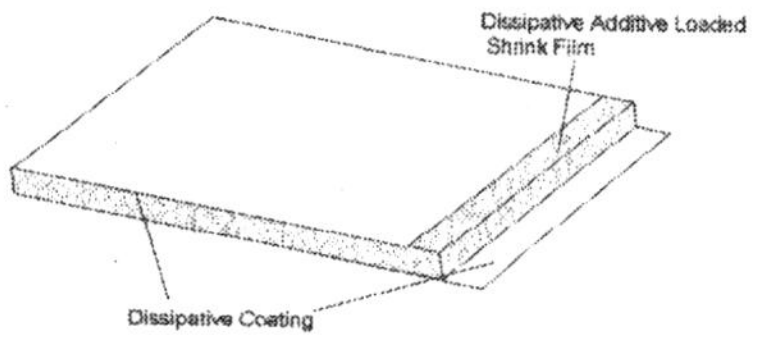

Figure 10. New shrink film.

In this case, the molecule is specifically an ethoxylated fatty amide. It should be noted that the commercial nomenclature "amine free" usually means that an amide additive is used instead of an amine. Other static dissipative additive systems for polyethylene exist with performance and properties very similar to amines and amides. Because these are not widely sold into the electronics industry, they are not mentioned in this paper. Either type of dissipative polymer starts as a homogeneous blend of additive, antiblock and polyolefin resin. However, two things must occur for static dissipation to take place. The amine or amide must first diffuse through the plastic volume to reach and wet the surface. This is commonly referred to as blooming. Good compatibility between the additive molecule and the resin results in an insufficient amount of amine or amide wetting the surface. Without enough additive on the surface, static dissipation will not occur. Too little additive and resin compatibility results in too much additive getting to the surface. In this case, the static dissipative property will work well but the surface will be excessively greasy with the additive contamination everything the film contacts. The additive must absorb at least a trace of atmospheric moisture with its two "claws." This hydrogen-bonded combination of additive and water is the surface that provides static dissipation. Essentially, the conductivity is provided by the layer of water attached to the bonds and helped by the ionic solution of the additive. It is an ionic soup. During certain seasons in some locations, it is possible that there is too little atmospheric moisture to be absorbed. It is not clearly understood the minimum relative humidity required for static dissipation. The best guide is probably the EIA's standard of 12.0±3.0% RH as a reasonable minimum limit.

ELECTRICAL PERMANENCE

Some have referred to questions about the permanence of dissipative polymers as one of the dirty little secrets of ESD. Many films lose their electrical properties when exposed to the Mil-Spec accelerated aging test at 160°F or after some period of time at storage conditions. The electrical performance lifetime of commercial dissipative polymers is often increased by increasing the film thickness and therefore the reservoir volume.

POLYCARBONATE COMPATIBILITY

The term "compatibility" when used in conjunction with polycarbonates has a different meaning than that used to describe additive and resin compatibility. A more accurate term would probably be polycarbonate "coexistence." Some antistatic additives such as tertiary amines cause polycarbonate parts to crack or "craze". A general rule for polycarbonate compatibility is that there be no serious crazing up to a 2000 psi stress level at 158°F.

SURFACE RESISTIVITY AND STATIC DECAY

Federal Test Standard 101C, Method 4046.1 requires a static protective material to have a static decay time of less than 2 seconds. EIA 541 defines a static dissipative material as one which has a surface resistivity of between IE5 and 1E12 Ohms per square. These two requirements are correlatable only for monolayer and homogeneous films. Multi-layer coextrusion technology as well as laminated materials separate these two parameters. The conception persists that a low surface resistivity is required to dissipate a static charge and that a good static decay is sufficient to characterize a dissipative material. The dissipation of static charges may be accomplished by several means: across the surface, by volumetric conduction through a relatively thin high resistivity skin, across a subcutaneous conductive or dissipative layer and out of the film again through the thin high resistivity skin. Metal laminates actually fool the static decay test. The reason for their seemingly instantaneous dissipation of charges is that the static decay time is dominated by voltage suppressed by the ground plane being capacitively coupled to the metal layer. For metal laminates with surface resistivities less than 1E12 Ohms/square, one can assume that the surface charges decay in less than 2 seconds. For those layers with resistivities above this level, no assumption can be made from the results of the static decay test as described in Method 4046.1. The application of laminations or coextrusion technology to ESD materials allows the separation of the parameters of surface resistivity and static decay for laminates that do not contain metal or metallized layers. Coextrusion is the process of forming multi-layer materials from several extruders directly out of a one special die. In a coextruded material, each layer can be made from different base polymers or a blend of polymers, each selected for design attributes such as moisture barrier, flame retardancy, sealability, stiffness or strength. Multi-ply materials allow an ESD protective packaging user to weigh the individual benefits of all material attributes without them being totally dependent on one another. Users of ESD protective materials need to weigh the cost/benefit parameters in selecting appropriate materials for specific applications. Probably no material can ever meet all the requirements of the ESD world. It would be nice to have a heat sealable transparent aluminum foil with a surface resistivity of 1E8 Ohms/square and a volume resistivity of 1E11 Ohms-cm and have an attenuation of at least 120 dB to all electromagnetic frequencies. However, this material is a dream. It is called: UNOBTAINIUM.

SURFACE RESISTIVITY AND TRIBOELECTRIFICATION

Triboelectric charge generation by plastic packaging materials is widely believed to be dependent on the surface resistivity of the materials in question. If a material has a low resistivity it is sometimes regarded as having a low propensity for charge generation. Surface resistivity and charge generation can not be correlated. However, the belief of a relation of these two parameters persists. For a material to be "antistatic" it must have a low propensity to generate triboelectric charges. As the following charts show, earlier surface resistivity scales listed an antistatic category. Presently the EIA, ESD Association and Military specifications have dropped any reference to such a relationship. Current standards recognize only three basic resistivities for non-shielding materials: CONDUCTIVE, DISSIPATIVE, INSULATIVE. Triboelectricity is "a positive or negative charge which is generated by friction." Triboelectricity is from the Greek, Tribein which means: "to rub." On the other hand, "contact charge" is the positive or negative charge generated by first the contact and then separation of two materials. Typically, in ESD work, these two mechanisms are lumped together in the term triboelectrification or just tribo. Early electrostatic work placed a great deal of emphasis on the relative position of materials in a tribo series. The relative polarity of charge acquired on contact between any material in the series with another was predicted by its location. There is little correlation between the series developed by different researchers due to very complex nature of the triboelectrification process. The question of whether or not materials at the positive end will always charge positive when rubbed with or contacted by materials lower in the series is not clear. If electron transfer was the only mechanism for charging, at least for certain material combinations, then such a series would certainly exist. However, instead of a uniform series of materials, some "rings" have been shown to exist. Silk charges glass negatively and glass charges zinc negatively, but zinc charges silk negatively. This is the case even though glass is higher than silk and silk is higher than zinc in most tribo series. One may not rely totally on a tribo series to determine the polarity of the charge for the contacting or rubbing together of two materials. No tribo series may be used to determine the actual quantity of charge resulting from the contacting or rubbing together of two materials. The mechanisms for determining the quantity of charge transfer are extremely complex. Surface resistivity does not play a role in the tribelectrification process. It does however, contribute to the material's ability to bleed off any charge which has been transferred. Materials with surface resistivities in the static dissipative range will not retain static charges accumulated by tribocharging if those materials are grounded. A total packaging system must be designed, one should not just chose a material which is "anti-static". All the requirements of the application must be taken into account before a material can be chosen. The user and the manufacturer must work together to design an appropriate static dissipative and low tribo generating packaging system.

NEW PACKAGING SYSTEM

SHRINK PACKAGING

Shrink films are usually thin biaxially oriented films which shrink back to the size when exposed to relatively high temperatures. They can provide 40% - 50% shrink characteristics. They are used in consumer packaging to bundle items or to provide tamper evident viewable packages. Despite several unique advantages for packaging electronics parts, subassemblies and finished goods that include overall package cost effectiveness, tamper evidence and immobilizing parts, antistatic shrink film has been generally considered novel in this industry. One reason for the novelty is that, while general consumer shrink films are well known, a genuine antistatic shrink film is a relatively recent development. In the late 1980's Cryovac pioneered the first anti-static (dissipative?) shrink film. The Cryovac shrink film was a three layer coextruded film with clean surfaces made of unloaded and uncoated EPC (ethylene propylene copolymer). This film used a dissipative core to provide a film with excellent surface cleanliness and shrink properties combined with suitable ESD properties. Even though its surface resistivity was on the order of IE13 Ohms/square, it had good static decay and tribo properties.

Several companies began to use the film using L-bar sealers and shrink tunnels to automate their packaging. Cryovac withdrew from the ESD market taking its EP films out of production. Two companies have entered the market with good replacement films and in some cases with better properties. The shrink film provides the ESD event wall against which any static discharges will occur when the product is handled. It is by its function separated from the product by some distance providing the air gap. In laboratory tests, an air gap of 1/8th inch or greater provides very good “shielding” attenuation to ESD events. With good package design, the use of shrink films can show effective costs savings in an automated packaging system. The use of air gaps in boxes and thermoformed packages is also effective as an attenuation technique or shield. Anytime the effective distance to a packaged product from the point of an ESD event can be increased, the protection goes up essentially by the square of the distance. This concept can be designed into any package whether it is retail or industrial. The idea of not always using metal shields may be uncomfortable with some end users. However, it is similar to one wearing a bullet proof vest all the time because he or she enjoys being in a hail of bullets. It is much better to put a little distance from the shooter and the body. No matter how good the vest, it always hurts. ESD shrink film has been shown to be an effective packaging system suitable for electronics components, subassemblies and finished products. For chips tested under lab and field conditions, the antistatic shrink films' static protection compares favorably with other well known materials. The costs are very favorable compared to conventional static protection films and bags. It has been shown that a few cents of shrink film automatically or semi-automatically applied can replace several tens of cents of shielding bags hand packaged.

AIR CUSHIONING/STATIC SHIELDING/MOISTURE BARRIER/DESICCATING PACKAGE

A recent development led to a protective packaging system which uses air channels surrounding the product to cushion and protect it, provide electrostatic shielding and moisture barrier both with and without desiccants. The system as it is conceived does away with the individual components used in the normal combination packaging of integrated circuits. Physical cushioning is provided by the air channels surrounding the product. Moisture barrier is provided by the multilayer materials in conjunction with the dry air in the channels. Since the air gap provides the majority of the electrostatic shielding the reuse of the system does not degrade its efficiency.

Conducting Polymers as Alignment Layers and Patterned Electrodes for Twisted Nematic Liquid Crystal Displays

Jerome B. Lando, J. Adin Mann, Jr., Andy Chang, Chin-Jen S. Tseng
Case Western Reserve University
David Johnson
Kent State University

INTRODUCTION

The construction and basic operation of TN displays is illustrated in Figure 1. The upper and lower substrate plates carry patterned, transparent conductive coatings of Indium-Tin Oxide (ITO) on their inner surfaces. In general, the transparent electrodes have a thin polyimide coating several hundred Å thick that is unidirectionally rubbed to align the local optic axis (director) of the liquid crystal at the surface parallel to the rubbing direction. The upper substrate is rubbed at right angles to the rubbing direction of the lower substrate. Thus, in the inactivate state (off), the local liquid crystal director undergoes a continuous 90° twist in the region between the substrates. Polarizers sheets are laminated on the outside of the plates so that the direction of vibration of the linear polarized light is parallel to the rubbing direction of the adjacent alignment layer of each substrate. The linear polarized light from the upper polarizer propagates through the layer, rotates its plane of polarization in step with the twisted structure, and emerges at the bottom of the layer polarized parallel to the transmission axis of the lower polarizer. Applying an electric field across the upper and lower electrodes orients the optic axis in the central portion of the LC layer predominantly parallel to the electric field and the twisted structure disappears (on). The polarization direction of the light is no longer rotated and light passing through the cell intersects the second polarizer in the crossed position where it is absorbed, causing the activated portion of the display to appear dark.

We have successfully prepared ultra thin polyimide alignment layers by utilizing the dipping direction in Langmuir-Blodgett film formation to orient a soluble low T_g precursor polyamic acid, which forms a monolayer at the gas-water interface. Subsequent to deposition of the precursor polymer the oriented multilayer is converted to oriented polyimide.[1,2] Such alignment layers eliminate the disadvantages inherent in the rubbing technique, which is used commercially at present.

In the present work we intend to eliminate the Indium, Tin Oxide (ITO) electrodes by utilizing dopable conducting polymers as alignment layers. In the present work precursor polymers, which can be converted to the desired conducting polymers after orientation during Langmuir-Blodgett deposition, are again utilized.

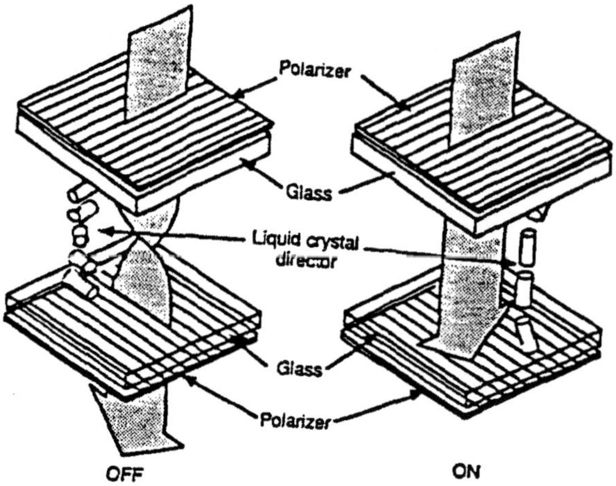

Figure 1. Construction and basic operation of a TN display.

EXPERIMENTAL

MATERIALS

The precursor polymer shown in Figure 2 poly(p-xylylene-α-diethylsulfonium chloride) was synthesized for this study by Kakimoto. 0.5 mM solutions of the precursor were prepared by mixing equal amounts of the polysulfonium salt with sodium perfluorononanoate in 1:1 EtOH:1,1,1 trichloroethane solution. After formation of precursor multilayers conversion to poly(p-phenylene vinylene) (PPV) was attempted by thermal treatment[2,3] for seven or more hours at 235°C. The precursor polymer for poly(p-phenylene) (PPP) poly(2,5-dicarboxyl-1,4-phenylene)(PDCP) was provided by Professor Virgil Percec. The tertiary amine o,o',o"-trihexadecanoyltriethanolamine used for salt formation was provided by Kakimoto.[1] See Figure 3.

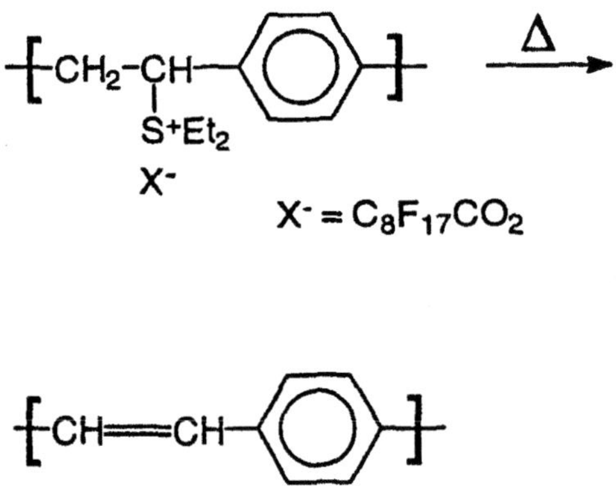

Figure 2. The conversion of precursor multilayers to PPV.

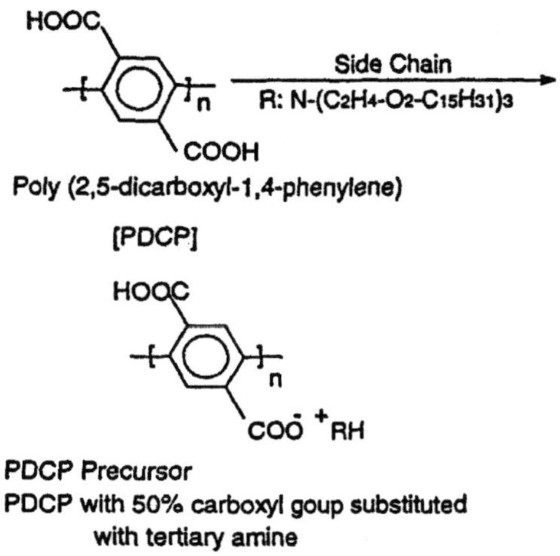

Figure 3. Salt formation.

PDCP was dissolved in DMSO (dimethyl sulfoxide) and tertiary amine was dissolved in DMAc (dimethyl acetamide). Mixing the above solutions with benzene, the PDCP precursor salt was obtained such that 50% of the carboxyl group are substituted with tertiary amine in the ideal case.

The spreading solution was mixed DMSO, DMAc and benzene in a volume ratio of 1:1:2, respectively. The PDCP precursor (salt) concentration was 0.770 mg/ml.

After multilayer formation side chain removal was attempted by a combination of chemical treatment followed by heat treatment.

Deposition substrates were fused quartz, needed for UV spectra, glass slides, Ge, ZnSe and Si.

DESCRIPTION OF METHODS

FORMATION OF MONOLAYERS AND MULTILAYERS

Monolayer manipulations were performed on a commercial Lauda film balance that employs the floating barrier method of measuring surface pressure. The brass trough was coated with Teflon. An IBM PC was interfaced with the film balance for data acquisition and processing. All the isotherm collection and deposition experiments were done in class 10 laminar flow areas inside a class 100 clean room. Subphase water was obtained from a Millipore water system. In addition, a “shake test” of subphase water in a clean volumetric flask was used to show the complete absence of any tendency to foam; the surface tension is always within the experimental uncertainty of the literature value for pure water. Capillary ripple damping of the pure water in also checked.

Initial spreading areas were greater than 70Å^2/molecule and dwell times of 10 min were used to ensure complete evaporation of the spreading solvent. Compression rates were 3.25 cm/min. The water temperature was controlled by circulating thermostated water underneath the brass trough. The temperature of the water in the trough was measured by a surface probe to a precision of $\pm 0.1^{\circ}$C

Surface pressure area isotherms and compressive creep tests were performed on each of precursor films. Creep tests were performed at the temperatures and surface pressures utilized for deposition of Langmuir-Blodgett multilayers.

Multilayer deposition was accomplished for the PPP precursor at 21°C and a surface pressure of 25 mN/M, while for the PPV precursor a temperature of 13.4°C and a surface pressure of 25 mN/M were used.

CHARACTERIZATION

Multilayers of the precursors were investigated before and after curing treatments as a function of the number of molecular layers by X-ray diffraction (glass substrate) using a Phillips diffractometer and by UV spectroscopy (fused quartz substrate) using a Varian Cary UV - vis spectrophotometer.

RESULTS AND DISCUSSION

Surface pressure area curves of the two precursor polymers are shown in Figure 4. The compressive creep curves of the two precursors are shown in Figure 5. It is clear that the PPV precursor at or near the deposition conditions (13.8°C and 25 mN/M) is far less stable at the gas-water interface than the PPP precursor at its deposition conditions (21°C and 25mN/m). This is attributed to the fact that ideal packing of the single stranded side chains of the PPV precursor can not accommodate in the area of the head group. However, between 10°C and 15°C PPV precursor is stable to compressive creep for 30 to 60 minutes. Thus stable deposition can be obtained for both materials although it is clear from the surface pressure area curves that the modulus of the PPP precursor monolayer is much higher than that of the PPV precursor. Note that the PPP precursor is stable to compressive creep for long times.

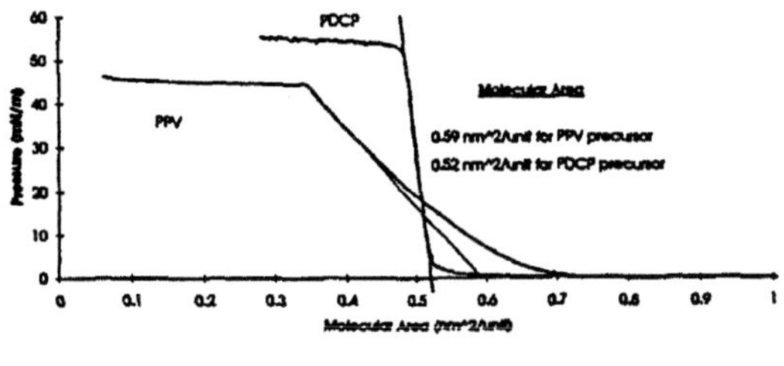

Figure 4. Surface pressure-area curves for the PPV-precursor (13.4°C and PDCP (21°C).

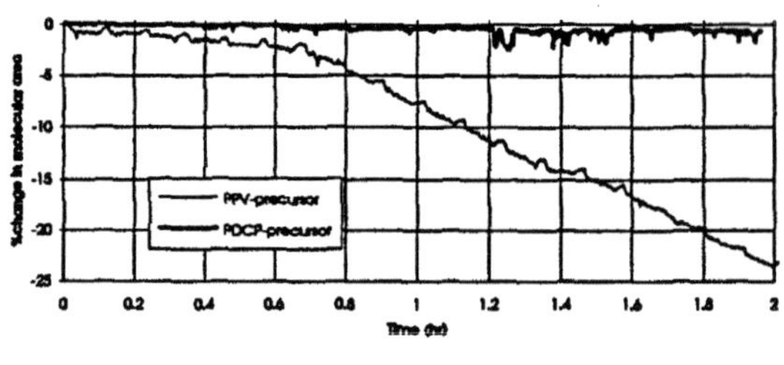

Figure 5. Compressive creep of PPV-precursor (13.8°C) and PDCP-precursor (21°C) at mN/m.

UV (190 - 640 nm) spectra of the precursors before and after curing treatment were obtained as a function of the number of layers on the fused quartz substrate. PPP precursor films at 9, 15 and 31 layers all have two peaks, one at 245 nm and the other at 315 nm. The peak at 245 is attributed to aromatic π electron to π * transition while the peak at 315 nm is an n electron to π * transition from the carbonyl group. The intensity changes with the number of layers follows Beer's law. A multilayer sample was tested after treatment. No new peaks were observed.

UV spectra for the PPV precursor as a function of the number of layers were also obtained before and after curing treatment. Absorbencies for the precursor were observed at 195, 225, 260 and 325 nm. After heat treatment there are absorbencies centered at 200, 250, 330, and 390 nm. This observed shift in absorbencies to lower energy radiation would support the hypothesis that significant side chain removal takes place after heating at 235°C for 7 hours. Further heat treatments at higher temperatures for longer durations did not yield changes in the observed spectra. A plot of peak absorbency versus the number of layers shows that layer uniformity was maintained after the heat treatment. Also, Beer's law is followed before and after treatment. We suspect the peak at 390 nm corresponds to $\pi - \pi$ * transition that is affected by the proximity of the polymer-substrate interface since thicker samples (15 or more layers) have similar intensities. As a result the 390 nm peak was not used for the absorbance versus layers plot.

X-ray diffraction plots of the PPV precursor before and after heat treatment both show a peak intensity at $2\theta=2.6°$ which corresponds to a layer thickness of 3.39 nm. This result is in agreement with the fact that deposition is occurring in well-ordered layers. In addition, the presence of the same peak after heat treatment, although reduced in intensity, shows that side chain removal is not complete. Kakimoto *et. al.* observed layer thickness to be 0.34 nm by ellipsometric techniques after heat treatment.[4] We do not observe the corresponding peak at $2\theta = 26.2°$, although that could be a consequence of low intensity.

X-ray diffractometry of 61 layers of the PPP precursor yielded indexing schemes as shown in Table 1. In scheme A two different layer structures are assumed and in scheme B a single structure is assumed with a doubled basic Y structure spacing of approximately 10 nm. An 11 layer sample had only the X-ray maximum at 2.56 degrees indicating a tilted structure as in scheme A. At some point (after 11 layers) in the deposition an orthogonal side chain packing occurs. Thus scheme A in Table 1 is correct. It should be noted that after heating the 61 layer sample at 280°C for four hours only the peak is at 2.56 degrees is present. This indicates that only layers near the multilayer upper surface had their side chains removed and/or a phase transformation occurred during treatment. Given the similar intensity of the peak at 2.56 degrees before and after treatment the former explanation is favored.

Table 1. Analysis data from X-ray diffraction

2θ, °	Relative intensity	d-spacing, nm	00l indices A	00l indices B
1.933	3705	4.570	001	002
2.56	993	3.451	001'	003
3.62	786	2.441	002	004
5.34	3770	1.655	003 and/or 002'	006
7.08	888	1.249	004	008

CONCLUSIONS

Langmuir-Blodgett films of both the PPV and PPP precursors have been successfully prepared. However, subsequent treatment indicates that the side chains have only been partially removed by chemical and/or thermal treatments. This work is continuing.

ACKNOWLEDGMENT

This work was supported by the National Science Foundation under the Science and Technology Center ALCOM DMR 89-20147 and by the Office of Naval Research under Grant No. N000 14-94-1-01270.

REFERENCES

1 M. Kakimoto, M. Suzuki, T. Konishi, Y. Imai, M. Iwamoto and T. Hino, *Chem. Lett.*, 823 (1986).
2 A. Albarici, Master of Science Thesis, "Langmuir-Blodgett Films as Alignment Layers for Twisted Nematic Liquid Crystal Displays," Case Western Reserve University, December 1994.
3 Y. Nishikata, M. Kakimoto, Y. Imai, *J. Chem. Soc., Chem. Commun.*, 1040 (1988).
4 Y. Nishikata, M. Kakimoto, Y. Imai, *J. Chem. Soc., Chem. Commun.*, 1042 (1988).

Flexible Conductive Coatings on Thermoformed Films for EMI/RFI Shielding

Bruce K. Bachman
Spraylat Corporation, Mount Vernon, New York

ELECTRONIC ENCLOSURE SHIELDING MATERIALS & METHODS OVERVIEW

Shielding of any plastic or composite enclosure for passing FCC, EU, VCCI or Tempest EMC emissions and susceptibility compliance regulations, with rare exceptions, has always been a secondary additive manufacturing process for an OEM. Shielding technology options can be placed in two distinct categories, add-on technology or conformal coatings technology, as noted in Table 1. As defined, an Add-On technology is one where the shielding technology is incorporated onto a second platform which is then set within the enclosure itself for shielding or ground-plane requirements. A Conformal Coating technology is one where the enclosure itself has the shielding technology integrated directly onto the surface - hence a conformal technology.

Table 1. Shielding technology options

Add-On-Technology	Conformal Coatings Technology
Coated Thermoformed Films	Conductive Coatings
Coated Woven Fiber Matrices	Electroless; Plating
Foils	Vacuum Deposition
Sheet Metal Inserts	
Thin-Film Laminates	

The merits or limitations of any specific technology are not addressed, but a specific new option - thermoformable films coated with a flexible Conductive Coating is discussed. The Table 1 technologies are further defined as follows:

Coated Thermoform Films - Flexible films such as PVC, polycarbonate, Kapton or PETG coated with in this case a Conductive Coating, utilizing a flexible polymer and treated coppers, silver or hybrid silver-coppers as the shielding metal.

Coated Woven Fiber Matrices - Graphite or woven polymers, plated with copper, nickel, tin or combinations, formed into a two dimensional lattice-work sheet.

Foils - Thin-rolled coppers, silver-plated coppers, tin-coated coppers, beryllium-copper or stainless steels which have foil thicknesses under 0.020" (500 μm).

Sheet Metal Inserts - Stainless steels, zinc or nickel plated, iron or zinc phosphated cold rolled steels of 0.030-0.060" (750-1,500 μm), laser or die cut, brake formed and welded.

Thin-Film Laminates - Thermoformed films to which thin-foils or vacuum deposited aluminum, silver or sputtered coppers have been added.

Conductive Coatings - Spray applied treated coppers, silvers or hybrid copper-silvers, usually in acrylic binders, solvent or water-based formulas, with other formulas in high solids epoxies, urethanes and alkaline strippable polymers.

Electroless-plating - Electroless copper, followed by a protective electroless phosphorous-nickel layer either by a conventional catalysis process, known as double-side plating, or by direct plating of a spray applied base-coat, and is commonly known as single-side plating, but more correctly is selective plating.

Vacuum deposition - Also known as Vacuum Plating, consists of aluminum directly applied within a vacuum chamber.

SHIELDING TECHNOLOGY MATERIALS VIABILITY CONSIDERATIONS

All the aforementioned shielding technologies are technically capable of providing for well under 0.050 ohms/square surface resistivity and can yield point-to-point resistance well below 0.25 ohms. All of these respective shielding technologies, when used in accordance to their supplier recommendations, can yield a variety of options for an OEM in both shielding and ground-plane capability for EMC compliance and susceptibility issues.

ELECTRONIC ENCLOSURE SHIELDING & GROUND-PLANE CONSIDERATIONS

Each of these shielding technologies represents an option for an OEM to resolve EMC compliance issues. Moreover, limitations come not from the technologies per se, but from their incorporation within an electronic enclosure under specific mechanical, design, functional or business considerations.

While a cellular telephone, notebook computer, desk top computer, modem, disk drive or PDA, each represent stand-alone EMC requirements, there are considerations common to all which influence the shielding and ground-plane options to an OEM. The following briefly covers some elements to consider.

HARDWARE & MECHANICAL

Weight consideration - critical in laptop or hand-held units
Dimensional concerns - close tolerance or tight fitting matting surfaces
Snap-fits, insert bosses, ribs - ground plane considerations
Internal support hardware - use of integrated plastic or added metal hardware
Keyboard or keypad support - ground plane
Accessory, hinge door covers - EMC leakage and ground plane
Vents & slots - EMC leakage and cooling support, cosmetics
Stress points - hardware or enclosure support for load bearing elements
Molding consideration - memory of plastic for possible stress relief
Complex internal details - increased limitations in EMC solutions

DESIGN

Wall thickness - dimensional stability
Ribbing - continuity of ground plane or EMC antenna slot
Seams - determine cosmetics and EMC leakage limitations
Edges & lips - determine ground plane and EMC leakage limitations
Molded in features - cosmetics

FUNCTIONAL

Line-of-sight - EMC leakage or possible ground plane effects
Blind or tight spots - EMC leakage or possible ground plane effects
Recyclability - EU compatibility

BUSINESS

OEM or Molder location - supplier location support limitations
Fixed asset requirements - program growth limitations
Supplier EPA restrictions of air, water or land disposal - added costs or reduced options
Selective tooling costs and lead times - program costs and burdens
Tooling complexities - increased costs through rework or rejects
World-wide programs - adequate supplier options and costs

Any specific electronic device requiring EMC compliance becomes subject to the limitations of the enclosure design, internal hardware and PCB radiated emissions, active ground planes and shielding technology enhancement combined.

An OEM designing a product subject to EMC compliance should be taking an active role with the supplier base at all levels, as early in a program as possible, to determine the widest choice of options available. Mechanical, design, and manufacturing engineers should take a pro-active role with outside suppliers to determine the EMC options, limitations and possibilities available on their specific program. This can only benefit an OEM through reducing the time and costs for shielding a specific program, as well as meeting manufacturing and marketing schedules.

THERMOFORMED FILM BUSINESS

Thermoformable plastics have been on the market for over 30 years, and represent a mature, well established technology and business base. As in any business market segment, it covers basic low-end markets to high-end value-added sophisticated market segments. It is most commonly seen in retail use, using blister packing of products for sales and marketing purposes of visually allowing a product to be seen, yet incorporates integration of a product packaging, cost and handling reductions and product security.

Some brief examples of thermoformed plastics is seen in trays separating fruits or components, sterile compartments for medical equipment use or disposable items, airline meal snack boxes, or outer shells of toys, to films directly competing with molded plastics. There literally are tens of thousands of applications utilizing thermoformed plastics on a world-wide basis throughout the service and industrial market segments.

The key element to note is that in the overwhelming majority of thermoformed plastics uses, the thermoformed plastic itself becomes the package for some specific item, and is thought-of as an outer shell, and except for some isolated cases, have not lent themselves as a platform for integrated ground-plane and EMC compliance prospects. There are several valid reasons for this point.

The use of a thermoformable film in packaging is a highly competitive business, and while as the business itself can be very sophisticated, the market perception is one of low end value - it represents a less expensive alternative in presenting and packaging a product.

The thermoforming market itself also has high and low value end suppliers, and a large minority of businesses offering thermoforming of films are neither capital intensive in nature nor offer engineering support staffs and expertise. No different than in companies offering plating, spray painting, sheet metal or laminated services, the suppliers actively participating in the EMC environment represent a sophisticated select number within their respective market as a rule.

The business of thermoforming plastics and EMC compliance for electronic enclosures have only recently become aware of their mutual business and technical prospects. Spraylat, through the introduction of its flexible Series 599 Conductive Coatings line, is to be a significant catalyst in bringing both business segments together and offering to an OEM, yet another EMC shielding solutions option for due consideration.

THERMOFORMED FILM TECHNOLOGY

In the business of thermoforming of films, a jig or tool is built to which the thermoformable sheet or web can be placed onto a platform over the tool or jig. The film has been dried for water removal and has been preheated to the manufacturers specified recommendations to allow elongation or thermoforming of the polymer over the tool.

The thermoforming process itself is accomplished by vacuum or pressure forming of the sheet or web-fed plastic film over the tool to conform to the geometries and complexities presented. The part is ejected, and goes to secondary operation for trimming from a die-cut pattern. Any additional decorative or add-on items are traditionally done at this point in production.

Briefly, in the vacuum forming process, vacuum draw-down of the heated film allows for part elongation and contouring. Pressure forming uses an additional platens above the film to pressure form the film into more complex contours, and tighter tolerances. Specific thermoforming business have developed sophisticated in-house engineering services and equipment to thermoform intricate details and geometries to tolerances almost approaching molded plastic tolerances.

Thermoformers with extensive experience and in-house tooling capabilities, can generate tools which will compensate for material dimensional changes, as well as relaxation changes during cool-down of the polymer film. Note that specific engineering films, as well as their respective film thicknesses will require changes in operating parameters, and possible tooling modifications for tight tolerance and conformal considerations.

THERMOFORM FILM CHARACTERISTICS

Since the thermoforming process reduces film thicknesses in draw-down areas, it is important to determine the elongation percentages and locations to establish the minimum required thermoform film thickness. Too thin a film, and structural and mechanical integrity will be lost. A minimum draw down such as used on a lap-top screen-back part, will have less impact than used on a cellular phone body, or the draw down on a desk-top computer base footprint.

Electronic component size is not a major issue, but does present limitations when considered being used on larger parts such as injected foam-molded housings, or computer screen monitors. Too small a foot print results in a less than desirable footprint to work with, but both represent insignificant market prospects to the mainstream electronics shielding business.

Since this is an add-on technology, rather than conformal, some method must be considered for mechanical securing or attachment and placement of the film into an electronic enclosure. Some enclosures will allow the boss-insert hardware to secure the film, while others will require added adhesives or tapes, compatible with the enclosure operating conditions or historical characteristics.

Some electronic enclosures may require conformal coatings tolerances, and would not be candidates for thermoformable films.

EMI/RFI THERMOFORMABLE FILMS

Traditionally, EMI/RFI technologies for thermoformable films have been limited in scope and use. Most commonly used have been:

- Conventional conductive coatings
- Vacuum deposited aluminum
- Sputtered silver or copper
- Metal foils
- Plated or coated woven fibers

With the exception of plated or random woven fiber materials, the process of thermoforming allows for shearing, tearing or cracking of the shielding technology, rendering it useless in all but most simple vertical draw applications. This presumes that the technology has been applied prior to thermoforming, which represents the greatest value for an OEM.

In cases where the shielding technology is applied after thermoforming, only conformal coatings will present any viable option. Foils and laminates will be economically unsuitable due to seams and matting surface problems. Handling of woven fiber matrixes is equally impractical and costly, again due to the two-dimensional nature of this technology.

However, while one is duplicating application of a conformal coating directly on the electronic enclosure, this still maintains some added value in reduction of lead time and cost savings as one is no longer shipping enclosures to the shielding source, manufacturing rejects and losses by the shielder, and multiple sourcing prospects for the thermoformed films on a world-wide basis for an OEM.

FLEXIBLE CONDUCTIVE COATINGS ON THERMOFORMABLE FILMS

Spraylat has developed a line of Series 599 Flexible Conductive Coatings which can be applied to a thermoformable film prior to vacuum or pressure thermoforming, and will maintain film integrity for conductivity and ground-plane throughout the thermoforming process. This technology is not new to Spraylat, having been used since 1990 on military applications of a far different nature.

The use of this unique technology solves a number of business and technical problems for thermoformed films to be viably considered for use within an electronic enclosure as another EMC option for EMI/RFI shielding.

Depending on the elongation percentage (drawn-down), specific film builds are applied to yield a minimum film build in the most critical drawn-down areas. For most applications, film builds of 1.0 - 2.0 mils (25-51 μm) will be adequate. For very deep-draw applications, films may be required up to 3+ mils (75 μm) in selected applications. Most portable and wireless devices will be in the 1.5 - 2.5 mil (39-63 μm) range.

Properly applied films for portable and wireless devices will have surface resistivities after thermoforming in the range of 0.015- 0.040 ohms/square, with point-to-point resistance of between 0. 15 and 0.5 ohms depending upon the electrical path of least resistance.

Incorporation of this technology within a thermoformed film offers to an OEM an EMC option, with a number of cost and manufacturing incentives, as follows:

- Costs due to cosmetic rejects and rework are removed from a program
- Shipping costs and lead times for transportation to and from shielding sites are removed
- Reliability issues of shielding technology continuity or thin-films are avoided
- Thermoformed films are readily removed for enclosure recyclability
- Reduction in tooling costs and lead times are probable
- EMC thermoformed shields are shipped directly to a location determined by the OEM
- Option of having the molder directly add the thermoformed parts
- Flexible film builds can be increased or decreased to meet EMC demands of a program

Like any shielding technology, the aspects of inter-component ground-plane activity, inherent leakage due to holes, doors and seams is a major EMC compliance concern. However, with this new Flexible Conductive Coatings technology, along with past experience in shielding designs, and working with leading thermoforming companies offers to an OEM, a new EMC option which may be favorable.

Nylon 6 in Thin-wall Housings for Portable Electronics

James F. Stevenson and Alan Dubin
AlliedSignal Plastics, Morristown NJ

INTRODUCTION

The semi-crystalline nature of nylon 6 results in an excellent balance of stiffness and impact, along with good performance at elevated temperature, and an outstanding resistance to most commonly encountered solvents. It also features a relatively wide processing-window due to its excellent flow characteristics. Nylon 6 can fill highly complex parts, down to wall thicknesses of 1 mm (.040 inch) or less, without the need for specialized molding equipment or extraordinary processing conditions. It normally yields a highly uniform surface, even when reinforced with mineral fillers and/or glass fiber, due to a thin film of pure resin that migrates to the surface during molding. This results in a higher level of surface gloss and more esthetically pleasing parts.

Nylon 6 undergoes dimensional growth and mechanical property change with absorption of moisture after molding. These effects are well-understood and can be readily predicted and compensated for in the design phase. In fact, this trait can often prove beneficial since the material actually becomes tougher and more resilient when it reaches equilibrium with ambient conditions. Subsequent dimensional variation and property change under normal circumstances are negligible. The widespread use of this versatile material for electrical devices and power tool housings, which contain high speed rotating mechanical components, demonstrate its ability to perform in critical tolerance parts.

Nylon 6 readily accepts glass fiber reinforcement and therefore can reach the higher modulus levels needed to compensate for the inherently low stiffness of thin-wall structures. This translates directly into less injection pressure and lower screw velocity, resulting in less wear and tear on molds and equipment. As a consequence of its high flow, nylon 6 will sometimes flash in a mold where other materials just barely fill. This tendency to flash can easily be avoided by designing the mold for the high flow characteristics of nylon. The fast fill and solidification capabilities of nylon 6 can, depending on the design, often reduce mold cycle times by up to 30%, yielding significant cost savings and throughput rates. This added value must be included in any calculation of total product cost, since machine cost and cycle time frequently represent the most significant component.

Small components such as portable electronic devices (cell phones, pagers, scanners, etc.) typically require wall thicknesses less than 1 mm (.040 inch), as the trend in these devices has been toward thinner and lighter packages. The high flow characteristics of nylon enable designers to avoid excessively large draft angles, which is more typical of products molded in amorphous thermoplastics. This feature also results in parts that are more easily extracted from the mold, without requiring special polishing of die-draw surfaces, and helps to avoid appearance problems such as sink marks. In addition, nylon 6 can be readily shielded by most of the techniques for EMI shielding currently used by the electronics industry.

When weld-lines are present due to required holes, slots or other openings in a part, nylon 6 enjoys a significant advantage in drop impact strength. Some plastics, particularly blends and alloys, exhibit weakness in these areas, often leading to unpredictable failure, either during assembly or when parts are dropped in use. One such example is the antenna opening in a cell phone housing, where parts made from PC/ABS have been known to fail during the insertion of a metal sleeve. This can take place either suddenly, as from impact, or over longer periods of time, due to creep. Nylon 6, on the other hand, actually gets tougher and less notch sensitive as it picks up moisture. Thus, if a nylon part passes impact testing in the dry-as-molded state, it will become even more resilient once it is assembled and placed into service, thereby assuring the designer of greater product reliability.

MATERIAL PROPERTY TESTING

MATERIALS

The materials evaluated for this study were typical nylon 6 compositions with impact-modifier systems designated as IM0, IM1, and IM2, and glass reinforcement levels ranging from 0% to 33%. The amorphous blends were a high flow and a commercial grade of PC/ABS, respectively designated as PC/ABS-1 HF and PC/ABS-2, both of which are currently used in thin-wall housing applications.

IMPACT-STIFFNESS DATA

Material selection for thin-wall housing applications usually involves a trade off between impact resistance and stiffness in the molded product. Stiffness for a given product thickness is proportional to the flexural modulus of the material. The correlation between impact properties obtained by standard testing of material samples and impact performance in actual molded parts is not well established. For the purposes of this comparison, we used total energy in an instrumented impact test (ASTM D3763) on discs that were molded with a central weldline. Our experience is that weldline strength offers a reasonable, although not perfect, prediction of part performance in housings with weldlines subjected to impact during drop testing. Nylon 6 compounds retain a relatively high proportion of their impact strength at weldlines compared to amorphous materials.

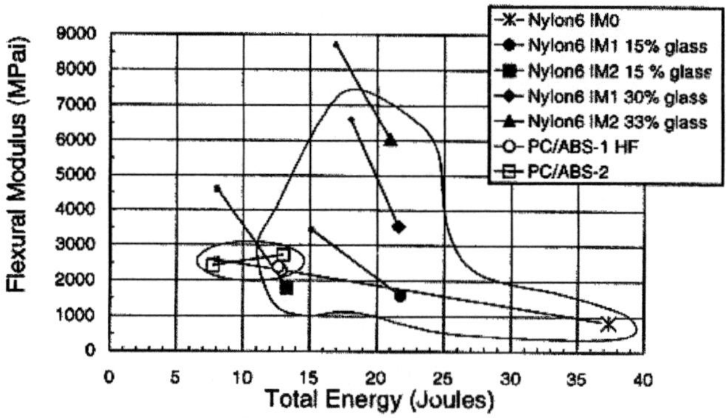

Figure 1. Flexural modulus vs. total energy from instrumented impact testing on discs with weldlines.

Exposure of nylon 6 to moisture at ambient temperature and humidity results in an increase in drop impact strength and a reduction in flexural modulus. The most appropriate testing conditions for housings are those most closely simulating actual end-use. This requires moisture conditioning the part at ambient temperature and relative humidity prior to testing, for varying time periods depending on the wall thickness.

A plot of flexural modulus (ASTM D790) vs. total energy for instrumented impact testing at weldlines is shown in Figure 1. For the nylon 6 materials, data are shown for the dry state immediately following molding (small symbols) and after 85 days in a 50% relative humidity environment (large symbols). The resulting moisture level of these 3.2 mm thick discs corresponds closely to the equilibrium moisture content for nylon 6 at room temperature and 30 % relative humidity.

The chart indicates a concentration of PC/ABS flexural modulus data in the range of 2200 - 2800 MPa, and total impact energy of 7 - 14 joules. By contrast, the use of glass fiber reinforcement greatly extends the modulus range for nylon 6. The total energy level for weldline impact in nylon 6 is equivalent or higher than that of PC/ABS. A balance of stiffness and impact strength, as required for a given application, can be obtained by selecting the right grade of nylon 6, with the appropriate type and quantity of reinforcement.

PRODUCT TESTING

DROP IMPACT TEST

Sample housings having a 1 mm wall thickness (Figure 2) were molded from several grades of nylon 6 and PC/ABS for drop impact testing according to a typical protocol used by the

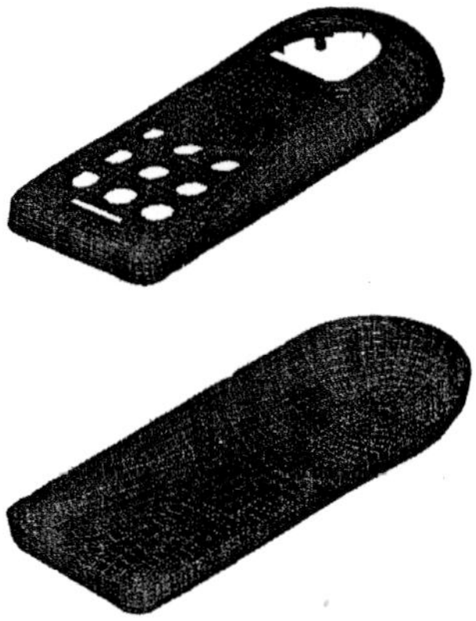

Figure 2. Cellular phone test housing with 1-mm-thick walls.

Table 1. Summary of results for drop impact testing

			Days
Nylon 6 IM0	pass	pass	14
Nylon 6 IM2 20% glass	pass		27
Nylon 6 IM2 15% glass	pass	pass	19,21
Nylon 6 IM1 15% glass	pass	pass	15,16
Nylon 6 IM2 33% glass	pass	fail	27
Nylon 6 IM2 30% glass	pass	fail	26
Nylon 6 IM1 30% glass	fail	pass	14,15
PC/ABS-1 HF	fail	fail	14+
PC/ABS-2	fail		27

telecommunications industry. These housings were dropped a total of 18 times from a height of 1.5 meters onto a concrete surface at room temperature. The 21.5 gram housings were weighted with a 160 gram internal steel plate secured on 6 bosses, to simulate the influence of internal components on impact. Any crack in the housing was considered to be a failure. All housings were preconditioned for a minimum of 2 weeks at 50% relative humidity. Results of this testing are summarized in Table 1. The data generally show that:

- The unfilled and 15% glass-filled nylon 6 materials consistently passed this test;
- Nylon with 30% or higher glass-filler levels passed this test about half the time;
- The two grades of PC/ABS did not pass this test.

Failures generally occurred at cracks along the weldline on the bottom of the part. These results are consistent with tests performed in our laboratory on commercial thin-wall housings.

PROCESSING

COMPARISON OF SIMULATION WITH EXPERIMENT

Filling of the 1-mm-thick housing shown in Figure 2 was simulated for the nylon 6 IM0 material and the PC/ABS-1 HF material. This simulation was performed using the Multilaminate Filling Analysis from Moldflow of Australia, Ltd. As shown in Figure 3, the predictions (lines) are in reasonable agreement with injection pressure data (solid symbols) obtained using a data acquisition system to measure pressure at the sprue.

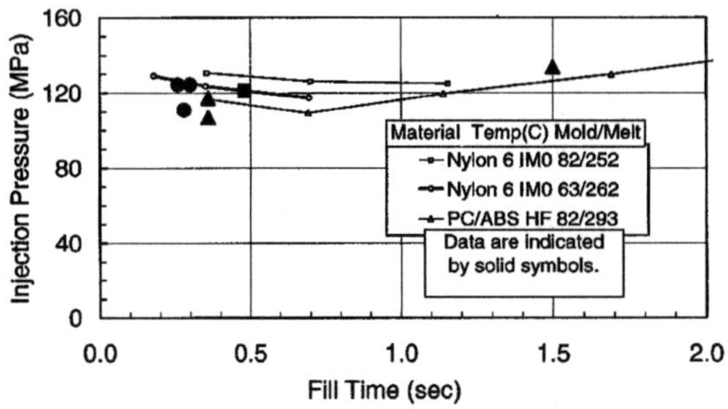

Figure 3. Comparison of simulation vs. data for injection pressure.

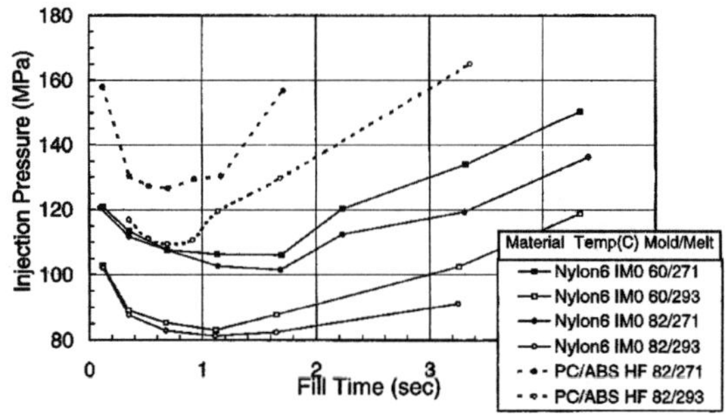

Figure 4. Simulation of injection pressure vs. fill time.

SIMULATION OVER A RANGE OF OPERATING CONDITIONS

A series of simulations was performed to obtain predictions for injection pressure and clamp force vs. fill time for the nylon 6 IM0 compound using recommended combinations of mold temperatures of 60°C and 82°C and melt temperatures of 271°C and 293°C. The injection pressure vs. fill-time predictions in Figure 4 show the expected U-shaped curve resulting from:

(a) Increase of injection pressure with injection rate at fill times that are too short for cooling to occur, and

(b) Increase of injection pressure with fill time at longer fill times due to cooling in the mold.[1]

The melt temperature has a much greater influence on injection pressure than does the mold temperature. Only at the longer fill times where cooling is more significant does mold temperature begin to influence the injection pressure. Similar curves are shown for PC/ABS-1 HF at its recommended processing temperature. The predictions show that this material requires anywhere from 25-35% greater injection pressure to fill than nylon 6 IM0 for the same temperatures. The processing window for nylon 6 as indicated by the simulation is fairly wide. For the highest mold/melt temperature combination 82°C/293°C, the injection pressure increases by no more than 10% from the minimum over injection times ranging from about 0.4 to 2.8 sec. The corresponding process window for PC/ABS- 1 HF ranges from about 0.3 to 1.2 sec.

Clamp force vs. fill-time predictions are shown in Figure 5. For nylon 6, the minimum in the clamp force profile occurs at fill times of about 0.35 sec, whereas the minimum in the pressure profile occurs at fill times greater than 1. 1 sec. An examination of pressure profiles (not shown) indicates that as fill time increases, a relatively larger portion of the pressure drop occurs within the cavity relative to the runners and gate. Consequently, clamp force starts to

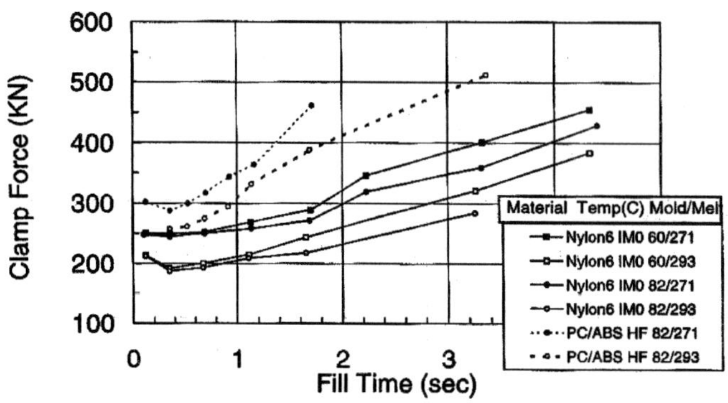

Figure 5. Simulation of clamp force vs. fill time.

increase with injection time at shorter fill times (0.35sec) than does injection pressure (> 1.1 sec).

An examination of temperature profiles in the mold cavity (not shown) for the mold/melt temperatures of 60°C/271°C shows the highest predicted melt temperature is at the end of the cavity for fill times of 0.15 and 0.35 sec. For these short fill times, shear heating dominates cooling during mold filling. For fill times of 1.1 sec and above, the lowest predicted material temperatures occur at the last location to fill, a result which indicates that cooling dominates over shear heating. These observations on temperature profiles are consistent with and help to explain the observations on the injection pressure and clamp force profiles.

CYCLE TIME

The mold for the 1-mm housing was used to evaluate minimum cycle times for nylon 6 IM0 and PC/ABS-1 HF using surface defects as the limiting factor, the cycle times thus obtained were 11.5 sec for nylon and 15.4 sec for PC/ABS. This 25% reduction in cycle time correlates well with actual observation in molding trials across a variety of different applications. Shorter cycle times are expected for nylon 6 materials because of its favorable crystallization rate, which accelerates the increase of rigidity during cooling.[2] This results in a shorter holding period with ejection from the mold able to occur much sooner than for amorphous materials.

AESTHETICS

The surface appearance of parts molded from nylon 6 compounds can be further enhanced by taking advantage of:

(a) the wider processing window of nylon 6 compared to amorphous materials, especially in thin-wall housings;

(b) the ability to vary the crystallization rate to create a resin-rich surface, particularly when glass-fiber reinforcement is present;

(c) the ability to obtain a non-glossy and uniform surface, without flow lines or other imperfections.

EMI SHIELDING

Nylon 6 compounds can be readily shielded by most of the common techniques currently in use by the electronics and telecommunications industry, as shown in Table 2.

CONCLUSIONS

In summary, nylon 6 compounds offer substantial processing and product-performance advantages over amorphous materials across a wide variety of applications. These benefits become more pronounced with decreasing wall thickness, as is frequently the case in thin-wall housings for electronics and telecommunications.

ACKNOWLEDGMENTS

The authors wish to acknowledge technical discussions with: Kris Akkapeddi, Sudhir Bhakuni, Geoff Burgeson, Al Chambers, Randy Fleck, Mark Minnichelli, Bill McMaster, Clark Smith, Bruce Van Buskirk, and Robert Welgos. Molding and mechanical testing results reported in this work were performed by Rowena McPherson, Igor Palley, Juan Ruiz, Roberto Sanchez, and Robert Seville. Computational assistance was provided by Prasanna Godbole, Christopher Roth, and Craig Scott.

REFERENCES

1 L. S. Turng, H. H. Chiang, J. F. Stevenson, Optimization Strategies for Injection Molding, SPE Technical Papers, 668, 41(1995).

2 R. H. Welgos, Nylon 6 and 6,6 aren't always the same, Machine Design, 55, Nov. 21, 1994.

Finite Element Analysis Aided Engineering of Elastomeric EMI Shielding Gaskets

Shu H. Peng and Kai Zhang

Chomerics Division, Parker Hannifin Corporation, 77 Dragon Court, Woburn, MA

INTRODUCTION

Electrically conductive elastomeric gaskets traditionally play a very important role in shielding military or commercial electronic devices from EMI and reducing electromagnetic emissions from such devices. Ever since they were first developed some thirty-five years ago, conductive elastomeric gasket technology has been well known for its complexity. As more digital electronic devices, using higher power and faster switching speeds, are produced and deployed into the commercial world, a high performance but cost effective shielding design involving electrically conductive gaskets has become a sophisticated and challenging engineering task.

A brief introduction to nonlinear FEA concepts and its application procedures is also presented. The Mooney-Rivlin model and the Ogden model are used to describe the highly filled electrically conductive silicone materials. FEA-assisted design examples are presented, which include deformation of a formed-in-place conductive gasket, a composite plastic/conductive-elastomer gasket with improved load-deflection characteristics, and a modified hollow-"D" extruded conductive gasket with an enhanced installation/attachment feature.

FINITE ELEMENT ANALYSIS

FEA has become an important part in the product design and prototyping processes. It allows engineers to assess the product performance before a prototype is built. Using FEA, the design can be modified quickly, with much more ease and much less cost than building another prototype for testing. The effective use of FEA serves to accelerate the design process, saving engineering time and cost.

In recent years, the use of FEA for the design of elastomeric products has increased substantially. Most applications of elastomeric gaskets, including EMI gaskets, involve large compressive deformation. Achieving an adequate simulation accuracy requires use of an advanced FEA program capable of tackling nonlinear problems, such as kinematic nonlinearity due to large deformation, material nonlinearity and changing boundary conditions. At the same time, software should also be user-friendly and efficient, which means that a FEA program should have a graphic user interface, efficient pre- and post-processors and an automatic mesh generator.

A nonlinear finite element program, MARC K6,[1] was used for the static analysis reported in this paper. Four-node plane strain Hermann elements were used to model the gasket cross-section. The compression of the gasket was simulated using the contact elements. The plastic spacer was considered as a rigid body since it is much stiffer than the elastomer material.

The Mooney-Rivlin strain energy function is used to model the gasket material. The Mooney-Rivlin model in MARC does not allow the input of bulk modulus. In order to take into account the near incompressibility of elastomeric materials, the Mooney-Rivlin constants are converted to the constants of the two-term Ogden strain energy function. MARC supplements the Ogden model by using the bulk modulus to account for the near incompressibility of elastomers.

The Mooney-Rivlin strain energy function,[2,3]

$$W = C_1(I_1 - 1) + C_2(I_2 - 1) \qquad [1]$$

where C_1, C_2 are material constants and I_1, I_2 are strain invariants, is a special case of the Ogden model,[2,4]

$$W = \sum_{i=1}^{m} \frac{\mu_i}{\alpha_i}\left(\lambda_1^{\alpha_i} + \lambda_2^{\alpha_i} + \lambda_3^{\alpha_i} - 3\right) \qquad [2]$$

where $\lambda_1, \lambda_2, \lambda_3$ are the stretch ratios and α_i, μ_i the material constants. For a two-term Ogden model (m=2) with $\alpha_1 = 2, \alpha_2 = -2$ and $\mu_1 = 2C_1$, $\mu_2 = -2C_2$, the Ogden and Mooney-Rivlin models become equivalent. The above functions are in their original incompressible forms. The actual functions employed

Table 1. Material constants of Cho-Seal 1310

Bulk modulus	K = 1380 MPa	
Ogden constants	μ_1 = -1.19 MPa	α_1 = 2
	μ_2 = -3.6 MPa	α_2 = -2

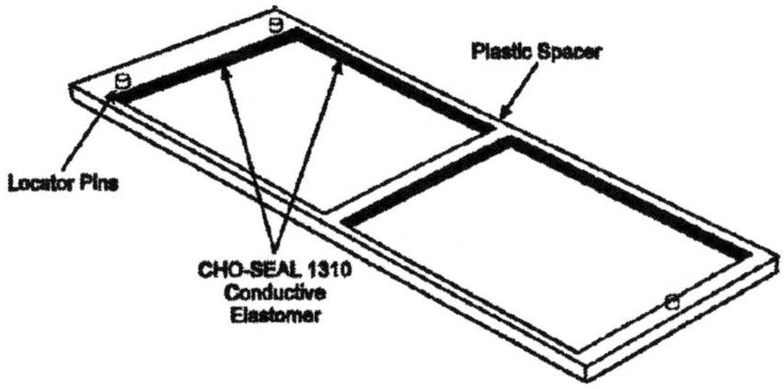

Figure 1. A schematic diagram of a plastic spacer gasket.

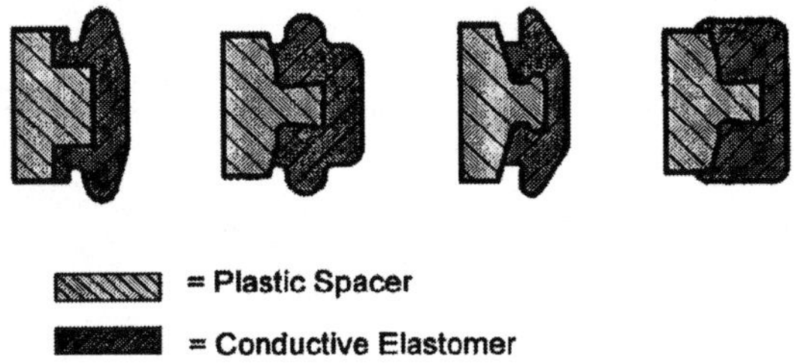

Figure 2. Typical spacer gasket cross-section profiles.

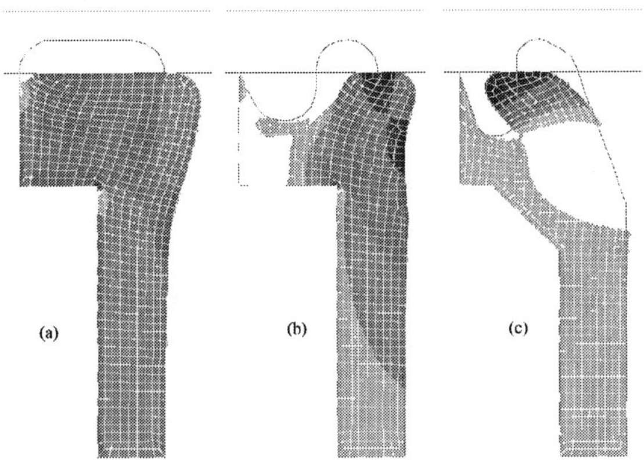

Figure 3. FEA simulated gasket shapes after the gaskets are installed. (a) Existing design (b) An improved design (c) The optimized design.

in FEA programs usually include an additional bulk term to account for the near incompressibility.[1-2]

Chomerics Cho-Seal 1310, a silicone-based compound filled with fine silver plated glass powder, is used as the gasket material. Table 1 lists the two-term Ogden constants and the bulk modulus of this compound, as obtained from material testing.

FEA AIDED DESIGN OF PLASTIC SPACER GASKET

The product designed using FEA in this work is a composite plastic/elastomer EMI spacer gasket. A spacer gasket features a thin plastic retainer frame onto which a conductive elastomer is molded. The elastomer can be located inside or outside the retainer frame, as well as on its top and bottom surfaces. The gasket is used as a grounding device between the EMI shielded housing of a cellular phone handset and its interior printed circuit board. It is a new approach to designing EMI gaskets into handheld electronics. A sketch of the gasket is shown in Figure 1 and its cross-section profiles in Figure 2. One of the design requirements in this type of application is that the gasket needs to deflect under a low compressive closure force. Using FEA, a sophisticated spacer gasket design was optimized to provide satisfactory deflection under low closure force, while also ensuring proper electrical/mechanical contact area to guarantee EMI performance.

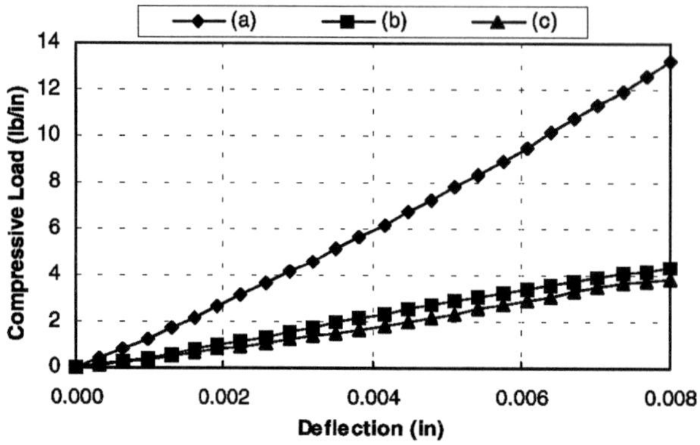

Figure 4. FEA predicted load-deflection curves for the initial design (a) an improved design (b) and the optimized design (c).

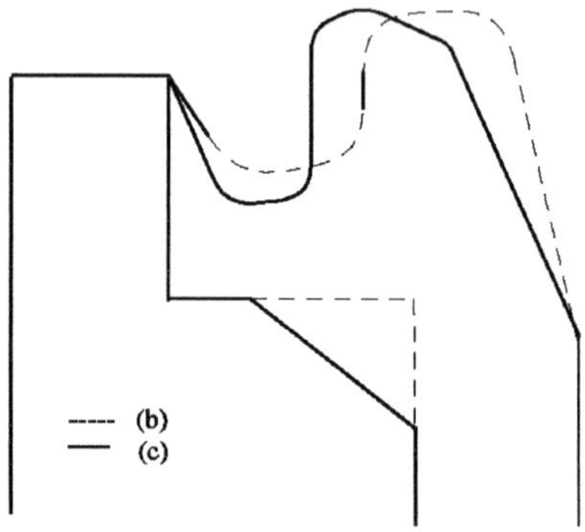

Figure 5. Comparison of the improved design (b) and the optimized design (c).

An existing design (a) was analyzed using FEA. The deformed shape after the compressive installation is shown as exhibit (a) of Figure 3. The straight horizontal lines at the top represent the flat mating surfaces, at positions before and after installation. The original gasket profiles before installation are indicated by solid curvature lines. The stress in the vertical direction is shown in colored contour plot. The mesh lines represent the finite elements used.

The predicted closure load as a function of deflection is illustrated in Figure 4. The closure force for this existing design was too high and needed to be decreased significantly to meet the application requirement.

After many design trials using FEA, only a limited number of design ideas proved to be effective in reducing the closure force. The best approach seemed to be modifying the gasket shape in such a way that the top portion of the gasket bends during installation. Exhibit (b) of Figure 3 shows the original and the deformed shapes of an improved design (b). A bending mechanism clearly existed after the design modification from the existing design. This bending mechanism led to a much reduced closure force, as indicated in Figure 4. Design (b) met the design requirement in terms of the closure force. However, the gasket top tilts away from the plastic spacer and into the interior and may interfere with the circuit board, which should be avoided.

Further efforts led to the final design as shown as Exhibit (c) of Figure 3. A detailed comparison of design (b) and design (c) is illustrated in Figure 5. The shape of the gasket top was modified to reverse the direction it tilts when compressed. The plastic spacer was also redesigned. The corner was cropped to increase the thickness of the elastomeric part on the gasket at that location, thereby further reducing the closure force, as indicated in Figure 4. Design (c) also feature some other advantages over design (b): more stable interface contact, larger contact area and less tearing of the elastomeric part.

Following the FEA-assisted design, prototype spacer gaskets were produced. Those parts met the application requirements of closure force and EMI shielding during performance trials. The design was approved for production without additional prototyping.

This example clearly demonstrates the value of advanced simulation technologies, such as FEA, in designing better products with reduced prototyping time and cost. Using finite element analysis, one can accurately predict and simulate a gasket in use, and reduce the possibility of a poor gasket design even before prototyping, which may cause poor EMI shielding performance.

REFERENCES

1 MARC K6.2, User Information, MARC Analysis Research Corp., Palo Alto, California, 1996.

2 S. H. Peng and W. V. Chang, A Compressible Approach in Finite Element Analysis of Rubber- Elastic Materials, *J. Comput. & Struct.*, **62**(3), 1997.

3 R. G. Treloar, **The Physics of Rubber Elasticity**, Third Edition, *Clarendon Press*, Oxford, 1975.

4. Ogden, **Nonlinear Elastic Deformation,** *Ellis Howard Limited*, New York, 1984.

Index